Student's Solutions Manual to Accompany

Trigonometry

David Cohen
Department of Mathematics
University of California
Los Angeles, California

Prepared by
Ross Rueger
Department of Mathematics
College of the Sequoias
Visalia, California

West Publishing Company
St. Paul   New York   Los Angeles   San Francisco

WEST'S COMMITMENT TO THE ENVIRONMENT
In 1906, West Publishing Company began recycling materials left over from the production of books. This began a tradition of efficient and responsible use of resources. Today, up to 95% of our legal books and 70% of our college texts are printed on recycled, acid-free stock. West also recycles nearly 22 million pounds of scrap paper annually—the equivalent of 181,717 trees. Since the 1960s, West has devised ways to capture and recycle waste inks, solvents, oils, and vapors created in the printing process. We also recycle plastics of all kinds, wood, glass, corrugated cardboard, and batteries, and have eliminated the use of styrofoam book packaging. We at West are proud of the longevity and the scope of our commitment to our environment.

Production, Prepress, Printing and Binding by West Publishing Company.

Contents

Chapter 7   Analytic Geometry

Preface

This <u>Student's Solutions Manual</u> contains complete solutions to all odd-numbered regular and graphing calculator exercises of <u>Trigonometry</u> by David Cohen. It also contains complete solutions to odd and even exercises for each chapter test. I have attempted to format solutions for readability and accuracy, and apologize to you for any errors that you may encounter. If you have any comments, suggestions, error corrections, or alternative solutions please feel free to drop me a note.

Please use this manual with some degree of caution. Be sure that you have attempted a solution, and re-attempted it, before you look it up in this manual. Mathematics can only be learned by **doing**, and not by observing! As you use this manual, do not just read the solution but work it along with the manual, using my solution to check your work. If you use this manual in that fashion then it should be helpful to you in your studying.

I would like to thank a number of people for their assistance in preparing this manual. Thanks go to Peter Marshall, Jane Bacon, and Mark Jacobsen at West Educational Publishing for their valuable assistance and support. Special thanks go to Chuck Heuer of Concordia College for his meticulous error-checking of my solutions. Finally, I wish to thank Joy Bishop of College of the Sequoias for her fabulous typing and editing of this entire manuscript.

I wish to express my deepest appreciation to David Cohen for continuing his tradition of excellence with this new textbook. I first became interested in trigonometry some ten years ago while in graduate school at UCLA, and David's unusual identities were the primary source of that interest. I hope as you work these problems that same curiosity arises in you.

Ross Rueger
College of the Sequoias
915 South Mooney Boulevard
Visalia, CA 93277

December, 1991

Chapter One
Coordinate Geometry and Functions

1.1 Sets of Real Numbers and Notation

1. (a) natural number, integer, rational number
    (b) integer, rational number

3. (a) rational number
    (b) irrational number

5. (a) natural number, integer, rational number
    (b) rational number

7. (a) rational number
    (b) rational number

9. irrational number

11. natural number, integer, rational number

13. Since $\frac{11}{4} = 2.75$, we have the following graph:

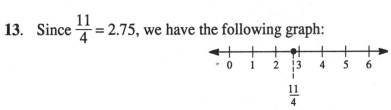

15. Since $1 + \sqrt{2} \approx 2.4$, we have the following graph:

$$0 \quad 1 \quad 2 \quad 3 \quad 4 \quad 5 \quad 6$$
$$1 + \sqrt{2}$$

17. Since $\sqrt{2} - 1 \approx 0.4$, we have the following graph:

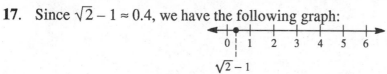

19. Since $\sqrt{2} + \sqrt{3} \approx 3.1$, we have the following graph:

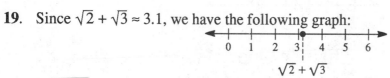

21. Since $\dfrac{1+\sqrt{2}}{2} \approx \dfrac{2.4}{2} \approx 1.2$, we have the following graph:

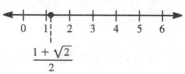

23. We draw the graph:

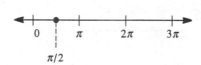

25. We draw the graph:

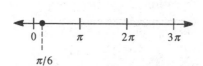

27. We draw the graph:

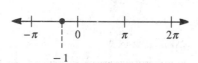

29. We draw the graph:

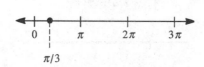

31. Since $1 \approx \dfrac{\pi}{3}$, we have the following graph:

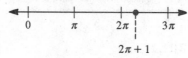

33. Since $\dfrac{\sqrt{139}-5}{3} \approx 2.26$, we have the following graph:

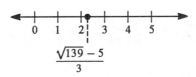

35. False

37. True (since $-2 = -2$, then it is also true that $-2 \le -2$)

39. False

41. False (since $2\pi \approx 6.2$)

43. True (since $2\sqrt{2} \approx 2.8$)

45. We graph the interval $(2,5)$:

47. We graph the interval $[1,4]$:

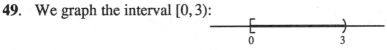

49. We graph the interval $[0,3)$:

51. We graph the interval $(-3,\infty)$:

53. We graph the interval $[-1,\infty)$:

55. We graph the interval $(-\infty,1)$:

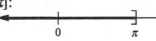

57. We graph the interval $(-\infty,\pi]$:

59. The domain is $(-\infty, \infty)$, the set of all real numbers.

61. In order for the denominator to be non-zero, we must be sure that $t - 4 \neq 0$, so $t \neq 4$. So the domain is the set of all real numbers except 4.

63. In order for the denominator to be non-zero, we must be sure that $y^2 - 16 \neq 0$, so $y^2 \neq 16$ and thus $y \neq \pm 4$. So the domain is the set of all real numbers except 4, –4.

65. We must be sure the quantity under the radical is non-negative, so:

$$21 - 7t \geq 0$$
$$-7t \geq -21$$
$$t \leq 3$$

So the domain is $(-\infty, 3]$.

67. We must be sure the quantity under the radical is non-negative, so $12 + t \geq 0$ and thus $t \geq -12$. So the domain is $[-12, \infty)$.

69. In order for the denominator to be non-zero, we must be sure that:

$$x^3 - 9x \neq 0$$
$$x(x^2 - 9) \neq 0$$
$$x(x + 3)(x - 3) \neq 0$$
$$x \neq 0, -3, 3$$

So the domain is the set of all real numbers except 0, –3, 3.

71. In order for the denominator to be non-zero, we must be sure that:

$$2t^3 - 13t^2 + 6t \neq 0$$
$$t(2t^2 - 13t + 6) \neq 0$$
$$t(2t - 1)(t - 6) \neq 0$$
$$t \neq 0, \tfrac{1}{2}, 6$$

So the domain is the set of all real numbers except $0, \tfrac{1}{2}, 6$.

1.2 Coordinate Geometry

1. We plot the points:

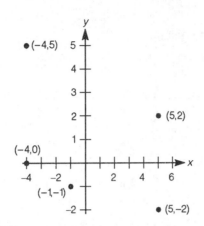

3. (a) We draw the right triangle PQR:

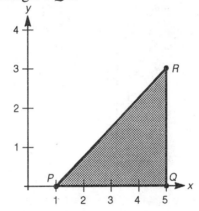

(b) Since the base is $b = 5 - 1 = 4$ and the height is $h = 3 - 0 = 3$, then the area is given by:
$$A = \tfrac{1}{2}bh = \tfrac{1}{2}(4)(3) = 6 \text{ square units}$$

5. (a) Here $(x_1, y_1) = (0,0)$ and $(x_2, y_2) = (-3, 4)$, so by the distance formula:
$$d = \sqrt{(-3-0)^2 + (4-0)^2} = \sqrt{9+16} = \sqrt{25} = 5$$

(b) Here $(x_1, y_1) = (2, 1)$ and $(x_2, y_2) = (7, 13)$, so:
$$d = \sqrt{(7-2)^2 + (13-1)^2} = \sqrt{25+144} = \sqrt{169} = 13$$

7. (a) Here $(x_1, y_1) = (-5, 0)$ and $(x_2, y_2) = (5, 0)$, so:
$$d = \sqrt{[5-(-5)]^2 + (0-0)^2} = \sqrt{100+0} = \sqrt{100} = 10$$

(b) Here $(x_1, y_1) = (0, -8)$ and $(x_2, y_2) = (0, 1)$, so:
$$d = \sqrt{(0-0)^2 + [1-(-8)]^2} = \sqrt{0+81} = \sqrt{81} = 9$$

Note that we really don't need to use the distance formula for either (a) or (b), since in each case one of the coordinates (either x or y) is the same. Draw quick graphs and you can find the distance by inspection:

(a)   This graph indicates the distance is 10:

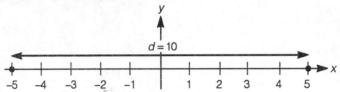

(b)   This graph indicates the distance is 9:

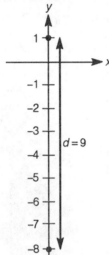

9.   Here $(x_1, y_1) = (1, \sqrt{3})$ and $(x_2, y_2) = (-1, -\sqrt{3})$, so:
$$d = \sqrt{(-1-1)^2 + \left(-\sqrt{3} - \sqrt{3}\right)^2} = \sqrt{(-2)^2 + \left(-2\sqrt{3}\right)^2} = \sqrt{4+12} = 4$$

11. (a)   We calculate the distance of each point from the origin:

$(3, -2)$: $d = \sqrt{(3-0)^2 + (-2-0)^2} = \sqrt{9+4} = \sqrt{13}$

$(4, \frac{1}{2})$: $d = \sqrt{(4-0)^2 + \left(\frac{1}{2} - 0\right)^2} = \sqrt{16 + \frac{1}{4}} = \sqrt{16.25}$

So $(4, \frac{1}{2})$ is farther from the origin.

(b)   We calculate the distance of each point from the origin:

$(-6, 7)$: $d = \sqrt{(-6-0)^2 + (7-0)^2} = \sqrt{36+49} = \sqrt{85}$

$(9, 0)$: $d = \sqrt{(9-0)^2 + (0-0)^2} = \sqrt{81+0} = \sqrt{81}$

So $(-6, 7)$ is farther from the origin.

13. We will graph each triangle and then determine (using the converse of the Pythagorean theorem) whether $a^2 + b^2 = c^2$.

    (a)   We graph the points:

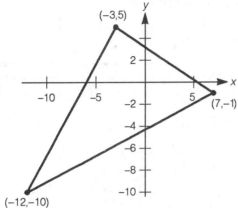

    Now calculate the distances:
$$a = \sqrt{(-3-7)^2 + [5-(-1)]^2} = \sqrt{100+36} = \sqrt{136}$$
$$b = \sqrt{[-12-(-3)]^2 + (-10-5)^2} = \sqrt{81+225} = \sqrt{306}$$
$$c = \sqrt{(-12-7)^2 + [-10-(-1)]^2} = \sqrt{361+81} = \sqrt{442}$$
    Now check the converse of the Pythagorean theorem:
$$a^2 + b^2 = 136 + 306 = 442 = c^2$$
    So the triangle is a right triangle.

    (b)   We graph the points:

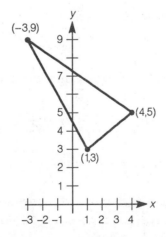

    Now calculate the distances:
$$a = \sqrt{(-3-1)^2 + (9-3)^2} = \sqrt{16+36} = \sqrt{52}$$
$$b = \sqrt{(4-1)^2 + (5-3)^2} = \sqrt{9+4} = \sqrt{13}$$
$$c = \sqrt{(-3-4)^2 + (9-5)^2} = \sqrt{49+16} = \sqrt{65}$$

Now check the converse of the Pythagorean theorem:
$$a^2 + b^2 = 52 + 13 = 65 = c^2$$
So the triangle is a right triangle.

(c)   We graph the points:

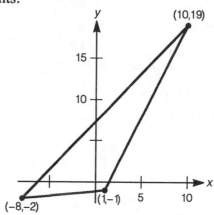

Now calculate the distances:
$$a = \sqrt{(-8-1)^2 + [-2-(-1)]^2} = \sqrt{81+1} = \sqrt{82}$$
$$b = \sqrt{(10-1)^2 + [19-(-1)]^2} = \sqrt{81+400} = \sqrt{481}$$
$$c = \sqrt{[10-(-8)]^2 + [19-(-2)]^2} = \sqrt{324+441} = \sqrt{765}$$
Now check the converse of the Pythagorean theorem:
$$a^2 + b^2 = 82 + 481 = 563 \neq 765 = c^2$$
So the triangle is not a right triangle.

15. (a)   Since $(h,k) = (4,6)$ and $r = 3$, the standard form $(x-h)^2 + (y-k)^2 = r^2$ becomes:
$$(x-4)^2 + (y-6)^2 = (3)^2, \text{ or } (x-4)^2 + (y-6)^2 = 9$$

(b)   Since $(h,k) = (-4,-6)$ and $r = \sqrt{3}$, the standard form $(x-h)^2 + (y-k)^2 = r^2$ becomes:
$$[x-(-4)]^2 + [y-(-6)]^2 = (\sqrt{3})^2, \text{ or } (x+4)^2 + (y+6)^2 = 3$$

17. (a)   Since $(h,k) = (-7,-3)$ and $r = 4$, the standard form $(x-h)^2 + (y-k)^2 = r^2$ becomes:
$$[x-(-7)]^2 + [y-(-3)]^2 = (4)^2, \text{ or } (x+7)^2 + (y+3)^2 = 16$$

(b)   Since $(h,k) = (7,-3)$ and $r = 16$, the standard form $(x-h)^2 + (y-k)^2 = r^2$ becomes:
$$(x-7)^2 + [y-(-3)]^2 = (16)^2, \text{ or } (x-7)^2 + (y+3)^2 = 256$$

19. (a)   Since $(h, k) = (0, 0)$ and $r = 1$, the standard form $(x - h)^2 + (y - k)^2 = r^2$ becomes:
$$(x - 0)^2 + (y - 0)^2 = (1)^2, \text{ or } x^2 + y^2 = 1$$

(b)   Since $(h, k) = (0, 0)$ and $r = 2$, the standard form $(x - h)^2 + (y - k)^2 = r^2$ becomes:
$$(x - 0)^2 + (y - 0)^2 = (2)^2, \text{ or } x^2 + y^2 = 4$$

21. (a)   Since P lies on the unit circle, the coordinates of P satisfy the equation $x^2 + y^2 = 1$. We are given that the x-coordinate of P is $\frac{3}{5}$. Substituting the value $x = \frac{3}{5}$ in the equation $x^2 + y^2 = 1$ gives us:
$$\left(\tfrac{3}{5}\right)^2 + y^2 = 1$$
$$\tfrac{9}{25} + y^2 = 1$$
$$y^2 = \tfrac{16}{25}$$
$$y = \pm\tfrac{4}{5}$$
We want to choose the positive root here because P lies in quadrant I (in which all y-coordinates are positive). Thus, the y-coordinate of P is $\frac{4}{5}$.

(b)   The solution for the y-coordinates are as in (a), where $y = \pm\frac{4}{5}$. We want to choose the negative root here because P lies in quadrant IV (in which all y-coordinates are negative). Thus, the y-coordinate of P is $-\frac{4}{5}$.

23. (a)   Since P lies on the unit circle, the coordinates of P satisfy the equation $x^2 + y^2 = 1$. We are given that the y-coordinate of P is $-\frac{\sqrt{2}}{2}$. Substituting the value $y = -\frac{\sqrt{2}}{2}$ in the equation $x^2 + y^2 = 1$ gives us:
$$x^2 + \left(-\tfrac{\sqrt{2}}{2}\right)^2 = 1$$
$$x^2 + \tfrac{1}{2} = 1$$
$$x^2 = \tfrac{1}{2}$$
$$x = \pm\tfrac{\sqrt{2}}{2}$$
We want to choose the negative root here because P lies in quadrant III (in which all x-coordinates are negative). Thus, the x-coordinate of P is $-\frac{\sqrt{2}}{2}$.

(b)   The solution for the x-coordinates are as in (a), where $x = \pm\frac{\sqrt{2}}{2}$. We want to choose the positive root here because P lies in quadrant IV (in which all x-coordinates are positive). Thus, the x-coordinate of P is $\frac{\sqrt{2}}{2}$.

25. (a) Since P lies on the unit circle, the coordinates of P satisfy the equation $x^2 + y^2 = 1$. We are given that the x-coordinate of P is $-\frac{2}{3}$. Substituting the value $x = -\frac{2}{3}$ in the equation $x^2 + y^2 = 1$ gives us:

$$\left(-\tfrac{2}{3}\right)^2 + y^2 = 1$$
$$\tfrac{4}{9} + y^2 = 1$$
$$y^2 = \tfrac{5}{9}$$
$$y = \pm\tfrac{\sqrt{5}}{3}$$

We want to choose the positive root here because P lies in quadrant II (in which all y-coordinates are positive). Thus, the y-coordinate of P is $\frac{\sqrt{5}}{3}$.

(b) The solution for the y-coordinates are as in (a), where $y = \pm\frac{\sqrt{5}}{3}$. We want to choose the negative root here because P lies in quadrant III (in which all y-coordinates are negative). Thus, the y-coordinate of P is $-\frac{\sqrt{5}}{3}$.

27. (a) Since P lies on the unit circle, the coordinates of P satisfy the equation $x^2 + y^2 = 1$. We are given that the x-coordinate of P is 0.652. Substituting the value $x = 0.652$ in the equation $x^2 + y^2 = 1$ gives us:

$$(0.652)^2 + y^2 = 1$$
$$0.425104 + y^2 = 1$$
$$y^2 = 0.574896$$
$$y \approx \pm 0.758$$

We want to choose the positive root here because P lies in quadrant I (in which all y-coordinates are positive). Thus, the y-coordinate of P is (approximately) 0.758.

(b) Substituting the value $y = 0.652$ in the equation $x^2 + y^2 = 1$ gives us:

$$x^2 + (0.652)^2 = 1$$
$$x^2 + 0.425104 = 1$$
$$x^2 = 0.574896$$
$$x \approx \pm 0.758$$

We want to choose the positive root here because P lies in quadrant I (in which all x-coordinates are positive). Thus, the x-coordinate of P is (approximately) 0.758.

29. Since the center is $(3, 2)$, we know the equation of the circle will take the form $(x - 3)^2 + (y - 2)^2 = r^2$. We can find r since the point $(-2, -10)$ must satisfy this equation:

$$(-2 - 3)^2 + (-10 - 2)^2 = r^2$$
$$25 + 144 = r^2$$
$$169 = r^2$$

So the equation of the circle is $(x - 3)^2 + (y - 2)^2 = 169$.

31. Since the circle is tangent to the y-axis and its center is $(3,5)$, its radius must be 3. Thus the equation of the circle is $(x-3)^2 + (y-5)^2 = 9$.

33. Using the Pythagorean theorem, we have:
$$a^2 = 1^2 + 1^2 = 2, \text{ so } a = \sqrt{2}$$
$$b^2 = 1^2 + \left(\sqrt{2}\right)^2 = 1+2 = 3, \text{ so } b = \sqrt{3}$$
$$c^2 = 1^2 + \left(\sqrt{3}\right)^2 = 1+3 = 4, \text{ so } c = 2$$
$$d^2 = 1^2 + 2^2 = 1+4 = 5, \text{ so } d = \sqrt{5}$$
$$e^2 = 1^2 + \left(\sqrt{5}\right)^2 = 1+5 = 6, \text{ so } e = \sqrt{6}$$
$$f^2 = 1^2 + \left(\sqrt{6}\right)^2 = 1+6 = 7, \text{ so } f = \sqrt{7}$$
$$g^2 = 1^2 + \left(\sqrt{7}\right)^2 = 1+7 = 8, \text{ so } g = \sqrt{8} = 2\sqrt{2}$$

35. (a) We sketch the parallelogram $ABCD$:

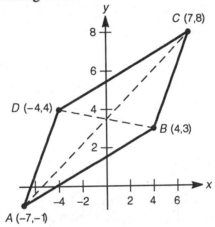

(b) For AC, let $(x_1, y_1) = (-7, -1)$ and $(x_2, y_2) = (7, 8)$, so the midpoint is:
$$M_1 = \left(\tfrac{-7+7}{2}, \tfrac{-1+8}{2}\right) = \left(0, \tfrac{7}{2}\right)$$
For BD, let $(x_1, y_1) = (4, 3)$ and $(x_2, y_2) = (-4, 4)$, so the midpoint is:
$$M_2 = \left(\tfrac{4-4}{2}, \tfrac{3+4}{2}\right) = \left(0, \tfrac{7}{2}\right)$$

(c) It appears that the midpoints of the two diagonals of a parallelogram are the same. Or, stated more concisely, the diagonals of a parallelogram bisect each other.

37. Since $A(-1, 6)$ and $B(3, -2)$ are the endpoints of a diameter, then their midpoint must be the center (h, k) of the circle:
$$(h, k) = \left(\tfrac{-1+3}{2}, \tfrac{6-2}{2}\right) = \left(\tfrac{2}{2}, \tfrac{4}{2}\right) = (1, 2)$$
Since the radius is the distance from this center $(1, 2)$ to $B(3, -2)$, using the distance formula:
$$r = \sqrt{(3-1)^2 + (-2-2)^2} = \sqrt{4+16} = \sqrt{20} = 2\sqrt{5}$$
The center of the circle is $(1, 2)$ and the radius is $2\sqrt{5}$.

1.3 Graphs and Symmetry

1. To find the x-intercept, we substitute $y = 0$ into the equation $3x + 4y = 12$:
$$3x + 4(0) = 12$$
$$3x = 12$$
$$x = 4$$
To find the y-intercept, we substitute $x = 0$ into the equation $3x + 4y = 12$:
$$3(0) + 4y = 12$$
$$4y = 12$$
$$y = 3$$
Now graph the line:

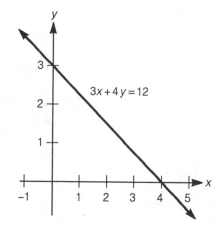

3. To find the x-intercept, we substitute $y = 0$ into the equation $y = 2x - 4$:
$$0 = 2x - 4$$
$$-2x = -4$$
$$x = 2$$
To find the y-intercept, we substitute $x = 0$ into the equation $y = 2x - 4$:
$$y = 2(0) - 4$$
$$y = -4$$
Now graph the line:

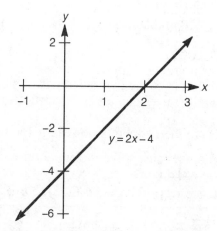

5. To find the x-intercept, we substitute $y = 0$ into the equation $x + y = 1$:
$$x + 0 = 1$$
$$x = 1$$
To find the y-intercept, we substitute $x = 0$ into the equation $x + y = 1$:
$$0 + y = 1$$
$$y = 1$$
Now graph the line:

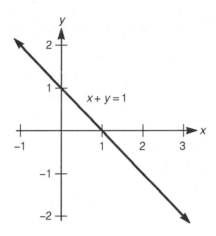

7. (a) To find the y-intercepts, let $x = 0$:
$$y = (0)^2 + 3(0) + 2 = 0 + 0 + 2 = 2$$

To find the x-intercepts, let $y = 0$:
$$x^2 + 3x + 2 = 0$$
$$(x + 2)(x + 1) = 0$$
$$x = -1, -2$$

(b) To find the y-intercepts, let $x = 0$:
$$y = (0)^2 + 2(0) + 3 = 0 + 0 + 3 = 3$$

To find the x-intercepts, we must solve the equation $x^2 + 2x + 3 = 0$. Using the quadratic formula, we obtain:
$$x = \frac{-2 \pm \sqrt{(2)^2 - 4(1)(3)}}{2(1)} = \frac{-2 \pm \sqrt{4 - 12}}{2} = \frac{-2 \pm \sqrt{-8}}{2}$$
Since this equation has no real solutions, there are no x-intercepts.

9. (a) To find the y-intercepts, let $x = 0$:
$$y = (0)^2 + 0 - 1 = 0 + 0 - 1 = -1$$

To find the x-intercepts, we must solve the equation $x^2 + x - 1 = 0$. Using the quadratric formula, we obtain:
$$x = \frac{-1 \pm \sqrt{(1)^2 - 4(1)(-1)}}{2(1)} = \frac{-1 \pm \sqrt{1 + 4}}{2} = \frac{-1 \pm \sqrt{5}}{2}$$

(b)   To find the y-intercepts, let $x = 0$:
$$y = (0)^2 + 0 + 1 = 0 + 0 + 1 = 1$$
To find the x-intercepts, we must solve the equation $x^2 + x + 1 = 0$. Using the quadratic formula, we obtain:
$$x = \frac{-1 \pm \sqrt{(1)^2 - 4(1)(1)}}{2(1)} = \frac{-1 \pm \sqrt{1-4}}{2} = \frac{-1 \pm \sqrt{-3}}{2}$$
Since this equation has no real solutions, there are no x-intercepts.

11.   To find the y-intercepts, let $x = 0$:
$$\tfrac{0}{2} + \tfrac{y}{3} = 1$$
$$\tfrac{y}{3} = 1$$
$$y = 3$$
To find the x-intercepts, let $y = 0$:
$$\tfrac{x}{2} + \tfrac{0}{3} = 1$$
$$\tfrac{x}{2} = 1$$
$$x = 2$$

13.   To find the y-intercepts, let $x = 0$:
$$3(0) - 5y = 10$$
$$-5y = 10$$
$$y = -2$$
To find the x-intercepts, let $y = 0$:
$$3x - 5(0) = 10$$
$$3x = 10$$
$$x = \tfrac{10}{3}$$

15.   To find the y-intercepts, let $x = 0$:
$$y = (0)^3 - 8 = 0 - 8 = -8$$
To find the x-intercepts, let $y = 0$:
$$x^3 - 8 = 0$$
$$x^3 = 8$$
$$x = 2$$

17.   (a)   To find the y-intercepts, let $x = 0$:
$$(-2)^2 + (y - 3)^2 = 1$$
$$4 + (y - 3)^2 = 1$$
$$(y - 3)^2 = -3$$
Since this equation has no real solutions, there are no y-intercepts for the graph.
To find the x-intercepts, let $y = 0$:
$$(x - 2)^2 + (-3)^2 = 1$$
$$(x - 2)^2 + 9 = 1$$
$$(x - 2)^2 = -8$$
Since this equation has no real solutions, there are no x-intercepts for the graph.

(b)   To find the y-intercepts, let $x = 0$:
$$(-2)^2 + (y-3)^2 = 4$$
$$4 + (y-3)^2 = 4$$
$$(y-3)^2 = 0$$
$$y = 3$$

To find the x-intercepts, let $y = 0$:
$$(x-2)^2 + (-3)^2 = 4$$
$$(x-2)^2 + 9 = 4$$
$$(x-2)^2 = -5$$
Since this equation has no real solutions, there are no x-intercepts for the graph.

(c)   To find the y-intercepts, let $x = 0$:
$$(-2)^2 + (y-3)^2 = 9$$
$$4 + (y-3)^2 = 9$$
$$(y-3)^2 = 5$$
$$y - 3 = \pm\sqrt{5}$$
$$y = 3 \pm \sqrt{5}$$

To find the x-intercepts, let $y = 0$:
$$(x-2)^2 + (-3)^2 = 9$$
$$(x-2)^2 + 9 = 9$$
$$(x-2)^2 = 0$$
$$x = 2$$

(d)   To find the y-intercepts, let $x = 0$:
$$(-2)^2 + (y-3)^2 = 16$$
$$4 + (y-3)^2 = 16$$
$$(y-3)^2 = 12$$
$$y - 3 = \pm\sqrt{12}$$
$$y - 3 = \pm 2\sqrt{3}$$
$$y = 3 \pm 2\sqrt{3}$$

To find the x-intercepts, let $y = 0$:
$$(x-2)^2 + (-3)^2 = 16$$
$$(x-2)^2 + 9 = 16$$
$$(x-2)^2 = 7$$
$$x - 2 = \pm\sqrt{7}$$
$$x = 2 \pm \sqrt{7}$$

19. To find the y-intercepts, let $x = 0$:
$$y = \sqrt{0-1} - 0 + 3 = \sqrt{-1} + 3$$
Since this is not a real number, there are no y-intercepts.
To find the x-intercepts, let $y = 0$:
$$0 = \sqrt{x-1} - x + 3$$
$$x - 3 = \sqrt{x-1}$$
$$(x-3)^2 = x - 1$$
$$x^2 - 6x + 9 = x - 1$$
$$x^2 - 7x + 10 = 0$$
$$(x-2)(x-5) = 0$$
$$x = 2, 5$$
Note that $x = 5$ satisfies the required equation, but $x = 2$ does not. So $x = 5$ is the only x-intercept.

21. To find the y-intercepts, let $x = 0$:
$$y = 11(0) - 2(0)^2 - (0)^3 = 0 - 0 - 0 = 0$$
To find the x-intercepts, let $y = 0$:
$$0 = 11x - 2x^2 - x^3$$
$$0 = x\left(11 - 2x - x^2\right)$$
So $x = 0$ is clearly one x-intercept. We find the other two using the quadratic formula:
$$x = \frac{2 \pm \sqrt{(-2)^2 - 4(11)(-1)}}{2(-1)} = \frac{2 \pm \sqrt{48}}{-2} = -1 \pm 2\sqrt{3} \approx 2.46, -4.46$$

23. To find the y-intercepts, let $x = 0$:
$$y = (0)^4 - 2(0)^2 - 3 = 0 - 0 - 3 = -3$$
To find the x-intercepts, let $y = 0$:
$$0 = x^4 - 2x^2 - 3$$
$$0 = \left(x^2 - 3\right)\left(x^2 + 1\right)$$
Setting the first factor equal to zero yields $x^2 - 3 = 0$, so $x^2 = 3$ and thus
$x = \pm\sqrt{3} \approx \pm 1.73$. Setting the second factor equal to zero yields $x^2 + 1 = 0$, so $x^2 = -1$, which has no real solutions.

25. (a) The reflection of $\overline{AB}$ about the x-axis is denoted by $\overline{A'B'}$:

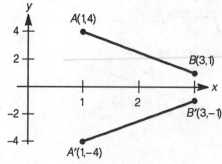

(b) The reflection of $\overline{AB}$ about the y-axis is denoted by $\overline{A'B'}$:

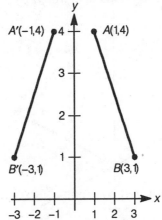

(c) The reflection of $\overline{AB}$ about the origin is denoted by $\overline{A'B'}$:

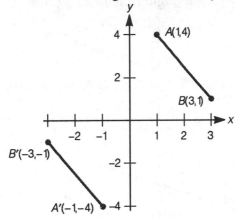

27. (a) The reflection of $\overline{AB}$ about the x-axis is denoted by $\overline{A'B'}$:

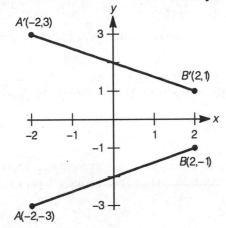

(b)   The reflection of $\overline{AB}$ about the y-axis is denoted by $\overline{A'B'}$:

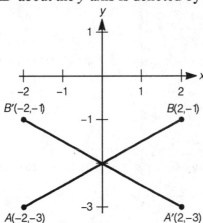

(c)   The reflection of $\overline{AB}$ about the origin is denoted by $\overline{A'B'}$:

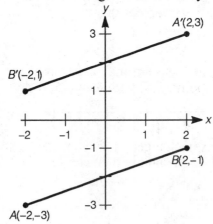

29.   (a)   To test the equation for symmetry about the x-axis, we replace the point (x, y) with the point $(x, -y)$:

$$3x^2 + (-y) = 16$$
$$3x^2 - y = 16$$

Since the equation is changed, it is not symmetric about the x-axis. To test the equation for symmetry about the y-axis, we replace the point (x, y) with the point $(-x, y)$:

$$3(-x)^2 + y = 16$$
$$3x^2 + y = 16$$

Since the equation remains unchanged, it is symmetric about the y-axis. To test the equation for symmetry about the origin, we replace the point (x, y) with the point $(-x, -y)$:

$$3(-x)^2 + (-y) = 16$$
$$3x^2 - y = 16$$

Since the equation is changed, it is not symmetric about the origin.

(b) To test the equation for symmetry about the x-axis, we replace the point (x, y) with the point $(x, -y)$:

$$3x + (-y)^2 = 16$$
$$3x + y^2 = 16$$

Since the equation remains unchanged, it is symmetric about the x-axis. To test the equation for symmetry about the y-axis, we replace the point (x, y) with the point $(-x, y)$:

$$3(-x) + y^2 = 16$$
$$-3x + y^2 = 16$$

Since the equation is changed, it is not symmetric about the y-axis. To test the equation for symmetry about the origin, we replace the point (x, y) with the point $(-x, -y)$:

$$3(-x) + (-y)^2 = 16$$
$$-3x + y^2 = 16$$

Since the equation is changed, it is not symmetric about the origin.

31. (a) To test for x-axis symmetry, we replace the point (x, y) with the point $(x, -y)$:

$$(-y) = x^4 - 3$$
$$y = -x^4 + 3$$

Since the equation is changed, there is no x-axis symmetry. To test for y-axis symmetry, we replace the point (x, y) with the point $(-x, y)$:

$$y = (-x)^4 - 3$$
$$y = x^4 - 3$$

Since the equation remains unchanged, it is symmetric about the y-axis. To test for origin symmetry, we replace the point (x, y) with the point $(-x, -y)$:

$$(-y) = (-x)^4 - 3$$
$$-y = x^4 - 3$$
$$y = -x^4 + 3$$

Since the equation is changed, there is no origin symmetry.

(b) To test for x-axis symmetry, we replace the point (x, y) with the point $(x, -y)$:

$$(-y) = x^3 - 3$$
$$y = -x^3 + 3$$

Since the equation is changed, there is no x-axis symmetry. To test for y-axis symmetry, we replace the point (x, y) with the point $(-x, y)$:

$$y = (-x)^3 - 3$$
$$y = -x^3 - 3$$

Since the equation is changed, there is no y-axis symmetry. To test for origin symmetry, we replace the point (x, y) with the point $(-x, -y)$:

$$(-y) = (-x)^3 - 3$$
$$-y = -x^3 - 3$$
$$y = x^3 + 3$$

Since the equation is changed, there is no origin symmetry.

(c)   To test for x-axis symmetry, we replace the point (x,y) with the point $(x,-y)$:
$$(-y)^2 = x^3 - 3$$
$$y^2 = x^3 - 3$$
Since the equation remains unchanged, it is symmetric about the x-axis. To test for y-axis symmetry, we replace the point (x,y) with the point $(-x,y)$:
$$y^2 = (-x)^3 - 3$$
$$y^2 = -x^3 - 3$$
Since the equation is changed, there is no y-axis symmetry. To test for origin symmetry, we replace the point (x,y) with the point $(-x,-y)$:
$$(-y)^2 = (-x)^3 - 3$$
$$y^2 = -x^3 - 3$$
Since the equation is changed, there is no origin symmetry.

(d)   To test for x-axis symmetry, we replace the point (x,y) with the point $(x,-y)$:
$$(-y)^3 = x$$
$$-y^3 = x$$
$$y^3 = -x$$
Since the equation is changed, there is no x-axis symmetry. To test for y-axis symmetry, we replace the point (x,y) with the point $(-x,y)$:
$$y^3 = -x$$
Since the equation is changed, there is no y-axis symmetry. To test for origin symmetry, we replace the point (x,y) with the point $(-x,-y)$:
$$(-y)^3 = -x$$
$$-y^3 = -x$$
$$y^3 = x$$
Since the equation remains unchanged, it is symmetric about the origin.

33. To test for x-axis symmetry, we replace the point (x,y) with the point $(x,-y)$:
$$x^2 + (-y)^2 = 16$$
$$x^2 + y^2 = 16$$
Since the equation remains unchanged, it is symmetric about the x-axis. To test for y-axis symmetry, we replace the point (x,y) with the point $(-x,y)$:
$$(-x)^2 + y^2 = 16$$
$$x^2 + y^2 = 16$$
Since the equation remains unchanged, it is symmetric about the y-axis. To test for origin symmetry, we replace the point (x,y) with the point $(-x,-y)$:
$$(-x)^2 + (-y)^2 = 16$$
$$x^2 + y^2 = 16$$
Since the equation remains unchanged, it is symmetric about the origin.

35. To test for x-axis symmetry, we replace the point (x, y) with the point $(x, -y)$:
$$(-y) = x^2 + x^3$$
$$y = -x^2 - x^3$$
Since the equation is changed, there is no x-axis symmetry. To test for y-axis symmetry, we replace the point (x, y) with the point $(-x, y)$:
$$y = (-x)^2 + (-x)^3$$
$$y = x^2 - x^3$$
Since the equation is changed, there is no y-axis symmetry. To test for origin symmetry, we replace the point (x, y) with the point $(-x, -y)$:
$$(-y) = (-x)^2 + (-x)^3$$
$$-y = x^2 - x^3$$
$$y = -x^2 + x^3$$
Since the equation is changed, there is no origin symmetry.

37. To test for x-axis symmetry, we replace the point (x, y) with the point $(x, -y)$:
$$x + (-y) = 1$$
$$x - y = 1$$
Since the equation is changed, there is no x-axis symmetry. To test for y-axis symmetry, we replace the point (x, y) with the point $(-x, y)$:
$$(-x) + y = 1$$
$$-x + y = 1$$
Since the equation is changed, there is no y-axis symmetry. To test for origin symmetry, we replace the point (x, y) with the point $(-x, -y)$:
$$(-x) + (-y) = 1$$
$$-x - y = 1$$
$$x + y = -1$$
Since the equation is changed, there is no origin symmetry.

39. The x- and y-intercepts are both 0. The graph is symmetric about the y-axis.

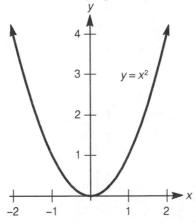

41. There are no x- or y-intercepts. The graph is symmetric about the origin.

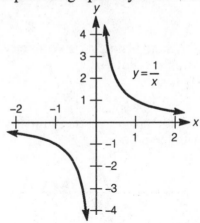

43. The x- and y-intercepts are both 0. The graph is symmetric about the y-axis.

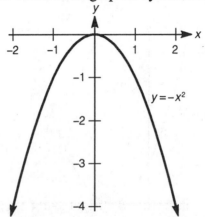

45. There are no x- or y-intercepts. The graph is symmetric about the origin.

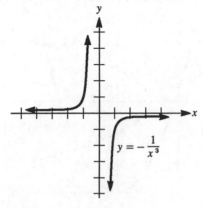

47. The x- and y-intercepts are both 0. The graph is symmetric about the y-axis.

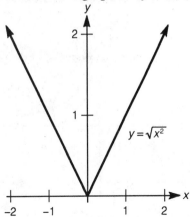

49. The x- and y-intercepts are both 1. The graph does not possess any of the three types of symmetry.

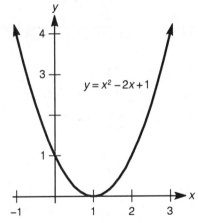

51. The x-intercept is 2 and there is no y-intercept. The graph is symmetric about the x-axis.

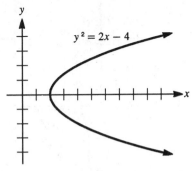

53. To find the x-intercepts, we use the quadratic formula:
$$x = \frac{-1 \pm \sqrt{(1)^2 - 4(2)(-4)}}{2(2)} = \frac{-1 \pm \sqrt{33}}{4}$$
The y-intercept is -4. The graph does not possess any of the three types of symmetry.

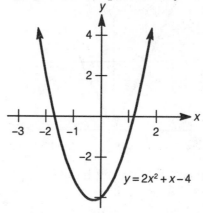

55. (a) The x-intercept is 2 and the y-intercept is -6. The graph does not possess any of the three types of symmetry.

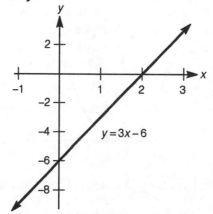

(b) The x-intercept is 2 and the y-intercept is 6. The graph does not possess any of the three types of symmetry.

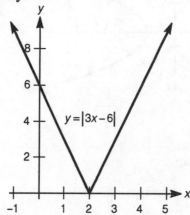

57. (a) Tracing up to the curve from $x = 2$, we get $y = \sqrt{2} \approx 1.4$.

(b) Tracing up to the curve from $x = 3$, we get $y = \sqrt{3} \approx 1.7$.

(c) Since $\sqrt{ab} = \sqrt{a} \bullet \sqrt{b}$, then $\sqrt{6} = \sqrt{2 \bullet 3} = \sqrt{2} \bullet \sqrt{3} \approx (1.4)(1.7) \approx 2.4$.

1.4 The Definition of a Function

1. (a) Since any value of x can be used to evaluate the expression, the domain is all real numbers.

(b) We must make sure that the quantity under the radical is non-negative, so $4x - 20 \geq 0$, and thus $4x \geq 20$ so $x \geq 5$. So the domain is $[5, \infty)$.

3. (a) We must make sure that the quantity under the radical is non-negative, so $x \geq 0$. So the domain is $[0, \infty)$.

(b) We must make sure that the quantity under the radical is non-negative, so $|x| \geq 0$. Since this is true for all values of x, the domain is all real numbers.

5. (a) We must make sure that the denominator is non-zero:
$$x^2 + 8x - 9 \neq 0$$
$$(x + 9)(x - 1) \neq 0$$
$$x \neq -9, 1$$
So the domain is all real numbers except -9 and 1.

(b) We must make sure that the denominator is non-zero. Using the quadratic formula to find the zeros for the denominator:
$$x = \frac{-7 \pm \sqrt{(7)^2 - 4(1)(-9)}}{2(1)} = \frac{-7 \pm \sqrt{85}}{2}$$

So the domain is all real numbers except $\dfrac{-7 \pm \sqrt{85}}{2}$.

7.   (a)   To find the range, we first solve for x:
$$y = \frac{x+3}{x-5}$$
$$y(x-5) = x+3$$
$$yx - 5y = x+3$$
$$yx - x = 5y+3$$
$$x(y-1) = 5y+3$$
$$x = \frac{5y+3}{y-1}$$

So the range is all real numbers except 1.

(b)   To find the range, we first solve for x:
$$y = \frac{x-5}{x+3}$$
$$y(x+3) = x-5$$
$$yx + 3y = x-5$$
$$yx - x = -3y-5$$
$$x(y-1) = -3y-5$$
$$x = \frac{-3y-5}{\cdot\, y-1}$$
$$x = \frac{3y+5}{1-y}$$

So the range is all real numbers except 1.

9.   (a)   To find the range, we first solve for x:
$$y = x^2 + 4$$
$$y - 4 = x^2$$
$$x = \pm\sqrt{y-4}$$

In order for the radical to be defined, we must have $y - 4 \geq 0$, so $y \geq 4$. So the range is $[4, \infty)$. Another approach is to note that since x^2 will range from 0 to ∞, then $x^2 + 4$ will range from 4 to ∞. So the range is $[4, \infty)$.

(b)   To find the range, we first solve for x:
$$y = x^3 + 4$$
$$y - 4 = x^3$$
$$x = \sqrt[3]{y-4}$$

Since the cube root is defined for all real numbers, the range is all real numbers. Another approach is to note that since x^3 will range from $-\infty$ to ∞, then so will $x^3 + 4$. So the range is all real numbers.

11.   Rules f, g, F and H are functions. Rule h is not a function since $h(x) = 1$ and $h(x) = 2$, which violates the definition of a function. Rule G is not a function, since $G(y)$ has not been assigned a value.

13. (a) The range of f is $\{1,2,3\}$, the range of g is $\{2,3\}$, the range of F is $\{1\}$, and the range of H is $\{1,2\}$.

(b) The range of g is $\{i,j\}$, the range of F is $\{i,j\}$, and the range of G is $\{k\}$.

15. (a) $y = (x-3)^2$
(b) $y = x^2 - 3$
(c) $y = (3x)^2$
(d) $y = 3x^2$

17. (a) We compute $f(1)$:
$$f(1) = (1)^2 - 3(1) + 1 = 1 - 3 + 1 = -1$$

(b) We compute $f(0)$:
$$f(0) = (0)^2 - 3(0) + 1 = 0 - 0 + 1 = 1$$

(c) We compute $f(-1)$:
$$f(-1) = (-1)^2 - 3(-1) + 1 = 1 + 3 + 1 = 5$$

(d) We compute $f\left(\frac{3}{2}\right)$:
$$f\left(\tfrac{3}{2}\right) = \left(\tfrac{3}{2}\right)^2 - 3\left(\tfrac{3}{2}\right) + 1 = \tfrac{9}{4} - \tfrac{9}{2} + 1 = -\tfrac{5}{4}$$

(e) We compute $f(z)$:
$$f(z) = (z)^2 - 3(z) + 1 = z^2 - 3z + 1$$

(f) We compute $f(x+1)$:
$$\begin{aligned} f(x+1) &= (x+1)^2 - 3(x+1) + 1 \\ &= x^2 + 2x + 1 - 3x - 3 + 1 \\ &= x^2 - x - 1 \end{aligned}$$

(g) We compute $f(a+1)$:
$$\begin{aligned} f(a+1) &= (a+1)^2 - 3(a+1) + 1 \\ &= a^2 + 2a + 1 - 3a - 3 + 1 \\ &= a^2 - a - 1 \end{aligned}$$

(h) We compute $f(-x)$:
$$f(-x) = (-x)^2 - 3(-x) + 1 = x^2 + 3x + 1$$

(i) Using our result from (a):
$$|f(1)| = |-1| = 1$$

19. (a) We compute $f(2)$:
$$f(2) = 4 - 3(2) = 4 - 6 = -2$$

(b) We compute $f(-3)$:
$$f(-3) = 4 - 3(-3) = 4 + 9 = 13$$

(c) We compute $f(2) + f(-3)$ using our answers from (a) and (b):
$$f(2) + f(-3) = -2 + 13 = 11$$

(d) We compute $f(2 + 3)$:
$$f(2 + 3) = f(5) = 4 - 3(5) = 4 - 15 = -11$$

(e) We compute $f(2x)$:
$$f(2x) = 4 - 3(2x) = 4 - 6x$$

(f) We compute $2f(x)$:
$$2f(x) = 2(4 - 3x) = 8 - 6x$$

(g) We compute $f(x^2)$:
$$f\left(x^2\right) = 4 - 3\left(x^2\right) = 4 - 3x^2$$

(h) We compute $[f(x)]^2$:
$$[f(x)]^2 = (4 - 3x)^2 = 16 - 24x + 9x^2$$

(i) We compute $\left[f\left(x^2\right)\right]^2$ using our answer from (g):
$$\left[f\left(x^2\right)\right]^2 = \left(4 - 3x^2\right)^2 = 16 - 24x^2 + 9x^4$$

21. (a) $H(0)$ is larger:
$$H(0) = 1 - 0 + (0)^2 - (0)^3 = 1 - 0 + 0 - 0 = 1$$
$$H(1) = 1 - 1 + (1)^2 - (1)^3 = 1 - 1 + 1 - 1 = 0$$

(b) We compute $H\left(\frac{1}{2}\right)$:
$$H\left(\tfrac{1}{2}\right) = 1 - \tfrac{1}{2} + \left(\tfrac{1}{2}\right)^2 - \left(\tfrac{1}{2}\right)^3 = 1 - \tfrac{1}{2} + \tfrac{1}{4} - \tfrac{1}{8} = \tfrac{5}{8}$$

Note that $H\left(\tfrac{1}{2}\right) + H\left(\tfrac{1}{2}\right) = \tfrac{5}{8} + \tfrac{5}{8} = \tfrac{5}{4}$, which does not equal $H(1) = 0$.

23. (a)   For the domain, we must exclude those values which make $x - 2 = 0$, or $x = 2$. So the domain is all real numbers except 2. For the range, we solve for x:

$$y = \frac{2x-1}{x-2}$$
$$y(x-2) = 2x-1$$
$$yx - 2y = 2x - 1$$
$$yx - 2x = 2y - 1$$
$$x(y-2) = 2y - 1$$
$$x = \frac{2y-1}{y-2}$$

Since the denominator cannot be zero, $y = 2$ is excluded. So the range is all real numbers except 2.

(b)   We compute $R(0)$:

$$R(0) = \frac{2(0)-1}{0-2} = \frac{-1}{-2} = \frac{1}{2}$$

(c)   We compute $R\left(\frac{1}{2}\right)$:

$$R\left(\tfrac{1}{2}\right) = \frac{2\left(\frac{1}{2}\right)-1}{\frac{1}{2}-2} = \frac{1-1}{-\frac{3}{2}} = 0$$

(d)   We compute $R(-1)$:

$$R(-1) = \frac{2(-1)-1}{-1-2} = \frac{-2-1}{-3} = \frac{-3}{-3} = 1$$

(e)   We compute $R\left(x^2\right)$:

$$R\left(x^2\right) = \frac{2\left(x^2\right)-1}{x^2-2} = \frac{2x^2-1}{x^2-2}$$

(f)   We compute $R\left(\frac{1}{x}\right)$:

$$R\left(\tfrac{1}{x}\right) = \frac{2\left(\frac{1}{x}\right)-1}{\frac{1}{x}-2} = \frac{\frac{2}{x}-1}{\frac{1}{x}-2} = \frac{2-x}{1-2x}$$

(g)   We compute $R(a)$:

$$R(a) = \frac{2(a)-1}{a-2} = \frac{2a-1}{a-2}$$

(h)   We compute $R(x-1)$:

$$R(x-1) = \frac{2(x-1)-1}{(x-1)-2} = \frac{2x-2-1}{x-3} = \frac{2x-3}{x-3}$$

25. (a) We compute $A(1)$ and $A(0)$:
$$A(1) = 1000\left(1+\tfrac{0.12}{4}\right)^{4(1)} \approx \$1125.51$$
$$A(0) = 1000\left(1+\tfrac{0.12}{4}\right)^{4(0)} = \$1000.00$$
So $A(1) - A(0) \approx 1125.51 - 1000 = \$125.51$.

(b) We compute $A(10)$ and $A(9)$:
$$A(10) = 1000\left(1+\tfrac{0.12}{4}\right)^{4(10)} \approx \$3262.04$$
$$A(9) = 1000\left(1+\tfrac{0.12}{4}\right)^{4(9)} \approx \$2898.28$$
So $A(10) - A(9) \approx 3262.04 - 2898.28 = \$363.76$.

27. (a) We compute $f(a)$, $f(2a)$ and $f(3a)$:
$$f(a) = \frac{a-a}{a+a} = \frac{0}{2a} = 0$$
$$f(2a) = \frac{2a-a}{2a+a} = \frac{a}{3a} = \frac{1}{3}$$
$$f(3a) = \frac{3a-a}{3a+a} = \frac{2a}{4a} = \frac{1}{2}$$
Since $\frac{1}{2} \neq 0 + \frac{1}{3}$, then $f(3a) \neq f(a) + f(2a)$.

(b) We compute $f(5a)$:
$$f(5a) = \frac{5a-a}{5a+a} = \frac{4a}{6a} = \frac{2}{3}$$
Since $f(2a) = \frac{1}{3}$, then $f(5a) = 2f(2a)$.

29. For the rule F, each input has exactly one output. That is, every person has a mother (so each value of x has been assigned), and no person has two (natural) mothers (so no value of x is assigned twice). The rule G, however, fails on both accounts. Not every person has an aunt (in the case that both parents are only children), and so not every value of x has been assigned. Furthermore, two aunts could be assigned to a person, which would violate the definition of a function.

1.5 The Graph of a Function

1. (a) positive
   (b) $f(-2) = 4$; $f(1) = 1$; $f(2) = 2$; $f(3) = 0$
   (c) $f(2)$, since $f(2) > 0$ and $f(4) < 0$
   (d) $f(4) - f(1) = -2 - 1 = -3$
   (e) $|f(4) - f(1)| = |-3| = 3$
   (f) domain $= [-2, 4]$; range $= [-2, 4]$

3. (a) $f(-2) = 0$ and $g(-2) = 1$, so $g(-2)$ is larger.
   (b) $f(0) - g(0) = 2 - (-3) = 2 + 3 = 5$
   (c) We compute the three values:
$$f(1) - g(1) = 1 - (-1) = 2$$
$$f(2) - g(2) = 1 - 0 = 1$$
$$f(3) - g(3) = 4 - 1 = 3$$
   So $f(2) - g(2)$ is the smallest.
   (d) Since $f(1) = 1$, we look for where $g(x) = 1$. This occurs at $x = -2$ or $x = 3$.
   (e) Since $(3, 4)$ is a point on the graph of f, then 4 is in the range of f.

5. The range of f is $[0, 4]$ and the range of g is $[-3, 3]$.

7. The completed table is given below:

Function	$|x|$	x^2	x^3
Turning Point	$(0,0)$	$(0,0)$	none
Maximum Value	none	none	none
Minimum Value	0	0	none
Interval(s) where Increasing	$[0,\infty)$	$[0,\infty)$	$(-\infty,\infty)$
Interval(s) where Decreasing	$(-\infty,0]$	$(-\infty,0]$	none

9. Since any vertical line drawn intersects the graph in at most one point, this is the graph of a function.

11. Since a vertical line can be drawn which intersects the graph in two points, this cannot be the graph of a function.

13. Since any vertical line drawn intersects the graph in at most one point, this is the graph of a function.

15. If a vertical line is drawn which hits the ends of the two "pieces", note that it will intersect the graph in two points. This cannot be the graph of a function.

17. The slope is given by:
$$m = \frac{f(4) - f(3)}{4 - 3} = \frac{4^2 - 3^2}{1} = \frac{16 - 9}{1} = 7$$

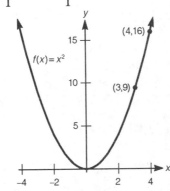

19. We find the two slopes:
$$m_1 = \frac{T(4)-T(1)}{4-1} = \frac{\sqrt{4}-\sqrt{1}}{3} = \frac{2-1}{3} = \frac{1}{3}$$
$$m_2 = \frac{T(9)-T(4)}{9-4} = \frac{\sqrt{9}-\sqrt{4}}{5} = \frac{3-2}{5} = \frac{1}{5}$$

Since $m_1 > m_2$, then the line between $(1, T(1))$ and $(4, T(4))$ has the larger slope.

21. (a) We draw the graph of $k(x)$ with domain $[0, 1]$:

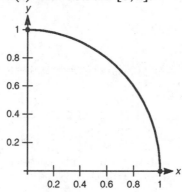

(b) This graph will be the same as in (a), except that the two endpoints $(0, 1)$ and $(1, 0)$ will be excluded. We draw the graph of $m(x)$ with domain $(0, 1)$:

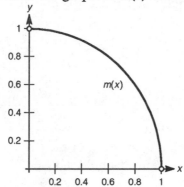

(c) Because the domain consists of just five values, this graph will consist of five points. We draw the graph of $n(x)$ with domain $\{0, 1, 2, 3, 4\}$:

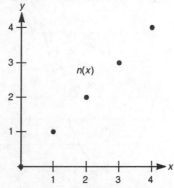

(d) We draw the graph of $z(x)$ with the required domain:

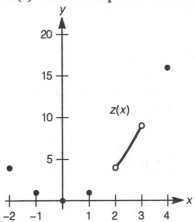

23. (a) The range is $[-1, 1]$.
     (b) The maximum value is 1 (occurring at $x = 1$).
     (c) The minimum value is -1 (occurring at $x = 3$).
     (d) The function is increasing on the intervals $[0, 1]$ and $[3, 4]$.
     (e) The function is decreasing on the interval $[1, 3]$.

25. (a) The range is $[-3, 0]$.
     (b) The maximum value is 0 (occurring at $x = 0$ and $x = 4$).
     (c) The minimum value is -3 (occurring at $x = 2$).
     (d) The function is increasing on the interval $[2, 4]$.
     (e) The function is decreasing on the interval $[0, 2]$.

27. (a) We complete the table:

t	0	0.25	0.5	0.75	1	1.25	1.5	1.75	2	2.25	2.5	2.75	3
$S(t)$	0	0.25	0.5	0.75	1	1	1	1	1	1	1	1	1

Now graph the function $S(t)$ on the interval $0 \le t \le 3$:

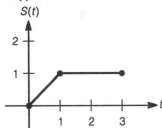

(b) We complete the table:

t	3	3.25	3.5	3.75	4	4.25	4.5	4.75	5
$S(t)$	1	0.75	0.5	0.25	0	−0.25	−0.5	−0.75	−1

Now graph the function $S(t)$ on the interval $3 \le t \le 5$:

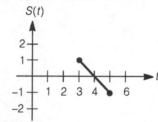

(c)   We draw the graph:

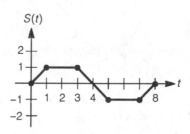

(d)   We draw the graph:

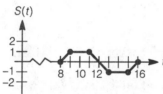

The graph is identical to that of part (c) except for the t-values. This is an example of a periodic function, that is, one which repeats its values over a specified period. In this example, the period is $t = 8$.

1.6 Techniques in Graphing

1.   (a) C   (b) F   (c) I   (d) A   (e) J   (f) K
      (g) D   (h) B   (i) E   (j) H   (k) G

3.   This graph will be that of $y = x^3$ translated down 3 units:

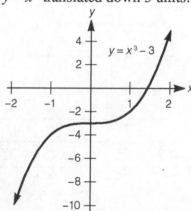

5. This graph will be that of $y = x^2$ translated to the left 4 units:

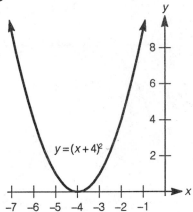

$y = (x + 4)^2$

7. This graph will be that of $y = x^2$ translated to the right 4 units:

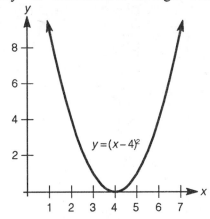

$y = (x - 4)^2$

9. This graph will be that of $y = x^2$ reflected across the x-axis:

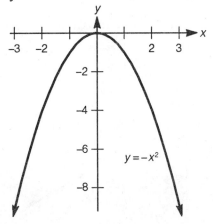

$y = -x^2$

11. This graph will be that of $y = x^2$ translated to the right 3 units, then reflected across the x-axis:

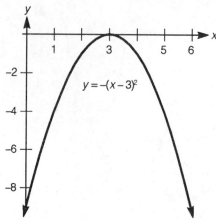

$y = -(x-3)^2$

13. This graph will be that of $y = \sqrt{x}$ translated to the right 3 units:

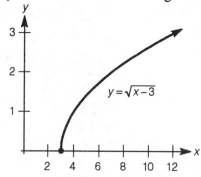

$y = \sqrt{x-3}$

15. This graph will be that of $y = \sqrt{x}$ translated to the left 1 unit, then reflected across the x-axis:

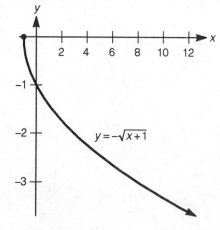

$y = -\sqrt{x+1}$

17. This graph will be that of $y = \dfrac{1}{x}$ translated to the left 2 units, then translated up 2 units:

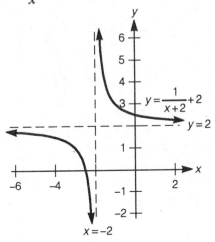

$y = \dfrac{1}{x+2} + 2$

$y = 2$

$x = -2$

19. This graph will be that of $y = x^3$ translated to the right 2 units:

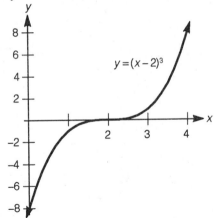

$y = (x-2)^3$

21. This graph will be that of $y = x^3$ reflected across the x-axis, then translated up 4 units:

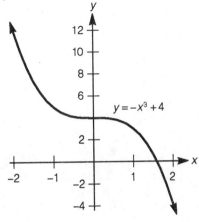

$y = -x^3 + 4$

23. **(a)**   This graph will be that of $y = |x|$ translated to the left 4 units:

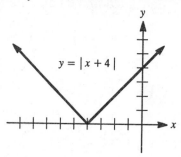

(b)   This graph will be that of $y = |x|$ reflected across the y-axis, then translated to the right 4 units. Note that the reflection has no effect on the graph, since $y = |x|$ is symmetric about the y-axis:

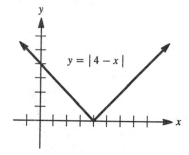

(c)   This graph will be that of $y = |x|$ reflected across the y-axis, translated to the right 4 units, reflected across the x-axis, then translated up 1 unit:

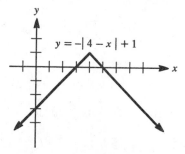

25.   This is $f(x) = |x|$ translated to the right 5 units:

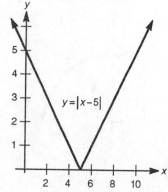

27. This is $f(x) = |x|$ reflected across the y-axis, then translated to the right 5 units. Note that the reflection has no effect on the graph, since $y = |x|$ is symmetric about the y-axis:

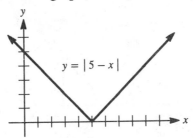

29. This is $f(x) = |x|$ translated to the right 5 units, reflected across the x-axis, then translated up 1 unit:

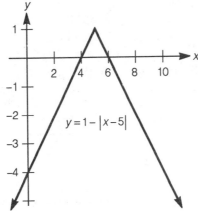

31. This is $F(x) = \dfrac{1}{x}$ translated to the left 3 units:

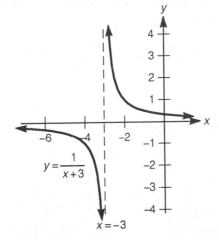

33. This is $F(x) = \dfrac{1}{x}$ translated to the left 3 units, then reflected across the x-axis:

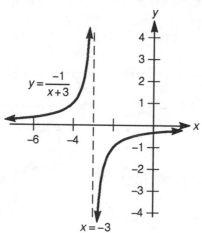

$$y = \frac{-1}{x+3}$$

$$x = -3$$

35. This is $g(x) = \sqrt{1-x^2}$ translated to the right 2 units:

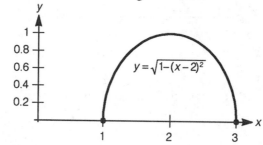

$$y = \sqrt{1-(x-2)^2}$$

37. This is $g(x) = \sqrt{1-x^2}$ translated to the right 2 units, reflected across the x-axis, then translated up 1 unit:

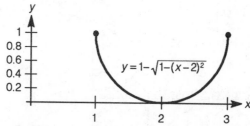

$$y = 1 - \sqrt{1-(x-2)^2}$$

39. This is $g(x) = \sqrt{1-x^2}$ reflected across the y-axis, then translated to the right 2 units:

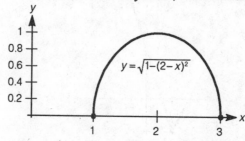

$$y = \sqrt{1-(2-x)^2}$$

41. (a) This is $y = f(x)$ reflected across the x-axis:

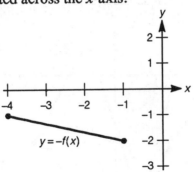

(b) This is $y = f(x)$ reflected across the y-axis:

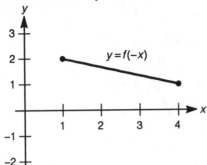

(c) This is $y = f(x)$ reflected across the y-axis, then reflected across the x-axis:

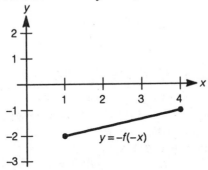

43. (a) This is $y = g(x)$ reflected across the y-axis:

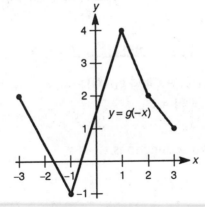

(b)   This is $y = g(x)$ reflected across the x-axis:

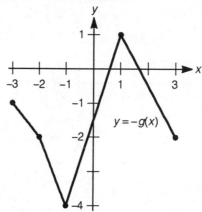

(c)   This is $y = g(x)$ reflected across the y-axis, then reflected across the x-axis:

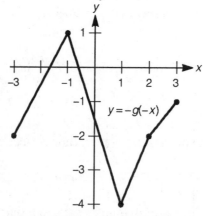

45.   (a)   The point is $(a + 3, b)$, since $f(a + 3-3) = f(a) = b$.
       (b)   The point is $(a, b - 3)$, since $f(a) - 3 = b - 3$.
       (c)   The point is $(a + 3, b - 3)$, since $f(a + 3 - 3) - 3 = f(a) - 3 = b - 3$.
       (d)   The point is $(a, -b)$, since $-f(a) = -b$.
       (e)   The point is $(-a, b)$, since $f[-(-a)] = f(a) = b$.
       (f)   The point is $(-a, -b)$, since $-f[-(-a)] = -f(a) = -b$.
       (g)   The point is $(-a + 3, b)$, since $f[3 - (-a + 3)] = f(3 + a - 3) = f(a) = b$.
       (h)   The point is $(-a + 3, -b + 1)$, since $-f[3 - (-a + 3)] + 1 = -f(a) + 1 = -b + 1$.

1.7 Inverse Functions

1.   (a)   We must show that $f[g(x)] = x$ and $g[f(x)] = x$:
$$f[g(x)] = f\left(\tfrac{x}{3}\right) = 3\left(\tfrac{x}{3}\right) = x$$
$$g[f(x)] = g(3x) = \tfrac{3x}{3} = x$$
So $f(x)$ and $g(x)$ are inverse functions.

(b) We must show that $f[g(x)] = x$ and $g[f(x)] = x$:

$$f[g(x)] = f\left(\tfrac{x+1}{4}\right) = 4\left(\tfrac{x+1}{4}\right) - 1 = x + 1 - 1 = x$$

$$g[f(x)] = g(4x - 1) = \tfrac{(4x-1)+1}{4} = \tfrac{4x}{4} = x$$

So $f(x)$ and $g(x)$ are inverse functions.

(c) We must show that $g[h(x)] = x$ and $h[g(x)] = x$:

$$g[h(x)] = g\left(x^2\right) = \sqrt{x^2} = x, \quad \text{since } x \geq 0$$

$$h[g(x)] = h(\sqrt{x}) = \left(\sqrt{x}\right)^2 = x$$

So $g(x)$ and $h(x)$ are inverse functions.

3. (a) Since $f[f^{-1}(x)] = x$ for all x in the domain of f^{-1}, then $f[f^{-1}(4)] = 4$ since the domain of f^{-1} is $(-\infty, \infty)$.

(b) Since $f^{-1}[f(x)] = x$ as long as $f(x)$ is in the domain of f^{-1}, then $f^{-1}[f(-1)] = -1$ since the domain of f^{-1} is $(-\infty, \infty)$.

(c) Since $f[f^{-1}(x)] = x$ for all x in the domain of f^{-1}, then $f[f^{-1}(\sqrt{2})] = \sqrt{2}$ since the domain of f^{-1} is $(-\infty, \infty)$.

(d) Since $f[f^{-1}(x)] = x$ for all x in the domain of f^{-1}, then $f[f^{-1}(t + 1)] = t + 1$ since the domain of f^{-1} is $(-\infty, \infty)$.

(e) We first compute $f(0)$:

$$f(0) = (0)^3 + 2(0) + 1 = 0 + 0 + 1 = 1$$

Since the inverse function interchanges inputs and outputs, and $f(0) = 1$, then $f^{-1}(1) = 0$.

(f) We first compute $f(-1)$:

$$f(-1) = (-1)^3 + 2(-1) + 1 = -1 - 2 + 1 = -2$$

Since the inverse function interchanges inputs and outputs, and $f(-1) = -2$, then $f^{-1}(-2) = -1$.

5. (a) Let $y = 3x - 1$. We switch the roles of x and y and solve the resulting equation for y:

$$x = 3y - 1$$
$$3y = x + 1$$
$$y = \tfrac{x+1}{3}$$

So the inverse is $f^{-1}(x) = \tfrac{x+1}{3}$.

(b)   We verify that $f[f^{-1}(x)] = x$ and $f^{-1}[f(x)] = x$:

$$f[f^{-1}(x)] = f\left(\tfrac{x+1}{3}\right) = 3\left(\tfrac{x+1}{3}\right) - 1 = x + 1 - 1 = x$$

$$f^{-1}[f(x)] = f^{-1}(3x - 1) = \frac{(3x - 1) + 1}{3} = \frac{3x}{3} = x$$

(c)   The graphs of each line are given below. Note the symmetry of the two lines about the line $y = x$:

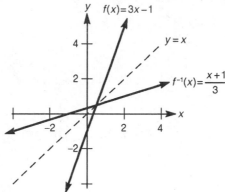

7.   (a)   Let $y = \sqrt{x - 1}$. We switch the roles of x and y and solve the resulting equation for y:

$$x = \sqrt{y - 1}$$
$$x^2 = y - 1$$
$$y = x^2 + 1$$

So the inverse is $f^{-1}(x) = x^2 + 1$ for $x \ge 0$.

(b)   We verify that $f[f^{-1}(x)] = x$ and $f^{-1}[f(x)] = x$:

$$f[f^{-1}(x)] = f(x^2 + 1) = \sqrt{(x^2 + 1) - 1} = \sqrt{x^2} = x, \quad \text{since } x \ge 0$$

$$f^{-1}[f(x)] = f^{-1}(\sqrt{x - 1}) = (\sqrt{x - 1})^2 + 1 = x - 1 + 1 = x$$

(c)   The graphs of each curve are given below. Note the symmetry of the curves about the line $y = x$:

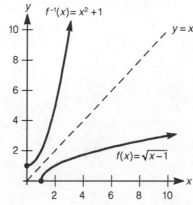

9.   (a)   Since the denominator is 0 when $x = 3$, the domain of f is the set of all real numbers except 3. To find the range of f, we first solve for x:

$$y = \frac{x+2}{x-3}$$
$$y(x-3) = x+2$$
$$yx - 3y = x+2$$
$$yx - x = 3y+2$$
$$x(y-1) = 3y+2$$
$$x = \frac{3y+2}{y-1}$$

Since the denominator is 0 when $y = 1$, the range of f is the set of all real numbers except 1.

(b)   Let $y = \frac{x+2}{x-3}$. We switch the roles of x and y and solve the resulting equation for y:

$$x = \frac{y+2}{y-3}$$
$$x(y-3) = y+2$$
$$xy - 3x = y+2$$
$$xy - y = 3x+2$$
$$y(x-1) = 3x+2$$
$$y = \frac{3x+2}{x-1}$$

So the inverse is $f^{-1}(x) = \frac{3x+2}{x-1}$.

(c)   Since the denominator is 0 when $x = 1$, the domain of f^{-1} is the set of all real numbers except 1. To find the range of f^{-1}, we first solve for x:

$$y = \frac{3x+2}{x-1}$$
$$y(x-1) = 3x+2$$
$$yx - y = 3x+2$$
$$yx - 3x = y+2$$
$$x(y-3) = y+2$$
$$x = \frac{y+2}{y-3}$$

Since the denominator is 0 when $y = 3$, the range of f^{-1} is the set of all real numbers except 3. We observe that the domain of f is equal to the range of f^{-1}, and that the range of f is equal to the domain of f^{-1}.

11. Let $y = 2x^3 + 1$. We switch the roles of x and y and solve the resulting equation for y:

$$x = 2y^3 + 1$$
$$2y^3 = x - 1$$
$$y^3 = \frac{x-1}{2}$$
$$y = \sqrt[3]{\frac{x-1}{2}}$$

So the inverse is $f^{-1}(x) = \sqrt[3]{\frac{x-1}{2}}$.

13. (a) Let $y = (x-3)^3 - 1$. We switch the roles of x and y and solve the resulting equation for y:

$$x = (y-3)^3 - 1$$
$$(y-3)^3 = x + 1$$
$$y - 3 = \sqrt[3]{x+1}$$
$$y = \sqrt[3]{x+1} + 3$$

So the inverse is $f^{-1}(x) = \sqrt[3]{x+1} + 3$.

(b) Note that $f(x)$ and $f^{-1}(x)$ are symmetric about the line $y = x$:

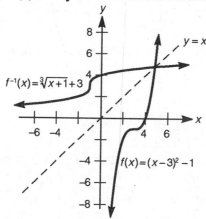

15. (a) The graph of $y = g^{-1}(x)$ is a reflection of $g(x)$ across the line $y = x$:

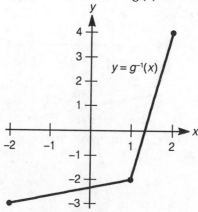

(b) The graph of $y = g^{-1}(x) - 1$ is a displacement of $g^{-1}(x)$ down 1 unit:

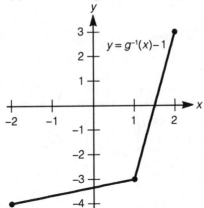

(c) The graph of $y = g^{-1}(x - 1)$ is a displacement of $g^{-1}(x)$ to the right 1 unit:

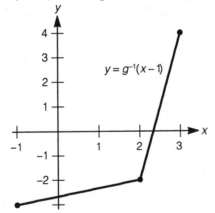

(d) The graph of $y = g^{-1}(-x)$ is a reflection of $g^{-1}(x)$ across the y-axis:

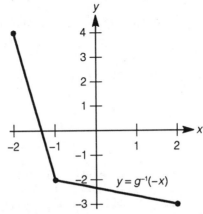

(e)   The graph of $y = -g^{-1}(x)$ is a reflection of $g^{-1}(x)$ across the x-axis:

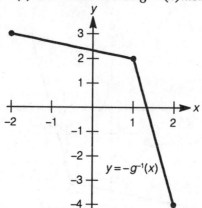

(f)   The graph of $y = -g^{-1}(-x)$ is a reflection of $g^{-1}(x)$ across the y-axis and across the x-axis:

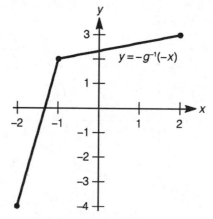

17. The graph of $y = x^2 + 1$ fails the horizontal line test, so it is not one-to-one:

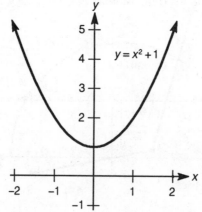

19. The graph of $y = \dfrac{1}{x}$ passes the horizontal line test, so it is one-to-one:

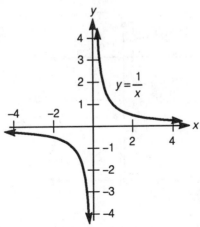

21. The graph of $y = x^3$ passes the horizontal line test, so it is one-to-one:

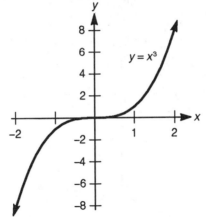

23. The graph of $y = \sqrt{1 - x^2}$ fails the horizontal line test, so it is not one-to-one:

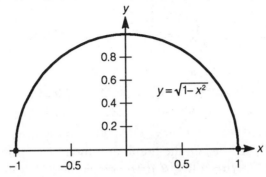

25. The graph of $g(x) = 5$ fails the horizontal line test (it *is* a horizontal line), so it is not one-to-one:

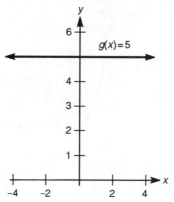

27. We find $f[f(x)]$:

$$f[f(x)] = f\left(\tfrac{3x-2}{5x-3}\right)$$

$$= \frac{3\left(\tfrac{3x-2}{5x-3}\right) - 2}{5\left(\tfrac{3x-2}{5x-3}\right) - 3}$$

$$= \frac{3(3x-2) - 2(5x-3)}{5(3x-2) - 3(5x-3)}$$

$$= \frac{9x - 6 - 10x + 6}{15x - 10 - 15x + 9}$$

$$= \frac{-x}{-1}$$

$$= x$$

Thus $f^{-1}(x) = f(x)$.

29. (a) Let $y = \sqrt{x}$. We switch the roles of x and y and solve the resulting equation for y:

$$x = \sqrt{y}$$
$$x^2 = y$$

So the inverse function is $f^{-1}(x) = x^2$. Since $x = \sqrt{y}$, then $x \geq 0$. So the domain of $f^{-1}(x)$ is $[0, \infty)$.

   (b) (i)   Since $2 = \sqrt{4}$, then $(4, 2)$ lies on the graph of f.
   (ii)   Since $(4, 2)$ lies on the graph of f, then $(2, 4)$ lies on the graph of f^{-1}.
   (iii)   Since $\sqrt{5} = \sqrt{5}$, then $(5, \sqrt{5})$ lies on the graph of f.
   (iv)   Since $(5, \sqrt{5})$ lies on the graph of f, then $(\sqrt{5}, 5)$ lies on the graph of f^{-1}.
   (v)   Since $f(a) = f(a)$, then $(a, f(a))$ lies on the graph of f.
   (vi)   Since $(a, f(a))$ lies on the graph of f, then $(f(a), a)$ lies on the graph of f^{-1}.
   (vii)   Since $(f^{-1}(b), b)$ lies on the graph of f, then $(b, f^{-1}(b))$ lies on the graph of f^{-1}.
   (viii)   Since $b = f[f^{-1}(b)]$, then $(f^{-1}(b), b)$ lies on the graph of f.

31. Let the points $P = (7, -1)$ and $Q = (-1, 7)$. We must first show that the line segment $\overline{PQ}$ is perpendicular to the line $y = x$. Line segment $\overline{PQ}$ has a slope of:

$$m = \frac{7 - (-1)}{-1 - 7} = \frac{8}{-8} = -1$$

Since $y = x$ has a slope of 1, and $1(-1) = -1$, then the two lines are perpendicular. Next, we must show that P and Q are equidistant from $y = x$. We find the midpoint of line segment $\overline{PQ}$:

$$M = \left(\frac{7-1}{2}, \frac{-1+7}{2}\right) = (3, 3)$$

Since $(3, 3)$ lies on the line $y = x$, and since $\overline{PQ}$ is perpendicular to $y = x$, and $PM = QM$, then by the definition of symmetry P and Q are symmetric about the line $y = x$.

33. Let the points $P = (a, b)$ and $Q = (b, a)$. We must first show that the line segment $\overline{PQ}$ is perpendicular to the line $y = x$. Line segment $\overline{PQ}$ has a slope of:

$$m = \frac{a - b}{b - a} = -1$$

Since $y = x$ has a slope of 1, and $1(-1) = -1$, then the two lines are perpendicular. Next, we must show that P and Q are equidistant from $y = x$. We find the midpoint of line segment $\overline{PQ}$:

$$M = \left(\frac{a+b}{2}, \frac{b+a}{2}\right) = \left(\frac{a+b}{2}, \frac{a+b}{2}\right)$$

Since $\left(\frac{a+b}{2}, \frac{a+b}{2}\right)$ lies on the line $y = x$, and since $\overline{PQ}$ is perpendicular to $y = x$, and $PM = QM$, then by the definition of symmetry P and Q are symmetric about the line $y = x$.

35. Let's pick the points $P(0, 3)$, $Q(2, 4)$, $R(4, 5)$. Then the reflected points will be $P'(3, 0)$, $Q'(4, 2)$, and $R'(5, 4)$. We will show that P', Q', and R' are collinear by computing the slopes:

$$m_{P'Q'} = \frac{2 - 0}{4 - 3} = \frac{2}{1} = 2 \qquad\qquad m_{Q'R'} = \frac{4 - 2}{5 - 4} = \frac{2}{1} = 2$$

So P', Q', and R' all are collinear (lie on the same line). There are two ways to find the equation of the line.

    1st way: We use the point $(3, 0)$ and $m = 2$ in the point-slope formula:

$$y - 0 = 2(x - 3)$$
$$y = 2x - 6$$

    2nd way: We realize that this line must be the inverse function of $y = \frac{1}{2}x + 3$.

        Exchange x and y and solve for y:

$$x = \frac{1}{2}y + 3$$
$$x - 3 = \frac{1}{2}y$$
$$y = 2x - 6$$

In either case, we obtain the same equation.

Chapter One Review Exercises

1. In inequality notation, we graph the interval $-3 \le x \le 2$:

3. In inequality notation, we graph the interval $-6 \le x < 0$:

5. In inequality notation, we graph the interval $x > -\sqrt{2}$:

    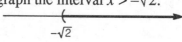

7. In inequality notation, we graph the interval $x < 1$:

9. We must make sure that the quantity inside the radical is non-negative, so:
    $$16 - 2t \ge 0$$
    $$-2t \ge -16$$
    $$t \le 8$$
    So the domain is $(-\infty, 8]$.

11. We must make sure that the denominator is non-zero, so:
    $$x^2 - 49 \ne 0$$
    $$x^2 \ne 49$$
    $$x \ne \pm 7$$
    So the domain is the set of all real numbers except -7 and 7.

13. We must make sure that the denominator is non-zero, so:
    $$t^2 - 12t - 13 \ne 0$$
    $$(t - 13)(t + 1) \ne 0$$
    $$t \ne -1, 13$$
    So the domain is the set of all real numbers except -1 and 13.

15. Since the expression $t^3 - 27$ has a value for every real number t, the domain is the set of all real numbers.

17. Since $x^2 + 1 \ge 0$ for all real values of x, then the radicand is always non-negative. So the domain is the set of all real numbers.

19. Using the distance formula with $(x_1, y_1) = (4, -7)$ and $(x_2, y_2) = (-2, 0)$:
    $$d = \sqrt{(x_2 - x_1)^2 + (y_2 - y_1)^2} = \sqrt{(-2 - 4)^2 + (0 + 7)^2} = \sqrt{36 + 49} = \sqrt{85}$$

21. Using the distance formula with $(x_1, y_1) = \left(\frac{\sqrt{3}}{2}, -\frac{1}{2}\right)$ and $(x_2, y_2) = (0, 0)$:

$$d = \sqrt{(x_2 - x_1)^2 + (y_2 - y_1)^2} = \sqrt{\left(0 - \frac{\sqrt{3}}{2}\right)^2 + \left(0 + \frac{1}{2}\right)^2} = \sqrt{\frac{3}{4} + \frac{1}{4}} = \sqrt{\frac{4}{4}} = 1$$

23. Since the point (x, y) lies on the unit circle, it must satisfy the equation $x^2 + y^2 = 1$. Also, since the point lies in quadrant II, we must have $y > 0$. Substituting $x = -\frac{3}{4}$ into the equation $x^2 + y^2 = 1$:

$$\left(-\frac{3}{4}\right)^2 + y^2 = 1$$
$$\frac{9}{16} + y^2 = 1$$
$$y^2 = \frac{7}{16}$$
$$y = \sqrt{\frac{7}{16}} = \frac{\sqrt{7}}{4} \quad \text{(since } y > 0\text{)}$$

25. Since the point (x, y) lies on the unit circle, it must satisfy the equation $x^2 + y^2 = 1$. Also, since the point lies in quadrant I, we must have $x > 0$. Substituting $y = \frac{7}{25}$ into the equation $x^2 + y^2 = 1$:

$$x^2 + \left(\frac{7}{25}\right)^2 = 1$$
$$x^2 + \frac{49}{625} = 1$$
$$x^2 = \frac{576}{625}$$
$$x = \sqrt{\frac{576}{625}} = \frac{24}{25} \quad \text{(since } x > 0\text{)}$$

27. (a) Since the graph of $y = f(x) + 1$ is obtained by translating the graph of $y = f(x)$ up one unit, then the point (a, b) is translated to the point $(a, b + 1)$. Thus (E) is the correct answer.

(b) Since the graph of $y = f(x + 1)$ is obtained by translating the graph of $y = f(x)$ to the left one unit, then the point (a, b) is translated to the point $(a - 1, b)$. Thus (C) is the correct answer.

(c) Since the graph of $y = f(x - 1) + 1$ is obtained by translating the graph of $y = f(x)$ to the right one unit and up one unit, then the point (a, b) is translated to the point $(a + 1, b + 1)$. Thus (L) is the correct answer.

(d) Since the graph of $y = f(-x)$ is obtained by reflecting the graph of $y = f(x)$ across the y-axis, then the point (a, b) is reflected to the point $(-a, b)$. Thus (A) is the correct answer.

(e) Since the graph of $y = -f(x)$ is obtained by reflecting the graph of $y = f(x)$ across the x-axis, then the point (a, b) is reflected to the point $(a, -b)$. Thus (J) is the correct answer.

(f)   Since the graph of $y = -f(-x)$ is obtained by reflecting the graph of $y = f(x)$ across the x-axis and across the y-axis, then the point (a, b) is reflected to the point $(-a, -b)$. Thus (G) is the correct answer.

(g)   Since the graph of $y = f^{-1}(x)$ is obtained by reflecting the graph of $y = f(x)$ across the line $y = x$, then the point (a, b) is reflected to the point (b, a). Thus (B) is the correct answer.

(h)   Since the graph of $y = f^{-1}(x) + 1$ is obtained by reflecting the graph of $y = f(x)$ across the line $y = x$, then translating up one unit, then the point (a, b) is reflected to the point (b, a) then translated to the point $(b, a + 1)$. Thus (M) is the correct answer.

(i)   Since the graph of $y = f^{-1}(x - 1)$ is obtained by reflecting the graph of $y = f(x)$ across the line $y = x$, then translating to the right one unit, then the point (a, b) is reflected to the point (b, a) then translated to the point $(b + 1, a)$. Thus (K) is the correct answer.

(j)   Since the graph of $y = f^{-1}(-x) + 1$ is obtained by reflecting the graph of $y = f(x)$ across the line $y = x$, then across the y-axis, then translating up one unit, then the point (a, b) is reflected to the point (b, a), then reflected to the point $(-b, a)$, then translated to the point $(-b, a + 1)$. Thus (D) is the correct answer.

(k)   Since the graph of $y = -f^{-1}(x)$ is obtained by reflecting the graph of $y = f(x)$ across the line $y = x$ then across the x-axis, then the point (a, b) is reflected to the point (b, a) then reflected to the point $(b, -a)$. Thus (I) is the correct answer.

(l)   Since the graph of $y = -f^{-1}(-x) + 1$ is obtained by reflecting the graph of $y = f(x)$ across the line $y = x$, the x-axis, and the y-axis, then translating up one unit, then the point (a, b) is reflected to the point (b, a), then reflected to the point $(-b, a)$ and then $(-b, -a)$, and finally translated to the point $(-b, -a + 1)$. Thus (H) is the correct answer.

(m)   Since the graph of $y = 1 - f^{-1}(x)$ is obtained by reflecting the graph of $y = f(x)$ across the line $y = x$ and across the x-axis, then translated up one unit, then the point (a, b) is reflected to the point (b, a), then reflected to the point $(b, -a)$, then translated to the point $(b, -a + 1)$. Thus (N) is the correct answer.

(n)   Since the graph of $y = f(1 - x)$ is obtained by reflecting the graph of $y = f(x)$ across the y-axis, then translating to the right one unit, then the point (a, b) is reflected to the point $(-a, b)$ then translated to the point $(-a + 1, b)$. Thus (F) is the correct answer.

29. To find the x-intercepts, we let $y = 0$:

$$-(x-1)^2 + 2 = 0$$
$$-(x-1)^2 = -2$$
$$x - 1 = \pm\sqrt{2}$$
$$x = 1 \pm \sqrt{2}$$

To find the y-intercept, we let $x = 0$:

$$y = -(0-1)^2 + 2 = -1 + 2 = 1$$

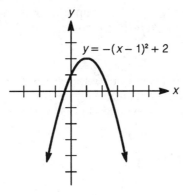

31. To find the x-intercepts, we let $y = 0$:

$$\frac{1}{x+1} = 0$$
$$1 = 0$$

Clearly this is impossible, so there are no x-intercepts. To find the y-intercept, we let $x = 0$:

$$y = \frac{1}{0+1} = 1$$

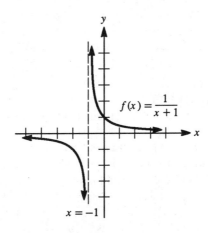

33. To find the x-intercepts, we let $y = 0$:
$$|x+3| = 0$$
$$x = -3$$

To find the y-intercept, we let $x = 0$:
$$y = |0+3| = 3$$

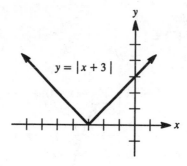

35. To find the x-intercepts, we let $y = 0$:
$$\sqrt{1-x^2} = 0$$
$$1-x^2 = 0$$
$$-x^2 = -1$$
$$x = \pm 1$$

To find the y-intercept, we let $x = 0$:
$$y = \sqrt{1-0} = 1$$

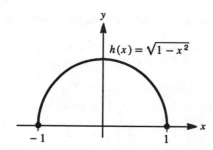

37. To find the x-intercepts, we let $y = 0$:
$$1 - (x+1)^3 = 0$$
$$-(x+1)^3 = -1$$
$$(x+1)^3 = 1$$
$$x + 1 = 1$$
$$x = 0$$

To find the y-intercept, we let $x = 0$:
$$y = 1 - (0+1)^3 = 1 - 1 = 0$$

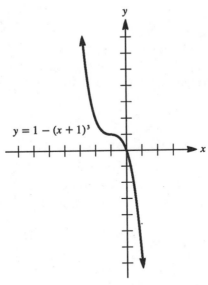

39. To find the inverse function, we switch the roles of x and y and solve the resulting equation for y:
$$x = \frac{y+1}{2}$$
$$2x = y+1$$
$$2x - 1 = y$$
So $f^{-1}(x) = 2x - 1$. To find the x-intercepts, we let $y = 0$:
$$2x - 1 = 0$$
$$2x = 1$$
$$x = \tfrac{1}{2}$$
To find the y-intercept, we let $x = 0$:
$$y = 2(0) - 1 = -1$$

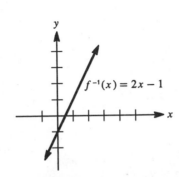

41. The graph is not symmetric in the x-axis or the origin, but it is symmetric in the y-axis.

43. The graph is not symmetric in either the x-axis, the y-axis, or the origin.

45. We must make sure that $x^2 - 9 \neq 0$. We find the points to exclude:
$$x^2 - 9 = 0$$
$$x^2 = 9$$
$$x = \pm 3$$
So the domain is all real numbers except 3 and -3.

47. We must be sure that the quantity inside the radical is non-negative:
$$8 - 2x \geq 0$$
$$-2x \geq -8$$
$$x \leq 4$$
So the domain is $(-\infty, 4]$.

49. We must be sure that the quantity inside the radical is non-negative, so $|2 - 5x| \geq 0$. But this is true for all real numbers, so the domain is all real numbers, or $(-\infty, \infty)$.

51. We solve for x:
$$y = \frac{x + 4}{3x - 1}$$
$$y(3x - 1) = x + 4$$
$$3xy - y = x + 4$$
$$3xy - x = y + 4$$
$$x(3y - 1) = y + 4$$
$$x = \frac{y + 4}{3y - 1}$$

Now $3y - 1 = 0$ when $y = \frac{1}{3}$, so the range is all real numbers except $\frac{1}{3}$.

53. Since the range of f^{-1} is the domain of f, we must exclude the values of x where $3x - 6 = 0$, or $x = 2$. So the range is all real numbers except 2.

55. Using $f(x)$, we compute $f(-3)$:
$$f(-3) = (-3)^2 - (-3) = 9 + 3 = 12$$

57. Using $F(x)$, we compute $F\left(\frac{3}{4}\right)$:
$$F\left(\tfrac{3}{4}\right) = \frac{\frac{3}{4} - 3}{\frac{3}{4} + 4} = \frac{3 - 3(4)}{3 + 4(4)} = \frac{3 - 12}{3 + 16} = -\frac{9}{19}$$

59. Using $f(x)$, we compute $f(-t)$:
$$f(-t) = (-t)^2 - (-t) = t^2 + t$$

61. Using $f(x)$, we compute $f(x - 2)$:
$$f(x - 2) = (x - 2)^2 - (x - 2) = x^2 - 4x + 4 - x + 2 = x^2 - 5x + 6$$

63. Using $g(x)$, we first compute $g(2)$ and $g(0)$:
$$g(2) = 1 - 2(2) = 1 - 4 = -3$$
$$g(0) = 1 - 2(0) = 1 - 0 = 1$$
So $g(2) - g(0) = -3 - 1 = -4$

65. Using $f(x)$, we first compute $f(1)$ and $f(3)$:
$$f(1) = (1)^2 - 1 = 1 - 1 = 0$$
$$f(3) = (3)^2 - 3 = 9 - 3 = 6$$
So $|f(1) - f(3)| = |0 - 6| = |-6| = 6$.

67. Using $f(x)$, we compute $f(x^2)$:
$$f(x^2) = (x^2)^2 - x^2 = x^4 - x^2$$

69. Let $y = \dfrac{x-3}{x+4}$. To find $F^{-1}(x)$, we switch the roles of x and y and solve the resulting equation for y:
$$x = \frac{y-3}{y+4}$$
$$x(y+4) = y - 3$$
$$xy + 4x = y - 3$$
$$xy - y = -4x - 3$$
$$y(x-1) = -4x - 3$$
$$y = \frac{-4x-3}{x-1}$$

So $F^{-1}(x) = \dfrac{-4x-3}{x-1} = \dfrac{4x+3}{1-x}$.

71. $F^{-1}[F(x)] = x$, by definition of $F^{-1}(x)$.

73. Let $y = 1 - 2x$. To find $g^{-1}(x)$, we switch the roles of x and y and solve the resulting equation for y:
$$x = 1 - 2y$$
$$2y = 1 - x$$
$$y = \frac{1-x}{2}$$

So $g^{-1}(x) = \dfrac{1-x}{2}$.

75. Since $\frac{22}{7}$ is in the domain of $F(x)$, then $F^{-1}\left[F\left(\frac{22}{7}\right)\right] = \frac{22}{7}$.

77. Since $f(0) = -2$, it is negative.

79. Since the point $(-3, -1)$ lies on the graph, then $f(-3) = -1$.

81. Since $f(0) = -2$ and $f(8) = -1$, then $f(0) - f(8) = -2 - (-1) = -2 + 1 = -1$.

83. The coordinates of the turning points are $(0, -2)$ and $(5, 1)$.

85. $f(x)$ is decreasing on the intervals $[-6, 0]$ and $[5, 8]$.

87. Since $|x| \leq 2$ corresponds to the interval $-2 \leq x \leq 2$, the largest value of $f(x)$ is 0, occurring at $x = 2$.

89. Since $f(x)$ is not a one-to-one function (it does not pass the horizontal line test) then it does not possess an inverse function.

Chapter One Test

1. (a) We must be sure the quantity under the radical is non-negative:
$$18 - 3x \geq 0$$
$$-3x \geq -18$$
$$x \leq 6$$
So the domain is $(-\infty, 6]$.

(b) We must be sure the quantity in the denominator is non-zero:
$$x^3 - 8x^2 - 9x = 0$$
$$x\left(x^2 - 8x - 9\right) = 0$$
$$x(x - 9)(x + 1) = 0$$
$$x = 0, -1, 9$$
So the domain is the set of all real numbers except 0, -1 and 9.

2. (a) We sketch the interval $(3, \infty)$:

(b) We sketch the interval $[-1, 2]$:

3. Using the distance formula, we have:
$$d = \sqrt{(-4 - 2)^2 + (-3 - 0)^2} = \sqrt{36 + 9} = \sqrt{45} = 3\sqrt{5}$$

4. Since P lies in the second quadrant, its x-coordinate is negative. Using the unit circle equation $x^2 + y^2 = 1$ and $y = \frac{1}{2}$, we have:

$$x^2 + \left(\frac{1}{2}\right)^2 = 1$$
$$x^2 + \frac{1}{4} = 1$$
$$x^2 = \frac{3}{4}$$
$$x = -\sqrt{\frac{3}{4}} = -\frac{\sqrt{3}}{2}$$

5. (a) We sketch the line segment L and its reflection about the x-axis:

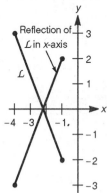

(b) We sketch the line segment L and its reflection about the y-axis:

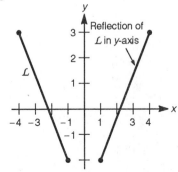

(c) We sketch the line segment L and its reflection about the origin:

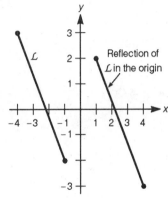

6.   (a)   To test for symmetry about the x-axis, we replace (x, y) with $(x, -y)$:

$$-y = x^3 - 2$$
$$y = -x^3 + 2$$

Since the equation is changed, there is no x-axis symmetry. To test for symmetry about the y-axis, we replace (x, y) with $(-x, y)$:

$$y = (-x)^3 - 2$$
$$y = -x^3 - 2$$

Since the equation is changed, there is no y-axis symmetry. To test for symmetry about the origin, we replace (x, y) with $(-x, -y)$:

$$-y = (-x)^3 - 2$$
$$-y = -x^3 - 2$$
$$y = x^3 + 2$$

Since the equation is changed, there is no origin symmetry.

(b)   To test for symmetry about the x-axis, we replace (x, y) with $(x, -y)$:

$$-y = |x^3| - 2$$
$$y = -|x^3| + 2$$

Since the equation is changed, there is no x-axis symmetry. To test for symmetry about the y-axis, we replace (x, y) with $(-x, y)$:

$$y = |(-x)^3| - 2$$
$$y = |-x^3| - 2$$
$$y = |x^3| - 2$$

Since the equation is unchanged, there is symmetry about the y-axis. To test for symmetry about the origin, we replace (x, y) with $(-x, -y)$:

$$-y = |(-x)^3| - 2$$
$$-y = |-x^3| - 2$$
$$-y = |x^3| - 2$$
$$y = -|x^3| + 2$$

Since the equation is changed, there is no origin symmetry.

7.   (a)   To find the x-intercepts, let $y = 0$:

$$0 = -\sqrt{x + 2}$$
$$0 = x + 2$$
$$-2 = x$$

To find the y-intercepts, let $x = 0$:

$$y = -\sqrt{0 + 2} = -\sqrt{2}$$

We graph the equation:

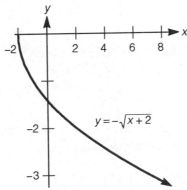

$$y=-\sqrt{x+2}$$

(b)   To find the *x*-intercepts, let $y = 0$:

$$0 = \frac{1}{x-1}+1$$
$$-1 = \frac{1}{x-1}$$
$$-x+1 = 1$$
$$-x = 0$$
$$x = 0$$

To find the *y*-intercepts, let $x = 0$:

$$y = \frac{1}{0-1}+1 = -1+1 = 0$$

We graph the equation:

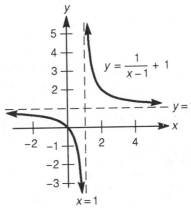

$$y = \frac{1}{x-1}+1$$

8.   We solve the equation for x:
$$y = \frac{3x-4}{2+x}$$
$$2y + xy = 3x - 4$$
$$xy - 3x = -2y - 4$$
$$x(y-3) = -2y - 4$$
$$x = \frac{-2y-4}{y-3}$$
Since $y \neq 3$, the range is all real numbers except 3.

9.   Let $y = 2x - 3$. Now we switch the roles of x and y, and solve the resulting equation for y:
$$x = 2y - 3$$
$$x + 3 = 2y$$
$$\frac{x+3}{2} = y$$

So $g^{-1}(x) = \frac{x+3}{2}$.

10.   As long as $\frac{2}{3}$ is in the domain of f^{-1}, then $f\left(f^{-1}\left(\frac{2}{3}\right)\right) = \frac{2}{3}$ since $f\left(f^{-1}(x)\right) = x$ for all values of x in the domain of f. To verify the domain of f^{-1}, which is the same as the range of f, we solve for x:
$$y = \frac{x+4}{3x-2}$$
$$3xy - 2y = x + 4$$
$$3xy - x = 2y + 4$$
$$x(3y - 1) = 2y + 4$$
$$x = \frac{2y+4}{3y-1}$$

So $\frac{2}{3}$ is in the range of f, and thus the domain of f^{-1}, therefore $f\left(f^{-1}\left(\frac{2}{3}\right)\right) = \frac{2}{3}$.

11.   (a)   We sketch the locations of 1 and 2:

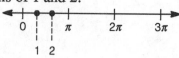

      (b)   We sketch the locations of $-\frac{\pi}{2}$ and $\frac{\pi}{4}$:

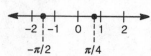

12. (a)   Using the equation $(x-h)^2 + (y-k)^2 = r^2$, we have:

$$(x-3)^2 + (y+1)^2 = 16$$

(b)   We replace $x = -\frac{1}{2}$ and $y = -\frac{11}{4}$ in the equation:

$$\left(-\tfrac{1}{2}-3\right)^2 + \left(-\tfrac{11}{4}+1\right)^2 = \left(-\tfrac{7}{2}\right)^2 + \left(-\tfrac{7}{4}\right)^2 = \tfrac{49}{4} + \tfrac{49}{16} = \tfrac{245}{16} = 15.3125$$

Since the expression did not simplify to 16, the point $\left(-\tfrac{1}{2}, -\tfrac{11}{4}\right)$ does not lie on this circle.

13. We evaluate and simplify $g(x-1)$:

$$g(x-1) = (x-1)^2 - 2(x-1) = x^2 - 2x + 1 - 2x + 2 = x^2 - 4x + 3$$

14. (a)   We compute $F(-x)$:

$$F(-x) = 5 - 2(-x) = 5 + 2x$$

(b)   We compute $-F(x)$:

$$-F(x) = -(5-2x) = -5 + 2x$$

(c)   We compute $-F(-x)$ using $F(-x) = 5 + 2x$ obtained in part (a):

$$-F(-x) = -(5+2x) = -5 - 2x$$

Chapter Two
Trigonometric Functions of Angles

2.1 Trigonometric Functions of Acute Angles

1. (a) We use the definitions:

$$\sin\theta = \frac{\text{opposite}}{\text{hypotenuse}} = \frac{15}{17} \qquad \cos\theta = \frac{\text{adjacent}}{\text{hypotenuse}} = \frac{8}{17}$$

$$\tan\theta = \frac{\text{opposite}}{\text{adjacent}} = \frac{15}{8} \qquad \cot\theta = \frac{\text{adjacent}}{\text{opposite}} = \frac{8}{15}$$

$$\sec\theta = \frac{\text{hypotenuse}}{\text{adjacent}} = \frac{17}{8} \qquad \csc\theta = \frac{\text{hypotenuse}}{\text{opposite}} = \frac{17}{15}$$

(b) We use the definitions:

$$\sin\beta = \frac{\text{opposite}}{\text{hypotenuse}} = \frac{8}{17} \qquad \cos\beta = \frac{\text{adjacent}}{\text{hypotenuse}} = \frac{15}{17}$$

$$\tan\beta = \frac{\text{opposite}}{\text{adjacent}} = \frac{8}{15} \qquad \cot\beta = \frac{\text{adjacent}}{\text{opposite}} = \frac{15}{8}$$

$$\sec\beta = \frac{\text{hypotenuse}}{\text{adjacent}} = \frac{17}{15} \qquad \csc\beta = \frac{\text{hypotenuse}}{\text{opposite}} = \frac{17}{8}$$

3.   **(a)**   We use the definitions:

$$\sin\theta = \frac{\text{opposite}}{\text{hypotenuse}} = \frac{3}{3\sqrt{5}} = \frac{1}{\sqrt{5}} = \frac{\sqrt{5}}{5}$$

$$\cos\theta = \frac{\text{adjacent}}{\text{hypotenuse}} = \frac{6}{3\sqrt{5}} = \frac{2}{\sqrt{5}} = \frac{2\sqrt{5}}{5}$$

$$\tan\theta = \frac{\text{opposite}}{\text{adjacent}} = \frac{3}{6} = \frac{1}{2}$$

$$\cot\theta = \frac{\text{adjacent}}{\text{opposite}} = \frac{6}{3} = 2$$

$$\sec\theta = \frac{\text{hypotenuse}}{\text{adjacent}} = \frac{3\sqrt{5}}{6} = \frac{\sqrt{5}}{2}$$

$$\csc\theta = \frac{\text{hypotenuse}}{\text{opposite}} = \frac{3\sqrt{5}}{3} = \sqrt{5}$$

(b)   We use the definitions:

$$\sin\beta = \frac{\text{opposite}}{\text{hypotenuse}} = \frac{6}{3\sqrt{5}} = \frac{2}{\sqrt{5}} = \frac{2\sqrt{5}}{5}$$

$$\cos\beta = \frac{\text{adjacent}}{\text{hypotenuse}} = \frac{3}{3\sqrt{5}} = \frac{1}{\sqrt{5}} = \frac{\sqrt{5}}{5}$$

$$\tan\beta = \frac{\text{opposite}}{\text{adjacent}} = \frac{6}{3} = 2$$

$$\cot\beta = \frac{\text{adjacent}}{\text{opposite}} = \frac{3}{6} = \frac{1}{2}$$

$$\sec\beta = \frac{\text{hypotenuse}}{\text{adjacent}} = \frac{3\sqrt{5}}{3} = \sqrt{5}$$

$$\csc\beta = \frac{\text{hypotenuse}}{\text{opposite}} = \frac{3\sqrt{5}}{6} = \frac{\sqrt{5}}{2}$$

5.   We first draw $\triangle ABC$ and label $AC = 3$ and $BC = 2$:

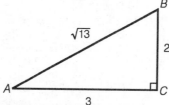

We can find AB by using the Pythagorean theorem:

$$(AC)^2 + (BC)^2 = (AB)^2$$
$$(3)^2 + (2)^2 = (AB)^2$$
$$9 + 4 = (AB)^2$$
$$\sqrt{13} = AB$$

(a) We find $\cos A$, $\sin A$ and $\tan A$:

$$\cos A = \frac{\text{adjacent}}{\text{hypotenuse}} = \frac{3}{\sqrt{13}} = \frac{3\sqrt{13}}{13}$$

$$\sin A = \frac{\text{opposite}}{\text{hypotenuse}} = \frac{2}{\sqrt{13}} = \frac{2\sqrt{13}}{13}$$

$$\tan A = \frac{\text{opposite}}{\text{adjacent}} = \frac{2}{3}$$

(b) We find $\sec B$, $\csc B$ and $\cot B$:

$$\sec B = \frac{\text{hypotenuse}}{\text{adjacent}} = \frac{\sqrt{13}}{2}$$

$$\csc B = \frac{\text{hypotenuse}}{\text{opposite}} = \frac{\sqrt{13}}{3}$$

$$\cot B = \frac{\text{adjacent}}{\text{opposite}} = \frac{2}{3}$$

7. We draw a sketch with $AB = 13$ and $BC = 5$:

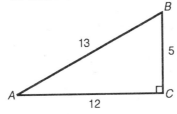

We can find AC by using the Pythagorean theorem:

$$(AC)^2 + (BC)^2 = (AB)^2$$
$$(AC)^2 + (5)^2 = (13)^2$$
$$(AC)^2 + 25 = 169$$
$$(AC)^2 = 144$$
$$AC = 12$$

Now we find the six trigonometric functions of angle B:

$$\sin B = \frac{\text{opposite}}{\text{hypotenuse}} = \frac{12}{13} \qquad \cos B = \frac{\text{adjacent}}{\text{hypotenuse}} = \frac{5}{13}$$

$$\tan B = \frac{\text{opposite}}{\text{adjacent}} = \frac{12}{5} \qquad \cot B = \frac{\text{adjacent}}{\text{opposite}} = \frac{5}{12}$$

$$\sec B = \frac{\text{hypotenuse}}{\text{adjacent}} = \frac{13}{5} \qquad \csc B = \frac{\text{hypotenuse}}{\text{opposite}} = \frac{13}{12}$$

9.   We draw a sketch where $AC = 1$ and $BC = \frac{3}{4}$:

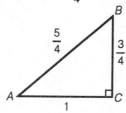

We can find AB by using the Pythagorean theorem:
$$(AC)^2 + (BC)^2 = (AB)^2$$
$$(1)^2 + \left(\tfrac{3}{4}\right)^2 = (AB)^2$$
$$1 + \tfrac{9}{16} = (AB)^2$$
$$\tfrac{25}{16} = (AB)^2$$
$$\tfrac{5}{4} = AB$$

(a)   We find $\sin B$ and $\cos A$:
$$\sin B = \frac{\text{opposite}}{\text{hypotenuse}} = \frac{1}{\frac{5}{4}} = \frac{4}{5}$$
$$\cos A = \frac{\text{adjacent}}{\text{hypotenuse}} = \frac{1}{\frac{5}{4}} = \frac{4}{5}$$

(b)   We find $\sin A$ and $\cos B$:
$$\sin A = \frac{\text{opposite}}{\text{hypotenuse}} = \frac{\frac{3}{4}}{\frac{5}{4}} = \frac{3}{5}$$
$$\cos B = \frac{\text{adjacent}}{\text{hypotenuse}} = \frac{\frac{3}{4}}{\frac{5}{4}} = \frac{3}{5}$$

(c)   We first find $\tan A$ and $\tan B$:
$$\tan A = \frac{\text{opposite}}{\text{adjacent}} = \frac{\frac{3}{4}}{1} = \frac{3}{4}$$
$$\tan B = \frac{\text{opposite}}{\text{adjacent}} = \frac{1}{\frac{3}{4}} = \frac{4}{3}$$

Now find $(\tan A)(\tan B)$:
$$(\tan A)(\tan B) = \tfrac{3}{4} \cdot \tfrac{4}{3} = 1$$

11. We draw a sketch where $AB = 25$ and $AC = 24$:

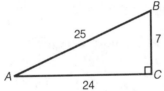

We can find BC by using the Pythagorean theorem:
$$(AC)^2 + (BC)^2 = (AB)^2$$
$$(24)^2 + (BC)^2 = (25)^2$$
$$576 + (BC)^2 = 625$$
$$(BC)^2 = 49$$
$$BC = 7$$

(a) We find $\cos A$, $\sin A$ and $\tan A$:
$$\cos A = \frac{\text{adjacent}}{\text{hypotenuse}} = \frac{24}{25}$$
$$\sin A = \frac{\text{opposite}}{\text{hypotenuse}} = \frac{7}{25}$$
$$\tan A = \frac{\text{opposite}}{\text{adjacent}} = \frac{7}{24}$$

(b) We find $\cos B$, $\sin B$ and $\tan B$:
$$\cos B = \frac{\text{adjacent}}{\text{hypotenuse}} = \frac{7}{25}$$
$$\sin B = \frac{\text{opposite}}{\text{hypotenuse}} = \frac{24}{25}$$
$$\tan B = \frac{\text{opposite}}{\text{adjacent}} = \frac{24}{7}$$

(c) Using the values obtained in parts (a) and (b):
$$(\tan A)(\tan B) = \tfrac{7}{24} \cdot \tfrac{24}{7} = 1$$

13. The calculated values for $\theta = 65°$ are approximately:

$\sin \theta \approx 0.906$ $\cos \theta \approx 0.423$ $\tan \theta \approx 2.145$

15. The calculated values for $\theta = 38.5°$ are approximately:

$\sin \theta \approx 0.623$ $\cos \theta \approx 0.783$ $\tan \theta \approx 0.795$

17. The calculated values for $\theta = 80.06°$ are approximately:

$\sin \theta \approx 0.985$ $\cos \theta \approx 0.173$ $\tan \theta \approx 5.706$

19. The calculated values for $\theta = 20°$ are approximately:

$\sec \theta \approx 1.064$ $\csc \theta \approx 2.924$ $\cot \theta \approx 2.747$

21. The calculated values for $\theta = 17.5°$ are approximately:

$\sec\theta \approx 1.049$ $\csc\theta \approx 3.326$ $\cot\theta \approx 3.172$

23. The calculated values for $\theta = 1°$ are approximately:

$\sec\theta \approx 1.000$ $\csc\theta \approx 57.299$ $\cot\theta \approx 57.290$

25. We will compute each side of the equation:

$\cos 60° = \frac{1}{2}$

$\cos^2 30° - \sin^2 30° = \left(\frac{\sqrt{3}}{2}\right)^2 - \left(\frac{1}{2}\right)^2 = \frac{3}{4} - \frac{1}{4} = \frac{2}{4} = \frac{1}{2}$

So $\cos 60° = \cos^2 30° - \sin^2 30°$.

27. We will compute the left-hand side of the equation:

$\sin^2 30° + \sin^2 45° + \sin^2 60° = \left(\frac{1}{2}\right)^2 + \left(\frac{\sqrt{2}}{2}\right)^2 + \left(\frac{\sqrt{3}}{2}\right)^2 = \frac{1}{4} + \frac{2}{4} + \frac{3}{4} = \frac{6}{4} = \frac{3}{2}$

So $\sin^2 30° + \sin^2 45° + \sin^2 60° = \frac{3}{2}$.

29. We will compute each side of the equation:

$2\sin 30° \cos 30° = 2\left(\frac{1}{2}\right)\left(\frac{\sqrt{3}}{2}\right) = \frac{2\sqrt{3}}{4} = \frac{\sqrt{3}}{2}$

$\sin 60° = \frac{\sqrt{3}}{2}$

So $2\sin 30° \cos 30° = \sin 60°$.

31. We will compute each side of the equation:

$\sin 30° = \dfrac{1}{2}$

$\sqrt{\dfrac{1 - \cos 60°}{2}} = \sqrt{\dfrac{1 - \frac{1}{2}}{2}} = \sqrt{\dfrac{\frac{1}{2}}{2}} = \sqrt{\dfrac{1}{4}} = \dfrac{1}{2}$

So $\sin 30° = \sqrt{\dfrac{1 - \cos 60°}{2}}$.

33. We will compute each side of the equation:

$\tan 30° = \dfrac{1}{\sqrt{3}} = \dfrac{\sqrt{3}}{3}$

$\dfrac{\sin 60°}{1 + \cos 60°} = \dfrac{\frac{\sqrt{3}}{2}}{1 + \frac{1}{2}} = \dfrac{\frac{\sqrt{3}}{2}}{\frac{3}{2}} = \dfrac{\sqrt{3}}{3}$

So $\tan 30° = \dfrac{\sin 60°}{1 + \cos 60°}$.

35. We will compute each side of the equation:
$$1 + \tan^2 45° = 1 + (1)^2 = 1 + 1 = 2$$
$$\sec^2 45° = \left(\frac{\sqrt{2}}{1}\right)^2 = 2$$
So $1 + \tan^2 45° = \sec^2 45°$.

37. (a) Using the Texas Instruments TI-81, the approximate values are:
$$\cos 30° \approx 0.8660254038$$
$$\cos 45° \approx 0.7071067812$$

(b) Using the values given in Table 1, the approximate values are:
$$\cos 30° = \frac{\sqrt{3}}{2} \approx 0.8660254038$$
$$\cos 45° = \frac{\sqrt{2}}{2} \approx 0.7071067812$$
Note that these results agree with those in part (a).

39. (a) $\cos\theta$ is larger
(b) $\sec\beta$ is larger

41. Extending AB to D an equal distance to AB guarantees $\triangle DBC$ congruent to $\triangle ABC$. Now $\triangle ADC$ is equilateral as each angle is 60°. By construction $AD = 2AB$ and since the triangle is equilateral, $AC = AD$, hence $AC = 2AB$.

43. The values are equal. Computed to ten decimal places, the values are:
$$\sin 3° \approx 0.0523359562 \qquad \sin 6° \approx 0.1045284633$$
$$\sin 9° \approx 0.1564344650 \qquad \sin 12° \approx 0.2079116908$$
$$\sin 15° \approx 0.2588190451 \qquad \sin 18° \approx 0.3090169944$$

2.2 Algebra and the Trigonometric Functions

1. (a) Combining like terms, we have:
$$-SC + 12SC = 11SC$$

(b) Combining as in (a), we have:
$$-\sin\theta\cos\theta + 12\sin\theta\cos\theta = 11\sin\theta\cos\theta$$

3. (a) Combining like terms, we have:
$$4C^3S - 12C^3S = -8C^3S$$

(b) Combining as in (a), we have:
$$4\cos^3\theta\sin\theta - 12\cos^3\theta\sin\theta = -8\cos^3\theta\sin\theta$$

5. (a) Squaring, we have:
$$(1+T)^2 = 1 + 2T + T^2$$

(b)   Squaring as in (a), we have:
$$(1+\tan\theta)^2 = 1+2\tan\theta+\tan^2\theta$$

7.   (a)   Using the distributive property, we have:
$$(T+3)(T-2)=T^2+3T-2T-6=T^2+T-6$$

(b)   Using the distributive property as in (a), we have:
$$(\tan\theta+3)(\tan\theta-2)=\tan^2\theta+3\tan\theta-2\tan\theta-6=\tan^2\theta+\tan\theta-6$$

9.   (a)   Factoring then simplifying, we have:
$$\frac{S-C}{C-S}=\frac{-1(C-S)}{C-S}=-1$$

(b)   Factoring then simplifying as in (a), we have:
$$\frac{\sin\theta-\cos\theta}{\cos\theta-\sin\theta}=\frac{-1(\cos\theta-\sin\theta)}{\cos\theta-\sin\theta}=-1$$

11.   (a)   Obtaining common denominators then adding, we have:
$$C+\frac{2}{S}=\frac{CS}{S}+\frac{2}{S}=\frac{CS+2}{S}$$

(b)   Obtaining common denominators then adding as in (a), we have:
$$\cos A+\frac{2}{\sin A}=\frac{\cos A\sin A}{\sin A}+\frac{2}{\sin A}=\frac{\cos A\sin A+2}{\sin A}$$

13.   (a)   The expression factors as:
$$T^2+8T-9=(T-1)(T+9)$$

(b)   Factoring as in (a):
$$\tan^2\beta+8\tan\beta-9=(\tan\beta-1)(\tan\beta+9)$$

15.   (a)   Factoring as a difference of squares:
$$4C^2-1=(2C+1)(2C-1)$$

(b)   Factoring as in (a):
$$4\cos^2 B-1=(2\cos B+1)(2\cos B-1)$$

17.   (a)   Factoring the greatest common factor:
$$9S^2T^3+6ST^2=3ST^2(3ST+2)$$

(b)   Factoring as in (a):
$$9\sec^2 B\tan^3 B+6\sec B\tan^2 B=3\sec B\tan^2 B(3\sec B\tan B+2)$$

19. Since $\sin\theta = \dfrac{\text{opposite}}{\text{hypotenuse}} = \dfrac{3}{4}$, we can construct the triangle:

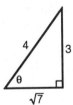

Using the Pythagorean theorem:

$$a^2 + 3^2 = 4^2$$
$$a^2 + 9 = 16$$
$$a^2 = 7$$
$$a = \sqrt{7}$$

So the six trigonometric functions are:

$$\sin\theta = \frac{\text{opposite}}{\text{hypotenuse}} = \frac{3}{4} \qquad\qquad \cos\theta = \frac{\text{adjacent}}{\text{hypotenuse}} = \frac{\sqrt{7}}{4}$$

$$\tan\theta = \frac{\text{opposite}}{\text{adjacent}} = \frac{3}{\sqrt{7}} = \frac{3\sqrt{7}}{7} \qquad\qquad \cot\theta = \frac{\text{adjacent}}{\text{opposite}} = \frac{\sqrt{7}}{3}$$

$$\sec\theta = \frac{\text{hypotenuse}}{\text{adjacent}} = \frac{4}{\sqrt{7}} = \frac{4\sqrt{7}}{7} \qquad\qquad \csc\theta = \frac{\text{hypotenuse}}{\text{opposite}} = \frac{4}{3}$$

21. Since $\cos\beta = \dfrac{\text{adjacent}}{\text{hypotenuse}} = \dfrac{\sqrt{3}}{5}$, we can construct the triangle:

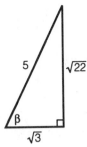

Using the Pythagorean theorem:. we have:

$$\left(\sqrt{3}\right)^2 + b^2 = 5^2$$
$$3 + b^2 = 25$$
$$b^2 = 22$$
$$b = \sqrt{22}$$

So the six trigonometric functions are:

$$\sin\beta = \frac{\text{opposite}}{\text{hypotenuse}} = \frac{\sqrt{22}}{5} \qquad \cos\beta = \frac{\text{adjacent}}{\text{hypotenuse}} = \frac{\sqrt{3}}{5}$$

$$\tan\beta = \frac{\text{opposite}}{\text{adjacent}} = \frac{\sqrt{22}}{\sqrt{3}} = \frac{\sqrt{66}}{3} \qquad \cot\beta = \frac{\text{adjacent}}{\text{opposite}} = \frac{\sqrt{3}}{\sqrt{22}} = \frac{\sqrt{66}}{22}$$

$$\sec\beta = \frac{\text{hypotenuse}}{\text{adjacent}} = \frac{5}{\sqrt{3}} = \frac{5\sqrt{3}}{3} \qquad \csc\beta = \frac{\text{hypotenuse}}{\text{opposite}} = \frac{5}{\sqrt{22}} = \frac{5\sqrt{22}}{22}$$

23. Since $\sin A = \dfrac{\text{opposite}}{\text{hypotenuse}} = \dfrac{5}{13}$, we can construct the triangle:

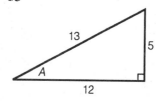

Using the Pythagorean theorem, we have:

$$a^2 + 5^2 = 13^2$$
$$a^2 + 25 = 169$$
$$a^2 = 144$$
$$a = 12$$

So the six trigonometric functions are:

$$\sin A = \frac{\text{opposite}}{\text{hypotenuse}} = \frac{5}{13} \qquad \cos A = \frac{\text{adjacent}}{\text{hypotenuse}} = \frac{12}{13}$$

$$\tan A = \frac{\text{opposite}}{\text{adjacent}} = \frac{5}{12} \qquad \cot A = \frac{\text{adjacent}}{\text{opposite}} = \frac{12}{5}$$

$$\sec A = \frac{\text{hypotenuse}}{\text{adjacent}} = \frac{13}{12} \qquad \csc A = \frac{\text{hypotenuse}}{\text{opposite}} = \frac{13}{5}$$

25. Since $\tan B = \dfrac{\text{opposite}}{\text{adjacent}} = \dfrac{4}{3}$, we can construct the triangle:

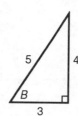

Using the Pythagorean theorem, we have:
$$3^2 + 4^2 = c^2$$
$$9 + 16 = c^2$$
$$25 = c^2$$
$$5 = c$$

So the six trigonometric functions are:

$$\sin B = \frac{\text{opposite}}{\text{hypotenuse}} = \frac{4}{5} \qquad\qquad \cos B = \frac{\text{adjacent}}{\text{hypotenuse}} = \frac{3}{5}$$

$$\tan B = \frac{\text{opposite}}{\text{adjacent}} = \frac{4}{3} \qquad\qquad \cot B = \frac{\text{adjacent}}{\text{opposite}} = \frac{3}{4}$$

$$\sec B = \frac{\text{hypotenuse}}{\text{adjacent}} = \frac{5}{3} \qquad\qquad \csc B = \frac{\text{hypotenuse}}{\text{opposite}} = \frac{5}{4}$$

27. Since $\sec C = \dfrac{\text{hypotenuse}}{\text{adjacent}} = \dfrac{3}{2}$, we can construct the triangle:

Using the Pythagorean theorem, we have:
$$2^2 + b^2 = 3^2$$
$$4 + b^2 = 9$$
$$b^2 = 5$$
$$b = \sqrt{5}$$

So the six trigonometric functions are:

$$\sin C = \frac{\text{opposite}}{\text{hypotenuse}} = \frac{\sqrt{5}}{3} \qquad\qquad \cos C = \frac{\text{adjacent}}{\text{hypotenuse}} = \frac{2}{3}$$

$$\tan C = \frac{\text{opposite}}{\text{adjacent}} = \frac{\sqrt{5}}{2} \qquad\qquad \cot C = \frac{\text{adjacent}}{\text{opposite}} = \frac{2}{\sqrt{5}} = \frac{2\sqrt{5}}{5}$$

$$\sec C = \frac{\text{hypotenuse}}{\text{adjacent}} = \frac{3}{2} \qquad\qquad \csc C = \frac{\text{hypotenuse}}{\text{opposite}} = \frac{3}{\sqrt{5}} = \frac{3\sqrt{5}}{5}$$

29. Since $\cot\alpha = \dfrac{\text{adjacent}}{\text{opposite}} = \dfrac{\sqrt{3}}{3}$, we can construct the triangle:

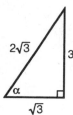

Using the Pythagorean theorem, we have:
$$\left(\sqrt{3}\right)^2 + (3)^2 = c^2$$
$$3 + 9 = c^2$$
$$12 = c^2$$
$$c = \sqrt{12} = 2\sqrt{3}$$

So the six trigonometric functions are:

$\sin\alpha = \dfrac{\text{opposite}}{\text{hypotenuse}} = \dfrac{3}{2\sqrt{3}} = \dfrac{3\sqrt{3}}{6} = \dfrac{\sqrt{3}}{2}$ $\cos\alpha = \dfrac{\text{adjacent}}{\text{hypotenuse}} = \dfrac{\sqrt{3}}{2\sqrt{3}} = \dfrac{1}{2}$

$\tan\alpha = \dfrac{\text{opposite}}{\text{adjacent}} = \dfrac{3}{\sqrt{3}} = \dfrac{3\sqrt{3}}{3} = \sqrt{3}$ $\cot\alpha = \dfrac{\text{adjacent}}{\text{opposite}} = \dfrac{\sqrt{3}}{3}$

$\sec\alpha = \dfrac{\text{hypotenuse}}{\text{adjacent}} = \dfrac{2\sqrt{3}}{\sqrt{3}} = 2$ $\csc\alpha = \dfrac{\text{hypotenuse}}{\text{opposite}} = \dfrac{2\sqrt{3}}{3}$

31. From the identity $\sin^2\theta + \cos^2\theta = 1$ we can find $\sin\theta$:
$$\sin^2\theta + (0.4626)^2 = 1$$
$$\sin^2\theta + 0.21399876 = 1$$
$$\sin^2\theta = 0.78600124$$
$$\sin\theta \approx 0.887$$

Using trigonometric identities, we compute:

$\tan\theta = \dfrac{\sin\theta}{\cos\theta} \approx \dfrac{0.887}{0.4626} \approx 1.916$

$\cot\theta = \dfrac{\cos\theta}{\sin\theta} \approx \dfrac{0.4626}{0.887} \approx 0.522$

$\sec\theta = \dfrac{1}{\cos\theta} = \dfrac{1}{0.4626} \approx 2.162$

$\csc\theta = \dfrac{1}{\sin\theta} \approx \dfrac{1}{0.887} \approx 1.128$

33. Since $\tan\theta = \dfrac{\text{opposite}}{\text{adjacent}} = \dfrac{1.1998}{1}$, we can construct the triangle:

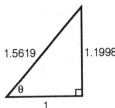

Using the Pythagorean theorem, we have:
$$1^2 + (1.1998)^2 = c^2$$
$$1 + 1.43952004 = c^2$$
$$2.43952004 = c^2$$
$$c \approx 1.5619$$

So the remaining trigonometric functions are:

$$\sin\theta = \frac{\text{opposite}}{\text{hypotenuse}} \approx \frac{1.1998}{1.5619} \approx 0.768$$

$$\cos\theta = \frac{\text{adjacent}}{\text{hypotenuse}} \approx \frac{1}{1.5619} \approx 0.640$$

$$\cot\theta = \frac{1}{\tan\theta} = \frac{1}{1.1998} \approx 0.833$$

$$\sec\theta = \frac{1}{\cos\theta} \approx \frac{1}{0.640} \approx 1.562$$

$$\csc\theta = \frac{1}{\sin\theta} \approx \frac{1}{0.768} \approx 1.302$$

35. By factoring the numerator, we have:
$$\frac{\sin^2 A - \cos^2 A}{\sin A - \cos A} = \frac{(\sin A - \cos A)(\sin A + \cos A)}{\sin A - \cos A} = \sin A + \cos A$$

37. By using the identities for $\csc\theta$ and $\sec\theta$, we have:
$$\sin^2\theta\cos\theta\csc^3\theta\sec\theta = \sin^2\theta \bullet \cos\theta \bullet \frac{1}{\sin^3\theta} \bullet \frac{1}{\cos\theta} = \frac{1}{\sin\theta} = \csc\theta$$

39. By using the identity for $\cot B$, we have:
$$\cot B\sin^2 B\cot B = \frac{\cos B}{\sin B} \bullet \sin^2 B \bullet \frac{\cos B}{\sin B} = \cos^2 B$$

41. By factoring the numerator, we have:
$$\frac{\cos^2 A + \cos A - 12}{\cos A - 3} = \frac{(\cos A - 3)(\cos A + 4)}{\cos A - 3} = \cos A + 4$$

43. By using the identities for $\tan\theta$ and $\sec\theta$, we have:

$$\frac{\tan\theta}{\sec\theta-1}+\frac{\tan\theta}{\sec\theta+1}=\frac{\frac{\sin\theta}{\cos\theta}}{\frac{1}{\cos\theta}-1}+\frac{\frac{\sin\theta}{\cos\theta}}{\frac{1}{\cos\theta}+1}$$

$$=\frac{\sin\theta}{1-\cos\theta}+\frac{\sin\theta}{1+\cos\theta}$$

$$=\frac{\sin\theta(1+\cos\theta)+\sin\theta(1-\cos\theta)}{1-\cos^2\theta}$$

$$=\frac{2\sin\theta}{\sin^2\theta}$$

$$=\frac{2}{\sin\theta}$$

$$=2\csc\theta$$

45. By using the identities for $\sec A$, $\csc A$, $\tan A$ and $\cot A$, we have:

$$\sec A\csc A-\tan A-\cot A=\frac{1}{\cos A}\bullet\frac{1}{\sin A}-\frac{\sin A}{\cos A}-\frac{\cos A}{\sin A}$$

$$=\frac{1-\sin^2 A-\cos^2 A}{\sin A\cos A}$$

$$=\frac{1-\left(\sin^2 A+\cos^2 A\right)}{\sin A\cos A}$$

$$=\frac{1-1}{\sin A\cos A}$$

$$=0$$

47. By using the identities for $\cot\theta$, $\tan\theta$, $\csc\theta$ and $\sec\theta$, we have:

$$\frac{\cot^2\theta}{\csc^2\theta}+\frac{\tan^2\theta}{\sec^2\theta}=\frac{\frac{\cos^2\theta}{\sin^2\theta}}{\frac{1}{\sin^2\theta}}+\frac{\frac{\sin^2\theta}{\cos^2\theta}}{\frac{1}{\cos^2\theta}}=\cos^2\theta+\sin^2\theta=1$$

49. By using the identity $\cos(90°-\theta)=\sin\theta$, we have:

$$\frac{\cos(90°-\theta)}{\cos\theta}=\frac{\sin\theta}{\cos\theta}=\tan\theta$$

51. By using the identities $\cos(90°-A)=\sin A$ and $\sin(90°-A)=\cos A$, we have:

$$\frac{\cos^2(90°-A)}{\sin^2(90°-A)}-\frac{1}{\cos^2 A}=\frac{\sin^2 A}{\cos^2 A}-\frac{1}{\cos^2 A}$$

$$=\frac{\sin^2 A-1}{\cos^2 A}$$

$$=-\frac{1-\sin^2 A}{\cos^2 A}$$

$$=-\frac{\cos^2 A}{\cos^2 A}$$

$$=-1$$

53. By using the identity $\sin(90°-\theta) = \cos\theta$, we have:
$$1 - \sin(90°-\theta)\cos\theta = 1 - \cos\theta \bullet \cos\theta = 1 - \cos^2\theta = \sin^2\theta$$

55. By using only algebraic simplification, we have:
$$\frac{\frac{\cos\theta+1}{\cos\theta}+1}{\frac{\cos\theta-1}{\cos\theta}-1} = \frac{\cos\theta+1+\cos\theta}{\cos\theta-1-\cos\theta} = \frac{2\cos\theta+1}{-1} = -2\cos\theta-1$$

57. We solve the first equation for A:
$$A\sin\theta + \cos\theta = 1$$
$$A\sin\theta = 1 - \cos\theta$$
$$A = \frac{1-\cos\theta}{\sin\theta}$$
Now solve the second equation for B:
$$B\sin\theta - \cos\theta = 1$$
$$B\sin\theta = 1 + \cos\theta$$
$$B = \frac{1+\cos\theta}{\sin\theta}$$
Now compute the product AB:
$$AB = \frac{1-\cos\theta}{\sin\theta} \bullet \frac{1+\cos\theta}{\sin\theta} = \frac{1-\cos^2\theta}{\sin^2\theta} = \frac{\sin^2\theta}{\sin^2\theta} = 1$$

59. Recall that $\sin^2\theta + \cos^2\theta = 1$, so we have the system of equations:
$$a\sin^2\theta + b\cos^2\theta = 1$$
$$\sin^2\theta + \cos^2\theta = 1$$
Multiply the second equation by $-b$:
$$a\sin^2\theta + b\cos^2\theta = 1$$
$$-b\sin^2\theta - b\cos^2\theta = -b$$
Adding, we get:
$$(a-b)\sin^2\theta = 1-b$$
$$\sin^2\theta = \frac{1-b}{a-b}$$
So, we have:
$$\cos^2\theta = 1 - \sin^2\theta = 1 - \frac{1-b}{a-b} = \frac{(a-b)-(1-b)}{a-b} = \frac{a-1}{a-b}$$
Using the identity for $\tan\theta$, we have:
$$\tan^2\theta = \frac{\sin^2\theta}{\cos^2\theta} = \frac{\frac{1-b}{a-b}}{\frac{a-1}{a-b}} = \frac{1-b}{a-1} = \frac{b-1}{1-a}$$
This proves the desired results.

61. Solving the identity $\sin^2\beta + \cos^2\beta = 1$ for $\cos\beta$ yields $\cos\beta = \sqrt{1 - \sin^2\beta}$, so:

$$\cos\beta = \sqrt{1 - \sin^2\beta}$$

$$= \sqrt{1 - \frac{m^4 - 2m^2n^2 + n^4}{\left(m^2 + n^2\right)^2}}$$

$$= \frac{\sqrt{m^4 + 2m^2n^2 + n^4 - m^4 + 2m^2n^2 - n^4}}{m^2 + n^2}$$

$$= \frac{\sqrt{4m^2n^2}}{m^2 + n^2}$$

$$= \frac{2mn}{m^2 + n^2} \quad \text{(since } m > 0, \, n > 0)$$

Now using the identity for $\tan\beta$, we have:

$$\tan\beta = \frac{\sin\beta}{\cos\beta} = \frac{\dfrac{m^2 - n^2}{m^2 + n^2}}{\dfrac{2mn}{m^2 + n^2}} = \frac{m^2 - n^2}{2mn}$$

63. First we compute $a^2 - b^2$:

$$a^2 - b^2 = (\tan\theta + \sin\theta)^2 - (\tan\theta - \sin\theta)^2$$
$$= \tan^2\theta + 2\sin\theta\tan\theta + \sin^2\theta - \tan^2\theta + 2\sin\theta\tan\theta - \sin^2\theta$$
$$= 4\sin\theta\tan\theta$$

Now compute $4\sqrt{ab}$:

$$4\sqrt{ab} = 4\sqrt{(\tan\theta + \sin\theta)(\tan\theta - \sin\theta)}$$

$$= 4\sqrt{\tan^2\theta - \sin^2\theta}$$

$$= 4\sqrt{\frac{\sin^2\theta}{\cos^2\theta} - \sin^2\theta}$$

$$= 4\sqrt{\frac{\sin^2\theta\left(1 - \cos^2\theta\right)}{\cos^2\theta}}$$

$$= 4\sqrt{\frac{\sin^4\theta}{\cos^2\theta}}$$

$$= 4\left(\frac{\sin^2\theta}{\cos\theta}\right)$$

$$= 4\left(\frac{\sin\theta}{\cos\theta}\right)\sin\theta$$

$$= 4\sin\theta\tan\theta$$

So $a^2 - b^2 = 4\sqrt{ab}$.

2.3 Right-Triangle Applications

1. We draw the figure:

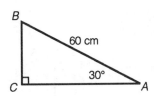

Since $\sin 30° = \dfrac{BC}{60}$, we have:

$\quad BC = 60 \sin 30° = 60 \bullet \frac{1}{2} = 30$ cm

Since $\cos 30° = \dfrac{AC}{60}$, we have:

$\quad AC = 60 \cos 30° = 60 \bullet \frac{\sqrt{3}}{2} = 30\sqrt{3}$ cm

3. We draw the figure:

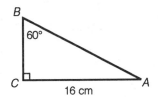

Since $\sin 60° = \dfrac{16}{AB}$, we have:

$\quad AB = \dfrac{16}{\sin 60°} = \dfrac{16}{\frac{\sqrt{3}}{2}} = \dfrac{32}{\sqrt{3}} = \dfrac{32\sqrt{3}}{3}$ cm

Since $\tan 60° = \dfrac{16}{BC}$, we have:

$\quad BC = \dfrac{16}{\tan 60°} = \dfrac{16}{\sqrt{3}} = \dfrac{16\sqrt{3}}{3}$ cm

5. We draw the figure:

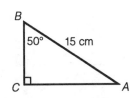

Since $\sin 50° = \dfrac{AC}{15}$, we have:

$\quad AC = 15 \sin 50° \approx 11.5$ cm

Since $\cos 50° = \dfrac{BC}{15}$, we have:

$BC = 15\cos 50° \approx 9.6$ cm

7.   We draw a figure:

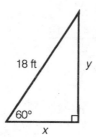

(a)   We are asked to find y:

$\sin 60° = \frac{y}{18}$

$y = 18\sin 60°$

$y = 18\left(\frac{\sqrt{3}}{2}\right)$

$y = 9\sqrt{3}$ ft

Using a calculator, this is approximately 15.59 ft.

(b)   We are asked to find x:

$\cos 60° = \frac{x}{18}$

$x = 18\cos 60°$

$x = 18\left(\frac{1}{2}\right)$

$x = 9$ ft

9.   Using the sine function, we have that $\sin(\angle SEM) = \dfrac{MS}{SE}$, so:

$MS = SE\sin(\angle SEM) = 93\sin 21.16° \approx 34$

Thus the distance MS is approximately 34 million miles.

11.   We draw a figure:

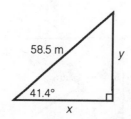

We first find x and y:
$$\cos 41.4° = \frac{x}{58.5}, \text{ so } x = 58.5\cos 41.4° \approx 43.9 \text{ m}$$
$$\sin 41.4° = \frac{y}{58.5}, \text{ so } y = 58.5\sin 41.4° \approx 38.7 \text{ m}$$
So the total length of fencing required is:
$$43.9 \text{ m} + 38.7 \text{ m} + 58.5 \text{ m} \approx 141.1 \text{ m}$$

13. (a) We first draw the triangle:

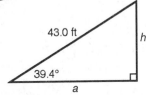

Now find h:
$$\sin 39.4° = \frac{h}{43.0}, \text{ so } h = 43.0\sin 39.4° \approx 27.3 \text{ ft}$$

(b) We first find a:
$$\cos 39.4° = \frac{a}{43.0}, \text{ so } a = 43.0\cos 39.4° \approx 33.2 \text{ ft}$$
Now the gable has a base of $2(33.2) = 66.4$ ft and a height of 27.3 ft, so its area is given by:
$$\tfrac{1}{2}(\text{base})(\text{height}) \approx \tfrac{1}{2}(66.4)(27.3) \approx 906.9 \text{ ft}^2$$
Note: This calculation was done with the full calculator approximation, not the values rounded to one decimal place.

15. Using the formula $A = \tfrac{1}{2}ab\sin\theta$, we have:
$$A = \tfrac{1}{2}(2)(3)\sin 30° = 1.5 \text{ in.}^2$$

17. We start by finding the measure of a central angle of a triangle drawn from the center out to two adjacent vertices:

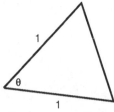

Since the polygon is 7-sided, we have $7\theta = 360°$, so $\theta = \frac{360°}{7}$. The area of this triangle is thus:
$$\tfrac{1}{2}ab\sin\theta = \tfrac{1}{2}(1)(1)\sin\frac{360°}{7} = \tfrac{1}{2}\sin\frac{360°}{7}$$
The area of the septagon (7-sided figure) is therefore:
$$7\left(\tfrac{1}{2}\sin\frac{360°}{7}\right) = \tfrac{7}{2}\sin\frac{360°}{7} \approx 2.736 \text{ square units}$$

19. We start by finding the measure of a central angle of a triangle drawn from the center out to two adjacent vertices:

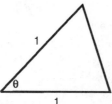

Since the polygon is 8-sided, we have $8\theta = 360°$, so $\theta = 45°$. The area of this triangle is thus:

$$\tfrac{1}{2}ab\sin\theta = \tfrac{1}{2}(1)(1)\sin 45° = \tfrac{\sqrt{2}}{4}$$

The area of the octagon (8-sided figure) is therefore:

$$8\left(\tfrac{\sqrt{2}}{4}\right) = 2\sqrt{2}$$

The shaded area is obtained by subtracting this area from the circular area, therefore:

$$\text{Area} = \pi(1)^2 - 2\sqrt{2} = \pi - 2\sqrt{2} \approx 0.313 \text{ square units}$$

21. We start by finding the measure of a central angle of a triangle drawn from the center out to two adjacent vertices:

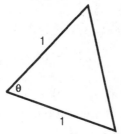

Since the polygon is 6-sided, we have $6\theta = 360°$, so $\theta = 60°$. The area of this triangle is thus:

$$\tfrac{1}{2}ab\sin\theta = \tfrac{1}{2}(1)(1)\sin 60° = \tfrac{\sqrt{3}}{4}$$

Since the shaded area is comprised of four such triangles, the shaded area is:

$$4\left(\tfrac{\sqrt{3}}{4}\right) = \sqrt{3} \approx 1.732 \text{ square units}$$

23. Using the hint, we draw the figure:

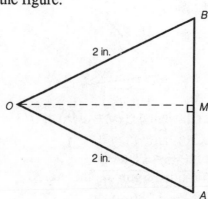

Now $\angle AOB = \frac{360°}{5} = 72°$, so $\angle AOM = 36°$. We find AM:

$$\sin 36° = \frac{AM}{2}$$

$$AM = 2\sin 36°$$

Thus $AB = 4\sin 36°$, and since there are five sides, the perimeter is:

$$5(AB) = 5(4\sin 36°) = 20\sin 36° \text{ inches}$$

25. We draw the figure:

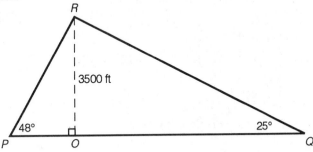

Now compute:

$$\tan 48° = \frac{3500}{PO} \quad \text{and} \quad \tan 25° = \frac{3500}{OQ}$$

So $PO = \dfrac{3500}{\tan 48°}$ and $OQ = \dfrac{3500}{\tan 25°}$. Thus $PQ = \dfrac{3500}{\tan 48°} + \dfrac{3500}{\tan 25°} \approx 10{,}660 \text{ ft.}$

27. We draw a figure:

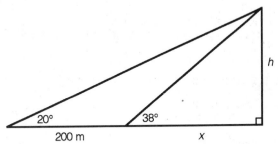

Now $\tan 38° = \dfrac{h}{x}$, so $x = \dfrac{h}{\tan 38°}$. Also $\tan 20° = \dfrac{h}{200 + x}$, so $h = (200 + x)\tan 20°$.
Substituting, we have:

$$h = \left(200 + \frac{h}{\tan 38°}\right)\tan 20°$$

$$h\tan 38° = 200\tan 38°\tan 20° + h\tan 20°$$

$$h(\tan 38° - \tan 20°) = 200\tan 38°\tan 20°$$

$$h = \frac{200\tan 38°\tan 20°}{\tan 38° - \tan 20°}$$

$$h \approx 136 \text{ m}$$

29. We draw the figure:

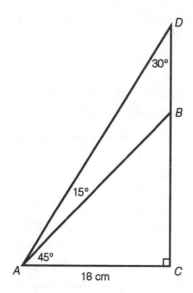

Now compute:

$$\tan 45° = \frac{BC}{18}, \text{ so } BC = 18\tan 45°$$

$$\tan 60° = \frac{CD}{18}, \text{ so } CD = 18\tan 60°$$

So:

$$BD = CD - BC = 18(\tan 60° - \tan 45°) = 18(\sqrt{3} - 1) \text{ cm}$$

31. (a) First, note that $\angle BOA = 90° - \theta$ since it forms a right angle with θ. Also, $\angle OAB = \theta$ since both angles are complemetary to the same angle ($\angle AOB$). Now, $\angle BAP = 90° - \theta$ since it forms a right angle with $\angle OAB$. Finally, $\angle BPA = \theta$ since $\angle BPA$ and $\angle OAB$ are both complementary to the same angle ($\angle BAP$).

(b) Using $\triangle AOP$, we have:

$$\sin\theta = \frac{AO}{OP} = \frac{AO}{1}, \text{ so } AO = \sin\theta$$

$$\cos\theta = \frac{AP}{OP} = \frac{AP}{1}, \text{ so } AP = \cos\theta$$

Using $\triangle AOB$, we have:

$$\sin\theta = \frac{OB}{OA} = \frac{OB}{\sin\theta}, \text{ so } OB = \sin^2\theta$$

Using $\triangle ABP$, we have:

$$\cos\theta = \frac{BP}{AP} = \frac{BP}{\cos\theta}, \text{ so } BP = \cos^2\theta$$

33. (a) Examine the similar triangles having $\overline{AB}$ and $\overline{BC}$ as hypotenuses, and notice that θ is also the angle at B in the smaller triangle. Thus:

$$\sin \theta = \frac{5}{BC}, \text{ so } BC = \frac{5}{\sin \theta}$$

(b) Using the smaller triangle:

$$\cos \theta = \frac{4}{AB}, \text{ so } AB = \frac{4}{\cos \theta}$$

(c) Since $AC = AB + BC$, then:

$$AC = \frac{4}{\cos \theta} + \frac{5}{\sin \theta} = 4\sec \theta + 5\csc \theta$$

35. First observe that the figure $x^2 + y^2 = 1$ is a circle with a radius of one. Since OA, OD and OF all represent the radius, they are each equal to 1. In each case, we will look for a trigonometric relationship involving the required segment:

(a) $\sin \theta = \dfrac{DE}{OD}$, so $DE = \sin \theta$

(b) $\cos \theta = \dfrac{OE}{OD}$, so $OE = \cos \theta$

(c) $\tan \theta = \dfrac{CF}{OF}$, so $CF = \tan \theta$

(d) $\sec \theta = \dfrac{OC}{OF}$, so $OC = \sec \theta$

Going to $\triangle OAB$, $\angle ABO = \theta$, thus:

(e) $\cot \theta = \dfrac{AB}{OA}$, so $AB = \cot \theta$

(f) $\csc \theta = \dfrac{OB}{OA}$, so $OB = \csc \theta$

37. (a) We can use $\sin\theta$ to set up a trigonometric relationship:

$$\sin\theta = \frac{r}{PS+r}$$

$$PS\sin\theta + r\sin\theta = r$$

$$PS\sin\theta = r - r\sin\theta$$

$$PS\sin\theta = r(1-\sin\theta)$$

$$r = \left(\frac{\sin\theta}{1-\sin\theta}\right)PS$$

(b) Using a calculator, we have:

$$r = \left(\frac{\sin 0.257°}{1-\sin 0.257°}\right)(238{,}857) \approx 1080 \text{ miles}$$

39. (a) We draw the figure:

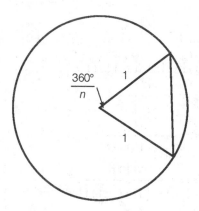

The area of the triangle is given by:

$$\tfrac{1}{2}ab\sin\theta = \tfrac{1}{2}(1)(1)\sin\tfrac{360°}{n} = \tfrac{1}{2}\sin\tfrac{360°}{n}$$

Since there are n congruent triangles in the n-gon, then its area is $\frac{n}{2}\sin\frac{360°}{n}$.

(b) We complete the table:

n	5	10	50	100	1,000	5,000	10,000
A_n	2.38	2.94	3.1333	3.1395	3.141572	3.1415918	3.1415924

(c) As n gets larger, A_n becomes closer to the area of the circle, which is π.

41. We draw the figure:

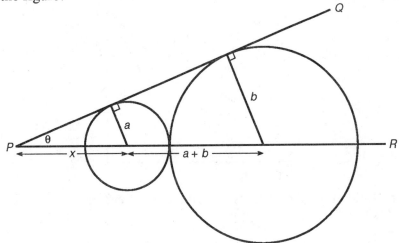

Using the smaller triangle:
$$\sin\theta = \frac{a}{x}, \text{ so } x = \frac{a}{\sin\theta}$$
Using the larger triangle and substituting for x:
$$\sin\theta = \frac{b}{x+a+b}$$
$$\sin\theta = \frac{b}{\frac{a}{\sin\theta}+a+b}$$
$$\sin\theta = \frac{b\sin\theta}{a+(a+b)\sin\theta}$$
$$a\sin\theta+(a+b)\sin^2\theta = b\sin\theta$$
$$a+(a+b)\sin\theta = b \quad (\text{since } \theta \neq 0 \text{ and } \theta \neq 180°, \text{ then } \sin\theta \neq 0)$$
$$(a+b)\sin\theta = b-a$$
$$\sin\theta = \frac{b-a}{a+b}$$
Since $\cos^2\theta + \sin^2\theta = 1$, then:
$$\cos\theta = \sqrt{1-\sin^2\theta}$$
$$= \sqrt{1-\frac{(b-a)^2}{(a+b)^2}}$$
$$= \sqrt{\frac{a^2+2ab+b^2-b^2+2ab-a^2}{(a+b)^2}}$$
$$= \sqrt{\frac{4ab}{(a+b)^2}}$$
$$= \frac{2\sqrt{ab}}{a+b}$$

43. We draw the figure:

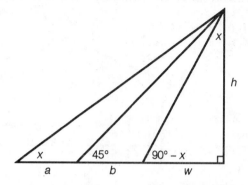

We have the following relationships from the three triangles:

small: $\tan x = \dfrac{w}{h}$

middle: $\tan 45° = \dfrac{h}{b+w}$, so $h = b + w$ and thus $w = h - b$

large: $\tan x = \dfrac{h}{a+b+w}$

Substituting, we have:
$$\frac{w}{h} = \frac{h}{a+b+w}$$

Since $w = h - b$, we have:
$$\frac{h-b}{h} = \frac{h}{a+b+h-b}$$
$$\frac{h-b}{h} = \frac{h}{h+a}$$
$$(h-b)(h+a) = h^2$$
$$h^2 - bh + ah - ab = h^2$$
$$ah - bh = ab$$
$$h(a-b) = ab$$
$$h = \frac{ab}{a-b}$$

So the tower is $\dfrac{ab}{a-b}$ feet high.

2.4 Trigonometric Functions of Angles

1. (a) The reference angle for 110° is 70°:

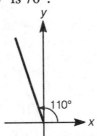

   (b) The reference angle for –110° is 70°:

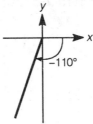

3. (a) The reference angle for 200° is 20°:

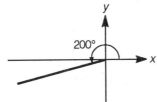

   (b) The reference angle for –200° is 20°:

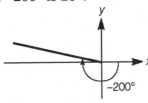

5. (a) The reference angle for 300° is 60°:

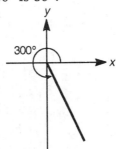

(b)   The reference angle for −300° is 60°:

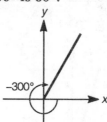

7.   (a)   The reference angle for 60° is 60°:

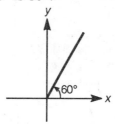

(b)   The reference angle for −60° is 60°:

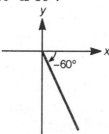

9.   Since 270° corresponds to the point $(0, -1)$ on the unit circle, then using $x = 0$ and $y = -1$ in the definitions yields:

$$\cos 270° = x = 0 \qquad\qquad \sec 270° = \frac{1}{x} = \frac{1}{0}, \text{ which is undefined}$$

$$\sin 270° = y = -1 \qquad\qquad \csc 270° = \frac{1}{y} = \frac{1}{-1} = -1$$

$$\tan 270° = \frac{y}{x} = \frac{-1}{0}, \text{ which is undefined} \qquad \cot 270° = \frac{x}{y} = \frac{0}{-1} = 0$$

11.   Since −270° corresponds to the point $(0, 1)$ on the unit circle, then using $x = 0$ and $y = 1$ in the definitions yields:

$$\cos(-270°) = x = 0 \qquad\qquad \sec(-270°) = \frac{1}{x} = \frac{1}{0}, \text{ which is undefined}$$

$$\sin(-270°) = y = 1 \qquad\qquad \csc(-270°) = \frac{1}{y} = \frac{1}{1} = 1$$

$$\tan(-270°) = \frac{y}{x} = \frac{1}{0}, \text{ which is undefined} \qquad \cot(-270°) = \frac{x}{y} = \frac{0}{1} = 0$$

13. Since $810°$ results in the same point as $90°$ on the unit circle, which corresponds to the point $(0, 1)$, then using $x = 0$ and $y = 1$ in the definitions yields:

$$\cos 810° = x = 0$$

$$\sec 810° = \frac{1}{x} = \frac{1}{0}, \text{ which is undefined}$$

$$\sin 810° = y = 1$$

$$\csc 810° = \frac{1}{y} = \frac{1}{1} = 1$$

$$\tan 810° = \frac{y}{x} = \frac{1}{0}, \text{ which is undefined}$$

$$\cot 810° = \frac{x}{y} = \frac{0}{1} = 0$$

15. Using Figure 8, we have $\sin 10° \approx 0.2$ and $\sin(-10°) \approx -0.2$. Using a calculator, we have $\sin 10° \approx 0.17$ and $\sin(-10°) \approx -0.17$.

17. Using Figure 8, we have $\cos 80° \approx 0.2$ and $\cos(-80°) \approx 0.2$. Using a calculator, we have $\cos 80° \approx 0.17$ and $\cos(-80°) \approx 0.17$.

19. Using Figure 8, we have $\sin 120° \approx 0.9$ and $\sin(-120°) \approx -0.9$. Using a calculator, we have $\sin 120° \approx 0.87$ and $\sin(-120°) \approx -0.87$.

21. Using Figure 8, we have $\sin 150° = 0.5$ and $\sin(-150°) = -0.5$. Using a calculator, we have $\sin 150° = 0.5$ and $\sin(-150°) = -0.5$.

23. Using Figure 8, we have $\cos 220° \approx -0.8$ and $\cos(-220°) \approx -0.8$. Using a calculator, we have $\cos 220° \approx -0.77$ and $\cos(-220°) \approx -0.77$.

25. Using Figure 8, we have $\cos 310° \approx 0.6$ and $\cos(-310°) \approx 0.6$. Using a calculator, we have $\cos 310° \approx 0.64$ and $\cos(-310°) \approx 0.64$.

27. Since $40° + 360°$ is one full revolution on the unit circle past $40°$, then $\sin(40° + 360°) = \sin 40°$. Using Figure 8, we have $\sin(40° + 360°) \approx 0.6$. Using a calculator, we have $\sin(40° + 360°) \approx 0.64$.

29. (a) The reference angle for $315°$ is $45°$, and $\cos 45° = \frac{\sqrt{2}}{2}$. Since $\cos \theta$ is the x-coordinate, and $\theta = 315°$ lies in the fourth quadrant where the x-coordinates are positive, then $\cos 315° = \frac{\sqrt{2}}{2}$.

(b) The reference angle for $-315°$ is $45°$, and $\cos 45° = \frac{\sqrt{2}}{2}$. Since $\cos \theta$ is the x-coordinate, and $\theta = -315°$ lies in the first quadrant where the x-coordinates are positive, then $\cos(-315°) = \frac{\sqrt{2}}{2}$.

(c) The reference angle for $315°$ is $45°$, and $\sin 45° = \frac{\sqrt{2}}{2}$. Since $\sin \theta$ is the y-coordinate, and $\theta = 315°$ lies in the fourth quadrant where the y-coordinates are negative, then $\sin 315° = -\frac{\sqrt{2}}{2}$.

(d)   The reference angle for $-315°$ is $45°$, and $\sin 45° = \frac{\sqrt{2}}{2}$. Since $\sin\theta$ is the y-coordinate, and $\theta = -315°$ lies in the first quadrant where the y-coordinates are positive, then $\sin(-315°) = \frac{\sqrt{2}}{2}$.

31. (a)   The reference angle for $300°$ is $60°$, and $\cos 60° = \frac{1}{2}$. Since $\cos\theta$ is the x-coordinate, and $\theta = 300°$ lies in the fourth quadrant where the x-coordinates are positive, then $\cos 300° = \frac{1}{2}$.

(b)   The reference angle for $-300°$ is $60°$, and $\cos 60° = \frac{1}{2}$. Since $\cos\theta$ is the x-coordinate, and $\theta = -300°$ lies in the first quadrant where the x-coordinates are positive, then $\cos(-300°) = \frac{1}{2}$.

(c)   The reference angle for $300°$ is $60°$, and $\sin 60° = \frac{\sqrt{3}}{2}$. Since $\sin\theta$ is the y-coordinate, and $\theta = 300°$ lies in the fourth quadrant where the y-coordinates are negative, then $\sin 300° = -\frac{\sqrt{3}}{2}$.

(d)   The reference angle for $-300°$ is $60°$, and $\sin 60° = \frac{\sqrt{3}}{2}$. Since $\sin\theta$ is the y-coordinate, and $\theta = -300°$ lies in the first quadrant where the y-coordinates are positive, then $\sin(-300°) = \frac{\sqrt{3}}{2}$.

33. (a)   The reference angle for $210°$ is $30°$, and $\cos 30° = \frac{\sqrt{3}}{2}$. Since $\cos\theta$ is the x-coordinate, and $\theta = 210°$ lies in the third quadrant where the x-coordinates are negative, then $\cos 210° = -\frac{\sqrt{3}}{2}$.

(b)   The reference angle for $-210°$ is $30°$, and $\cos 30° = \frac{\sqrt{3}}{2}$. Since $\cos\theta$ is the x-coordinate, and $\theta = -210°$ lies in the second quadrant where the x-coordinates are negative, then $\cos(-210°) = -\frac{\sqrt{3}}{2}$.

(c)   The reference angle for $210°$ is $30°$, and $\sin 30° = \frac{1}{2}$. Since $\sin\theta$ is the y-coordinate, and $\theta = 210°$ lies in the third quadrant where the y-coordinates are negative, then $\sin 210° = -\frac{1}{2}$.

(d) The reference angle for $-210°$ is $30°$, and $\sin 30° = \frac{1}{2}$. Since $\sin \theta$ is the y-coordinate, and $\theta = -210°$ lies in the second quadrant where the y-coordinates are positive, then $\sin(-210°) = \frac{1}{2}$.

35. (a) The reference angle for $390°$ is $30°$ ($390° = 30° + 360°$), and $\cos 30° = \frac{\sqrt{3}}{2}$. Since $\cos \theta$ is the x-coordinate, and $\theta = 390°$ lies in the first quadrant where the x-coordinates are positive, then $\cos 390° = \frac{\sqrt{3}}{2}$.

(b) The reference angle for $-390°$ is $30°$ ($-390° = -30° - 360°$), and $\cos 30° = \frac{\sqrt{3}}{2}$. Since $\cos \theta$ is the x-coordinate, and $\theta = -390°$ lies in the fourth quadrant where the x-coordinates are positive, then $\cos(-390°) = \frac{\sqrt{3}}{2}$.

(c) The reference angle for $390°$ is $30°$, and $\sin 30° = \frac{1}{2}$. Since $\sin \theta$ is the y-coordinate, and $\theta = 390°$ lies in the first quadrant where the y-coordinates are positive, then $\sin 390° = \frac{1}{2}$.

(d) The reference angle for $-390°$ is $30°$, and $\sin 30° = \frac{1}{2}$. Since $\sin \theta$ is the y-coordinate, and $\theta = -390°$ lies in the fourth quadrant where the y-coordinates are negative, then $\sin(-390°) = -\frac{1}{2}$.

37. (a) The reference angle for $600°$ is $60°$ ($600° = 240° + 360°$), and $\sec 60° = 2$. Since $\sec \theta$ is the reciprocal of the x-coordinate, and $\theta = 600°$ lies in the third quadrant where the x-coordinates are negative, then $\sec 600° = -2$.

(b) The reference angle for $-600°$ is $60°$ ($-600° = -240° - 360°$), and $\csc 60° = \frac{2\sqrt{3}}{3}$. Since $\csc \theta$ is the reciprocal of the y-coordinate, and $\theta = -600°$ lies in the second quadrant where the y-coordinates are positive, then $\csc(-600°) = \frac{2\sqrt{3}}{3}$.

(c) The reference angle for $600°$ is $60°$, and $\tan 60° = \sqrt{3}$. Since $\tan \theta = \frac{y}{x}$, and $\theta = 600°$ lies in the third quadrant where both the x- and y-coordinates are negative, then $\tan 600° = \sqrt{3}$.

(d)   The reference angle for $-600°$ is $60°$, and $\cot 60° = \frac{\sqrt{3}}{3}$. Since $\cot\theta = \frac{x}{y}$ and $\theta = -600°$ lies in the second quadrant where the y-coordinates are positive and the x-coordinates are negative, then $\cot(-600°) = -\frac{\sqrt{3}}{3}$.

39. We complete the table:

θ	$\sin\theta$	$\cos\theta$	$\tan\theta$
$0°$	0	1	0
$30°$	$\frac{1}{2}$	$\frac{\sqrt{3}}{2}$	$\frac{\sqrt{3}}{3}$
$45°$	$\frac{\sqrt{2}}{2}$	$\frac{\sqrt{2}}{2}$	1
$60°$	$\frac{\sqrt{3}}{2}$	$\frac{1}{2}$	$\sqrt{3}$
$90°$	1	0	undefined
$120°$	$\frac{\sqrt{3}}{2}$	$-\frac{1}{2}$	$-\sqrt{3}$
$135°$	$\frac{\sqrt{2}}{2}$	$-\frac{\sqrt{2}}{2}$	-1
$150°$	$\frac{1}{2}$	$-\frac{\sqrt{3}}{2}$	$-\frac{\sqrt{3}}{3}$
$180°$	0	-1	0

41. We draw the figure:

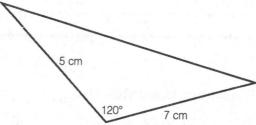

Now using the area formula, we have:

$$\text{Area} = \tfrac{1}{2}ab\sin\theta = \tfrac{1}{2}(7)(5)\sin 120° = \left(\tfrac{35}{2}\right)\left(\tfrac{\sqrt{3}}{2}\right) = \tfrac{35\sqrt{3}}{4} \text{ cm}^2$$

43. We draw the figure:

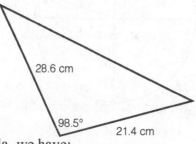

Now using the area formula, we have:

$$\text{Area} = \tfrac{1}{2}ab\sin\theta = \tfrac{1}{2}(21.4)(28.6)\sin 98.5° \approx 302.7 \text{ cm}^2$$

45. We start by finding the measure of a central angle of a triangle drawn from the center out to two adjacent vertices:

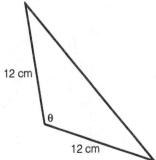

Since the triangle is 3-sided, we have $3\theta = 360°$, so $\theta = 120°$. The area of this triangle is thus:

$$\tfrac{1}{2}ab\sin\theta = \tfrac{1}{2}(12)(12)\sin 120° = 72 \bullet \tfrac{\sqrt{3}}{2} = 36\sqrt{3}$$

The area of the shaded figure (3 triangles) is therefore:

$$3(36\sqrt{3}) = 108\sqrt{3} \approx 187.06 \text{ cm}^2$$

47. (a)   We complete the table:

Terminal side of angle θ lies in

	Quadrant I	Quadrant II	Quadrant III	Quadrant IV
$\sin\theta$	positive	positive	negative	negative
$\cos\theta$	positive	negative	negative	positive
$\tan\theta$	positive	negative	positive	negative

(b)   It works!

49. (a)   A sketch of $\sin 10°$ indicates that $\sin 10° \approx 0.2$:

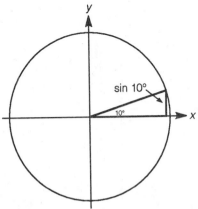

A sketch of $\sin 70°$ indicates that $\sin 70° \approx 0.9$:

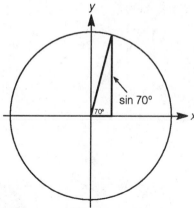

Clearly $\sin 70°$ is larger.

(b)   A sketch of $\cos 10°$ indicates that $\cos 10° \approx 1.0$:

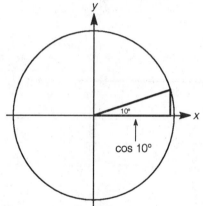

A sketch of $\cos 70°$ indicates that $\cos 70° \approx 0.3$:

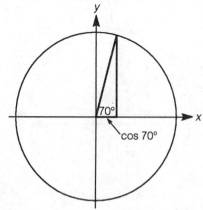

Clearly $\cos 10°$ is larger.

51. (a)   A sketch of cos 140° indicates that cos 140° ≈ −0.8:

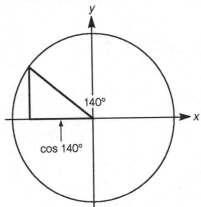

A sketch of cos 100° indicates that cos 100° ≈ −0.2:

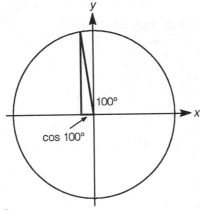

Clearly cos 100° is larger.

(b)   A sketch of sin 140° indicates that sin 140° ≈ 0.6:

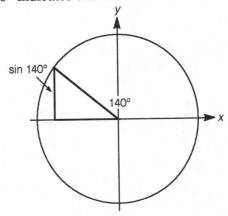

A sketch of sin 100° indicates that sin 100° ≈ 1.0:

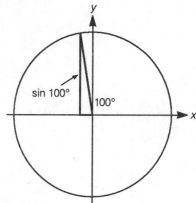

Clearly sin 100° is larger.

53. (a)   We sketch the figure showing the obtuse angle 180° − θ in standard position:

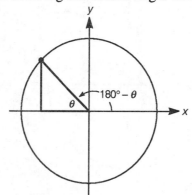

Note that the terminal side of the angle 180° − θ lies in the second quadrant.

(b)   The reference angle for 180° − θ is θ.

(c)   Since 180° − θ lies in the second quadrant where all y-coordinates are positive, then
$\sin(180° - \theta) = \sin\theta$.

55. (a)   In $\triangle ADC$, $\sec\alpha = \dfrac{AC}{AD} = \dfrac{AC}{1} = AC$.

Similarly, in $\triangle ADB$ we have $\sec\beta = \dfrac{AB}{AD} = \dfrac{AB}{1} = AB$.

(b)   We compute the areas:

$\text{Area } \triangle ADC = \tfrac{1}{2}(AC)(AD)\sin\alpha = \tfrac{1}{2}\sec\alpha\sin\alpha$

$\text{Area } \triangle ADB = \tfrac{1}{2}(AD)(AB)\sin\beta = \tfrac{1}{2}\sec\beta\sin\beta$

$\text{Area } \triangle ABC = \tfrac{1}{2}(AC)(AB)\sin(\alpha + \beta) = \tfrac{1}{2}\sec\alpha\sec\beta\sin(\alpha + \beta)$

(c) Since the sum of the areas of the two smaller triangles equals the area of $\triangle ABC$, we have:

$$\tfrac{1}{2}\sec\alpha\sec\beta\sin(\alpha+\beta) = \tfrac{1}{2}\sec\alpha\sin\alpha + \tfrac{1}{2}\sec\beta\sin\beta$$

Now multiply both sides of this equation by the quantity $2\cos\alpha\cos\beta$. This yields:

$$\sin(\alpha+\beta) = \sin\alpha\cos\beta + \cos\alpha\sin\beta$$

(d) Using the hint, we have:

$$\begin{aligned}
\sin 75° &= \sin(30°+45°) \\
&= \sin 30°\cos 45° + \cos 30°\sin 45° \\
&= \frac{1}{2}\cdot\frac{\sqrt{2}}{2} + \frac{\sqrt{3}}{2}\cdot\frac{\sqrt{2}}{2} \\
&= \frac{\sqrt{2}+\sqrt{6}}{4}
\end{aligned}$$

(e) We compute $\sin 75° = \dfrac{\sqrt{2}+\sqrt{6}}{4} \approx \dfrac{1.4+2.4}{4} = \dfrac{3.8}{4} < 1.$

But $\sin 30° + \sin 45° \approx 0.5 + 0.7 = 1.2 > 1$. Since $\sin 75°$ is less than 1, whereas $\sin 30° + \sin 45°$ is greater than 1, the quantities cannot be equal. We can also argue with exact values (not approximations):

$$\sin 30° + \sin 45° = \frac{1}{2} + \frac{\sqrt{2}}{2} = \frac{2+2\sqrt{2}}{4}$$

$$\sin 75° = \frac{\sqrt{2}+\sqrt{6}}{4}$$

So, if these were to be equal then $2 + 2\sqrt{2} = \sqrt{2} + \sqrt{6}$, thus $2 + \sqrt{2} = \sqrt{6}$. But $2 + \sqrt{2} > 3$ while $\sqrt{6} < 3$, thus they cannot be equal.

(f) We compute $\sin 105° \approx 0.9659$, $\sin 45° \approx 0.7071$, and $\sin 60° \approx 0.8660$. So $\sin 45° + \sin 60° \approx 1.5731 \neq \sin 105°$.
Note: An argument similar to (e) using exact values can be made here.

57. (a) $h(60°) = \tan 60° = \sqrt{3}$

(b) $h(2\cdot 60°) = \tan 120° = -\sqrt{3}$

(c) $h\!\left(\dfrac{60°}{2}\right) = \tan 30° = \dfrac{\sqrt{3}}{3}$

59. (a) Using a calculator, we have:

$$\frac{f(27°)}{g(27°)} = \frac{\sin 27°}{\cos 27°} \approx 0.510$$

(b) $h(27°) = \tan 27° \approx 0.510$

(c) $[f(27°)]^2 + [g(27°)]^2 = \sin^2 27° + \cos^2 27° = 1$

Note that a calculator is not necessary, since $\sin^2 \theta + \cos^2 \theta = 1$ for all angles θ.

2.5 Trigonometric Identities

1. (a) Substituting $\sin \theta = \frac{1}{5}$ into $\cos^2 \theta = 1 - \sin^2 \theta$ yields:

$$\cos^2 \theta = 1 - \left(\tfrac{1}{5}\right)^2 = 1 - \tfrac{1}{25} = \tfrac{24}{25}$$

Since the terminal side of θ lies in quadrant 2, $\cos \theta < 0$ and thus:

$$\cos \theta = -\sqrt{\tfrac{24}{25}} = -\tfrac{2\sqrt{6}}{5}$$

Now compute the remaining four trigonometric values:

$$\tan \theta = \frac{\sin \theta}{\cos \theta} = \frac{\tfrac{1}{5}}{-\tfrac{2\sqrt{6}}{5}} = -\frac{1}{2\sqrt{6}} = -\frac{\sqrt{6}}{12}$$

$$\cot \theta = \frac{\cos \theta}{\sin \theta} = \frac{-\tfrac{2\sqrt{6}}{5}}{\tfrac{1}{5}} = -2\sqrt{6}$$

$$\sec \theta = \frac{1}{\cos \theta} = \frac{1}{-\tfrac{2\sqrt{6}}{5}} = -\frac{5}{2\sqrt{6}} = -\frac{5\sqrt{6}}{12}$$

$$\csc \theta = \frac{1}{\sin \theta} = \frac{1}{\tfrac{1}{5}} = 5$$

(b) Substituting $\sin \theta = -\frac{1}{5}$ into $\cos^2 \theta = 1 - \sin^2 \theta$ yields:

$$\cos^2 \theta = 1 - \left(\tfrac{1}{5}\right)^2 = 1 - \tfrac{1}{25} = \tfrac{24}{25}$$

Since the terminal side of θ lies in quadrant 3, $\cos \theta < 0$ and thus:

$$\cos \theta = -\sqrt{\tfrac{24}{25}} = -\tfrac{2\sqrt{6}}{5}$$

Now compute the remaining four trigonometric values:

$$\tan \theta = \frac{\sin \theta}{\cos \theta} = \frac{-\tfrac{1}{5}}{-\tfrac{2\sqrt{6}}{5}} = \frac{1}{2\sqrt{6}} = \frac{\sqrt{6}}{12}$$

$$\cot \theta = \frac{\cos \theta}{\sin \theta} = \frac{-\tfrac{2\sqrt{6}}{5}}{-\tfrac{1}{5}} = 2\sqrt{6}$$

$$\sec \theta = \frac{1}{\cos \theta} = \frac{1}{-\tfrac{2\sqrt{6}}{5}} = -\frac{5}{2\sqrt{6}} = -\frac{5\sqrt{6}}{12}$$

$$\csc \theta = \frac{1}{\sin \theta} = \frac{1}{-\tfrac{1}{5}} = -5$$

3. (a) Substituting $\cos\theta = \frac{5}{13}$ into $\sin^2\theta = 1 - \cos^2\theta$ yields:

$$\sin^2\theta = 1 - \left(\tfrac{5}{13}\right)^2 = 1 - \tfrac{25}{169} = \tfrac{144}{169}$$

Since the terminal side of θ lies in quadrant 1, $\sin\theta > 0$ and thus:

$$\sin\theta = \sqrt{\tfrac{144}{169}} = \tfrac{12}{13}$$

Now compute the remaining four trigonometric values:

$$\tan\theta = \frac{\sin\theta}{\cos\theta} = \frac{\frac{12}{13}}{\frac{5}{13}} = \frac{12}{5} \qquad \cot\theta = \frac{\cos\theta}{\sin\theta} = \frac{\frac{5}{13}}{\frac{12}{13}} = \frac{5}{12}$$

$$\sec\theta = \frac{1}{\cos\theta} = \frac{1}{\frac{5}{13}} = \frac{13}{5} \qquad \csc\theta = \frac{1}{\sin\theta} = \frac{1}{\frac{12}{13}} = \frac{13}{12}$$

(b) Substituting $\cos\theta = -\frac{5}{13}$ into $\sin^2\theta = 1 - \cos^2\theta$ yields:

$$\sin^2\theta = 1 - \left(-\tfrac{5}{13}\right)^2 = 1 - \tfrac{25}{169} = \tfrac{144}{169}$$

Since the terminal side of θ lies in quadrant 3, $\sin\theta < 0$ and thus:

$$\sin\theta = -\sqrt{\tfrac{144}{169}} = -\tfrac{12}{13}$$

Now compute the remaining four trigonometric values:

$$\tan\theta = \frac{\sin\theta}{\cos\theta} = \frac{-\frac{12}{13}}{-\frac{5}{13}} = \frac{12}{5} \qquad \cot\theta = \frac{\cos\theta}{\sin\theta} = \frac{-\frac{5}{13}}{-\frac{12}{13}} = \frac{5}{12}$$

$$\sec\theta = \frac{1}{\cos\theta} = \frac{1}{-\frac{5}{13}} = -\frac{13}{5} \qquad \csc\theta = \frac{1}{\sin\theta} = \frac{1}{-\frac{12}{13}} = -\frac{13}{12}$$

5. Since $\csc A = -3$, then $\sin A = -\frac{1}{3}$. Substituting into $\cos^2 A = 1 - \sin^2 A$ yields:

$$\cos^2 A = 1 - \left(-\tfrac{1}{3}\right)^2 = 1 - \tfrac{1}{9} = \tfrac{8}{9}$$

Since the terminal side of A lies in quadrant 4, $\cos A > 0$ and thus:

$$\cos A = \sqrt{\tfrac{8}{9}} = \tfrac{2\sqrt{2}}{3}$$

Now compute the remaining three trigonometric values:

$$\tan A = \frac{\sin A}{\cos A} = \frac{-\frac{1}{3}}{\frac{2\sqrt{2}}{3}} = -\frac{1}{2\sqrt{2}} = -\frac{\sqrt{2}}{4}$$

$$\cot A = \frac{\cos A}{\sin A} = \frac{\frac{2\sqrt{2}}{3}}{-\frac{1}{3}} = -2\sqrt{2}$$

$$\sec A = \frac{1}{\cos A} = \frac{1}{\frac{2\sqrt{2}}{3}} = \frac{3}{2\sqrt{2}} = \frac{3\sqrt{2}}{4}$$

7.   Since $\sec B = -\frac{3}{2}$, then $\cos B = -\frac{2}{3}$. Substituting into $\sin^2 B = 1 - \cos^2 B$ yields:

$$\sin^2 B = 1 - \left(-\frac{2}{3}\right)^2 = 1 - \frac{4}{9} = \frac{5}{9}$$

Since the terminal side of B lies in quadrant 3, $\sin B < 0$ and thus:

$$\sin B = -\sqrt{\frac{5}{9}} = -\frac{\sqrt{5}}{3}$$

Now compute the remaining three trigonometric values:

$$\tan B = \frac{\sin B}{\cos B} = \frac{-\frac{\sqrt{5}}{3}}{-\frac{2}{3}} = \frac{\sqrt{5}}{2}$$

$$\cot B = \frac{\cos B}{\sin B} = \frac{-\frac{2}{3}}{-\frac{\sqrt{5}}{3}} = \frac{2}{\sqrt{5}} = \frac{2\sqrt{5}}{5}$$

$$\csc B = \frac{1}{\sin B} = \frac{1}{-\frac{\sqrt{5}}{3}} = -\frac{3}{\sqrt{5}} = -\frac{3\sqrt{5}}{5}$$

9.   Substituting $\cos\theta = \frac{t}{3}$ into $\sin^2\theta = 1 - \cos^2\theta$ yields:

$$\sin^2\theta = 1 - \left(\frac{t}{3}\right)^2 = 1 - \frac{t^2}{9} = \frac{9 - t^2}{9}$$

Since the terminal side of θ lies in quadrant 4, $\sin\theta < 0$ and thus:

$$\sin\theta = -\sqrt{\frac{9 - t^2}{9}} = -\frac{\sqrt{9 - t^2}}{3}$$

Now compute the remaining four trigonometric values:

$$\tan\theta = \frac{\sin\theta}{\cos\theta} = \frac{-\frac{\sqrt{9-t^2}}{3}}{\frac{t}{3}} = -\frac{\sqrt{9 - t^2}}{t}$$

$$\cot\theta = \frac{\cos\theta}{\sin\theta} = \frac{\frac{t}{3}}{-\frac{\sqrt{9-t^2}}{3}} = -\frac{t}{\sqrt{9 - t^2}} = -\frac{t\sqrt{9 - t^2}}{9 - t^2}$$

$$\sec\theta = \frac{1}{\cos\theta} = \frac{1}{\frac{t}{3}} = \frac{3}{t}$$

$$\csc\theta = \frac{1}{\sin\theta} = \frac{1}{-\frac{\sqrt{9-t^2}}{3}} = -\frac{3}{\sqrt{9 - t^2}} = -\frac{3\sqrt{9 - t^2}}{9 - t^2}$$

11.   Substituting $\sin\theta = -3u$ into $\cos^2\theta = 1 - \sin^2\theta$ yields:

$$\cos^2\theta = 1 - (-3u)^2 = 1 - 9u^2$$

Since the terminal side of θ lies in quadrant 3, $\cos\theta < 0$ and thus:

$$\cos\theta = -\sqrt{1 - 9u^2}$$

Now compute the remaining four trigonometric values:

$$\tan\theta = \frac{\sin\theta}{\cos\theta} = \frac{-3u}{-\sqrt{1-9u^2}} = \frac{3u\sqrt{1-9u^2}}{1-9u^2}$$

$$\cot\theta = \frac{\cos\theta}{\sin\theta} = \frac{-\sqrt{1-9u^2}}{-3u} = \frac{\sqrt{1-9u^2}}{3u}$$

$$\sec\theta = \frac{1}{\cos\theta} = \frac{1}{-\sqrt{1-9u^2}} = -\frac{\sqrt{1-9u^2}}{1-9u^2}$$

$$\csc\theta = \frac{1}{\sin\theta} = \frac{1}{-3u} = -\frac{1}{3u}$$

13. Substituting $\cos\theta = \dfrac{u}{\sqrt{3}}$ into $\sin^2\theta = 1 - \cos^2\theta$ yields:

$$\sin^2\theta = 1 - \left(\frac{u}{\sqrt{3}}\right)^2 = 1 - \frac{u^2}{3} = \frac{3-u^2}{3}$$

Since the terminal side of θ lies in quadrant 1, $\sin\theta > 0$ and thus:

$$\sin\theta = \sqrt{\frac{3-u^2}{3}} = \frac{\sqrt{3-u^2}}{\sqrt{3}} = \frac{\sqrt{9-3u^2}}{3}$$

Now compute the remaining four trigonometric values:

$$\tan\theta = \frac{\sin\theta}{\cos\theta} = \frac{\frac{\sqrt{9-3u^2}}{3}}{\frac{u}{\sqrt{3}}} = \frac{\sqrt{9-3u^2}}{u\sqrt{3}} = \frac{\sqrt{3-u^2}}{u}$$

$$\cot\theta = \frac{\cos\theta}{\sin\theta} = \frac{\frac{u}{\sqrt{3}}}{\frac{\sqrt{9-3u^2}}{3}} = \frac{u\sqrt{3}}{\sqrt{9-3u^2}} = \frac{u}{\sqrt{3-u^2}} = \frac{u\sqrt{3-u^2}}{3-u^2}$$

$$\sec\theta = \frac{1}{\cos\theta} = \frac{1}{\frac{u}{\sqrt{3}}} = \frac{\sqrt{3}}{u}$$

$$\csc\theta = \frac{1}{\sin\theta} = \frac{1}{\frac{\sqrt{9-3u^2}}{3}} = \frac{3}{\sqrt{9-3u^2}} = \frac{3\sqrt{9-3u^2}}{9-3u^2} = \frac{\sqrt{9-3u^2}}{3-u^2}$$

15. Using the identities for $\sec\theta$ and $\csc\theta$, we have:

$$\sin\theta\cos\theta\sec\theta\csc\theta = \frac{\sin\theta\cos\theta}{\sin\theta\cos\theta} = 1$$

17. Using the identities for $\sec\theta$ and $\tan\theta$, we have:

$$\frac{\sin\theta\sec\theta}{\tan\theta} = \frac{\sin\theta\bullet\frac{1}{\cos\theta}}{\frac{\sin\theta}{\cos\theta}} = \frac{\frac{\sin\theta}{\cos\theta}}{\frac{\sin\theta}{\cos\theta}} = 1$$

19. Working from the right-hand side and using the identities for $\sec x$ and $\tan x$, we have:
$$\sec x - 5\tan x = \frac{1}{\cos x} - \frac{5\sin x}{\cos x} = \frac{1-5\sin x}{\cos x}$$

21. Using the identity for $\sec A$, we have:
$$\cos A(\sec A - \cos A) = \cos A\left(\frac{1}{\cos A} - \cos A\right) = 1 - \cos^2 A = \sin^2 A$$

23. Multiplying out parentheses and using the identities for $\sec\theta$ and $\tan\theta$, we have:
$$(1-\sin\theta)(\sec\theta + \tan\theta) = \sec\theta + \tan\theta - \sin\theta\sec\theta - \sin\theta\tan\theta$$
$$= \frac{1}{\cos\theta} + \frac{\sin\theta}{\cos\theta} - \sin\theta\bullet\frac{1}{\cos\theta} - \sin\theta\bullet\frac{\sin\theta}{\cos\theta}$$
$$= \frac{1+\sin\theta - \sin\theta - \sin^2\theta}{\cos\theta}$$
$$= \frac{1-\sin^2\theta}{\cos\theta}$$
$$= \frac{\cos^2\theta}{\cos\theta}$$
$$= \cos\theta$$

25. Multiplying out parentheses and using the identities for $\sec\alpha$ and $\tan\alpha$, we have:
$$(\sec\alpha - \tan\alpha)^2 = \left(\frac{1}{\cos\alpha} - \frac{\sin\alpha}{\cos\alpha}\right)^2$$
$$= \frac{(1-\sin\alpha)^2}{\cos^2\alpha}$$
$$= \frac{(1-\sin\alpha)^2}{1-\sin^2\alpha}$$
$$= \frac{(1-\sin\alpha)^2}{(1+\sin\alpha)(1-\sin\alpha)}$$
$$= \frac{1-\sin\alpha}{1+\sin\alpha}$$

27. Working from the right-hand side and using identities for $\cot A$ and $\tan A$, we have:
$$\frac{\sin A}{1-\cot A} - \frac{\cos A}{\tan A - 1} = \frac{\sin A}{1-\frac{\cos A}{\sin A}} - \frac{\cos A}{\frac{\sin A}{\cos A} - 1}$$
$$= \frac{\sin^2 A}{\sin A - \cos A} - \frac{\cos^2 A}{\sin A - \cos A}$$
$$= \frac{(\sin A - \cos A)(\sin A + \cos A)}{\sin A - \cos A}$$
$$= \sin A + \cos A$$

29. Working from the left-hand side and using identities for $\csc \theta$ and $\sec \theta$, we have:

$$\csc^2 \theta + \sec^2 \theta = \frac{1}{\sin^2 \theta} + \frac{1}{\cos^2 \theta}$$

$$= \frac{\cos^2 \theta + \sin^2 \theta}{\sin^2 \theta \cos^2 \theta}$$

$$= \frac{1}{\sin^2 \theta \cos^2 \theta}$$

$$= \frac{1}{\sin^2 \theta} \bullet \frac{1}{\cos^2 \theta}$$

$$= \csc^2 \theta \sec^2 \theta$$

31. Using the identity for $\tan A$ and $\sin^2 A = 1 - \cos^2 A$, we have:

$$\sin A \tan A = \sin A \bullet \frac{\sin A}{\cos A} = \frac{\sin^2 A}{\cos A} = \frac{1 - \cos^2 A}{\cos A}$$

33. Working from the right-hand side, we have:

$$-\cot^4 A + \csc^4 A = \frac{-\cos^4 A}{\sin^4 A} + \frac{1}{\sin^4 A}$$

$$= \frac{\left(1 - \cos^2 A\right)\left(1 + \cos^2 A\right)}{\sin^4 A}$$

$$= \frac{\sin^2 A\left(1 + \cos^2 A\right)}{\sin^4 A}$$

$$= \frac{1 + \cos^2 A}{\sin^2 A}$$

$$= \frac{1}{\sin^2 A} + \frac{\cos^2 A}{\sin^2 A}$$

$$= \csc^2 A + \cot^2 A$$

35. Working from the left-hand side, we have:

$$\frac{\sin A - \cos A}{\sin A} + \frac{\cos A - \sin A}{\cos A}$$

$$= \frac{\cos A(\sin A - \cos A) + \sin A(\cos A - \sin A)}{\sin A \cos A}$$

$$= \frac{\cos A \sin A - \cos^2 A + \sin A \cos A - \sin^2 A}{\sin A \cos A}$$

$$= \frac{2\cos A \sin A - \left(\cos^2 A + \sin^2 A\right)}{\sin A \cos A}$$

$$= 2 - \frac{1}{\sin A \cos A}$$

$$= 2 - \sec A \csc A$$

37. (a) We start by testing the value $\alpha = 30°$. Since $\cos 30° = \frac{\sqrt{3}}{2}$ and $\csc 30° = 2$, the equation states that:

$$\frac{2^2 - 1}{2^2} = \frac{\sqrt{3}}{2}$$

$$\frac{3}{4} = \frac{\sqrt{3}}{2}$$

Since the left-hand side is the square of the right-hand side, this is not an identity.

(b) Proceeding as in (a), we test the value $\alpha = 30°$. Since $\sec 30° = \frac{2}{\sqrt{3}}$ and $\csc 30° = 2$, the equation states that:

$$\left[\left(\tfrac{2}{\sqrt{3}}\right)^2 - 1\right]\left[(2)^2 - 1\right] = 1$$

$$\left[\tfrac{4}{3} - 1\right][4 - 1] = 1$$

$$\tfrac{1}{3} \bullet 3 = 1$$

Since this is true, we proceed by proving the identity:

$$\left(\sec^2 \alpha - 1\right)\left(\csc^2 \alpha - 1\right) = \left(\frac{1}{\cos^2 \alpha} - 1\right)\left(\frac{1}{\sin^2 \alpha} - 1\right)$$

$$= \frac{1 - \cos^2 \alpha}{\cos^2 \alpha} \bullet \frac{1 - \sin^2 \alpha}{\sin^2 \alpha}$$

$$= \frac{\sin^2 \alpha}{\cos^2 \alpha} \bullet \frac{\cos^2 \alpha}{\sin^2 \alpha}$$

$$= 1$$

39. (a) Starting with the left-hand side, we multiply the numerator and denominator by $1 + \cos \theta$ to obtain:

$$\frac{\sin \theta}{1 - \cos \theta} \bullet \frac{1 + \cos \theta}{1 + \cos \theta} = \frac{\sin \theta(1 + \cos \theta)}{1 - \cos^2 \theta} = \frac{\sin \theta(1 + \cos \theta)}{\sin^2 \theta} = \frac{1 + \cos \theta}{\sin \theta}$$

(b) Starting with the left-hand side, we multiply the numerator and denominator by $\sin \theta$ to obtain:

$$\frac{\sin \theta}{1 - \cos \theta} \bullet \frac{\sin \theta}{\sin \theta} = \frac{\sin^2 \theta}{\sin \theta(1 - \cos \theta)}$$

$$= \frac{1 - \cos^2 \theta}{\sin \theta(1 - \cos \theta)}$$

$$= \frac{(1 + \cos \theta)(1 - \cos \theta)}{\sin \theta(1 - \cos \theta)}$$

$$= \frac{1 + \cos \theta}{\sin \theta}$$

41. Starting with the left-hand side, we use the identities for $\sec\theta$ and $\csc\theta$, then multiply the resulting fraction by $\sin\theta$ to obtain:

$$\frac{\sec\theta - \csc\theta}{\sec\theta + \csc\theta} = \frac{\frac{1}{\cos\theta} - \frac{1}{\sin\theta}}{\frac{1}{\cos\theta} + \frac{1}{\sin\theta}} = \frac{\frac{\sin\theta}{\cos\theta} - \frac{\sin\theta}{\sin\theta}}{\frac{\sin\theta}{\cos\theta} + \frac{\sin\theta}{\sin\theta}} = \frac{\tan\theta - 1}{\tan\theta + 1}$$

43. Starting with the left-hand side, we use the identity for $\cot\theta$ to obtain:

$$\sin^2\theta\left(1 + n\cot^2\theta\right) = \sin^2\theta\left(1 + n\cdot\frac{\cos^2\theta}{\sin^2\theta}\right)$$

$$= \sin^2\theta + n\cos^2\theta$$

$$= \cos^2\theta\left(\frac{\sin^2\theta}{\cos^2\theta} + n\right)$$

$$= \cos^2\theta\left(n + \tan^2\theta\right)$$

45. (a) Using the difference of cubes formula $A^3 - B^3 = (A - B)\left(A^2 + AB + B^2\right)$, we obtain:

$$\cos^3\theta - \sin^3\theta = (\cos\theta - \sin\theta)\left(\cos^2\theta + \cos\theta\sin\theta + \sin^2\theta\right)$$

$$= (\cos\theta - \sin\theta)(1 + \cos\theta\sin\theta)$$

(b) Starting with the left-hand side, we use the identities for $\cot\phi$, $\tan\phi$, $\sec\phi$ and $\csc\phi$ to obtain:

$$\frac{\cos\phi\cot\phi - \sin\phi\tan\phi}{\csc\phi - \sec\phi} = \frac{\cos\phi\cdot\frac{\cos\phi}{\sin\phi} - \sin\phi\cdot\frac{\sin\phi}{\cos\phi}}{\frac{1}{\sin\phi} - \frac{1}{\cos\phi}}$$

$$= \frac{\frac{\cos^2\phi}{\sin\phi} - \frac{\sin^2\phi}{\cos\phi}}{\frac{\cos\phi - \sin\phi}{\sin\phi\cos\phi}}$$

$$= \frac{\cos^3\phi - \sin^3\phi}{\cos\phi - \sin\phi}$$

$$= \frac{(\cos\phi - \sin\phi)(1 + \cos\phi\sin\phi)}{\cos\phi - \sin\phi}$$

$$= 1 + \sin\phi\cos\phi$$

47. Since $\tan\alpha\tan\beta = 1$, then $\tan\alpha = \dfrac{1}{\tan\beta}$. Using the identities from Exercise 46, we have:

$$\sec^2\alpha = 1 + \tan^2\alpha = 1 + \frac{1}{\tan^2\beta} = 1 + \cot^2\beta = \csc^2\beta$$

Since α and β are acute angles, then $\sec\alpha > 0$ and $\csc\beta > 0$, so $\sec^2\alpha = \csc^2\beta$ implies $\sec\alpha = \csc\beta$.

49. Using the identity $\cos^2\theta = 1 - \sin^2\theta$, we have:

$$\cos^2\theta = 1 - \frac{(p-q)^2}{(p+q)^2}$$

$$= 1 - \frac{p^2 - 2pq + q^2}{p^2 + 2pq + q^2}$$

$$= \frac{p^2 + 2pq + q^2 - p^2 + 2pq - q^2}{p^2 + 2pq + q^2}$$

$$= \frac{4pq}{(p+q)^2}$$

Since the terminal side of θ lies in quadrant 2, then $\cos\theta < 0$ and thus:

$$\cos\theta = -\sqrt{\frac{4pq}{(p+q)^2}} = -\frac{2\sqrt{pq}}{p+q} \qquad \text{(since } p > 0,\, q > 0\text{)}$$

Now we find $\tan\theta$:

$$\tan\theta = \frac{\sin\theta}{\cos\theta} = \frac{\dfrac{p-q}{p+q}}{-\dfrac{2\sqrt{pq}}{p+q}} = \frac{p-q}{-2\sqrt{pq}} = \frac{q-p}{2\sqrt{qp}}$$

51. Working from the left-hand side, we have:

$$\frac{\tan\theta + \sec\theta - 1}{\tan\theta - \sec\theta + 1} = \frac{\frac{\sin\theta}{\cos\theta} + \frac{1}{\cos\theta} - 1}{\frac{\sin\theta}{\cos\theta} - \frac{1}{\cos\theta} + 1}$$

$$= \frac{\sin\theta + 1 - \cos\theta}{\sin\theta - 1 + \cos\theta}$$

$$= \frac{(\sin\theta + 1) - \cos\theta}{(\sin\theta - 1) + \cos\theta} \cdot \frac{(\sin\theta + 1) - \cos\theta}{(\sin\theta + 1) - \cos\theta}$$

$$= \frac{(\sin\theta + 1)^2 - 2\cos\theta(\sin\theta + 1) + \cos^2\theta}{(\sin^2\theta - 1) + \cos\theta(\sin\theta + 1) - \cos\theta(\sin\theta - 1) - \cos^2\theta}$$

$$= \frac{\sin^2\theta + 2\sin\theta + 1 - 2\sin\theta\cos\theta - 2\cos\theta + \cos^2\theta}{\sin^2\theta - 1 + \sin\theta\cos\theta + \cos\theta - \sin\theta\cos\theta + \cos\theta - \cos^2\theta}$$

$$= \frac{2 + 2\sin\theta - 2\sin\theta\cos\theta - 2\cos\theta}{\sin^2\theta - \cos^2\theta - 1 + 2\cos\theta}$$

$$= \frac{2(1 - \cos\theta) + 2\sin\theta(1 - \cos\theta)}{1 - \cos^2\theta - \cos^2\theta - 1 + 2\cos\theta}$$

$$= \frac{2(1 - \cos\theta)(1 + \sin\theta)}{-2\cos^2\theta + 2\cos\theta}$$

$$= \frac{(1 - \cos\theta)(1 + \sin\theta)}{\cos\theta(1 - \cos\theta)}$$

$$= \frac{1 + \sin\theta}{\cos\theta}$$

Chapter Two Review Exercises

1. Since the terminal side of 135° lies in quadrant 2, then sin 135° > 0. Since the reference angle for 135° is 45°, then:
    $$\sin 135° = \sin 45° = \frac{\sqrt{2}}{2}$$

3. Since the terminal side of –240° lies in quadrant 2, then tan (–240°) < 0. Since the reference angle for –240° is 60°, then:
    $$\tan(-240°) = -\tan 60° = -\sqrt{3}$$

5. Since the terminal side of 210° lies in quadrant 3, then csc (210°) < 0. Since the reference angle for 210° is 30°, then:
    $$\csc 210° = -\csc 30° = -2$$

7. Since 270° corresponds to the point $(0,-1)$ on the unit circle, then sin 270° = –1, which is the y-coordinate.

9. Since the terminal side of –315° lies in quadrant 1, then cos (–315°) > 0. Since the reference angle for –315° is 45°, then:
    $$\cos(-315°) = \cos 45° = \frac{\sqrt{2}}{2}$$

11. Since 1800° corresponds to the point $(1,0)$ on the unit circle, then cos 1800° = 1, which is the x-coordinate.

13. Since the terminal side of 240° lies in quadrant 3, then csc 240° < 0. Since the reference angle for 240° is 60°, then:
    $$\csc 240° = -\csc 60° = -\frac{2\sqrt{3}}{3}$$

15. Since the terminal side of 780° lies in quadrant 1, then sec 780° > 0. Since the reference angle for 780° is 60°, then:
    $$\sec 780° = \sec 60° = 2$$

17. (a) Since $\sin^2(A+B) \neq \sin^2 A + \sin^2 B$, then $\sin^2 33° + \sin^2 57° \neq \sin^2(33°+57°)$.

    (b) Since $\sin\theta = \cos(90°-\theta)$, we compute the quantity as follows:
    $$\sin^2 33° + \sin^2 57° = \sin^2 33° + \cos^2(90°-57°) = \sin^2 33° + \cos^2 33° = 1$$
    The correct answer is 1, but not for the reasons the student gave.

19. We draw the triangle:

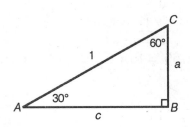

Now compute a and c:

$$\sin 30° = \frac{a}{1}, \text{ so } a = \sin 30° = \frac{1}{2}$$

$$\cos 30° = \frac{c}{1}, \text{ so } c = \cos 30° = \frac{\sqrt{3}}{2}$$

21. We draw a triangle:

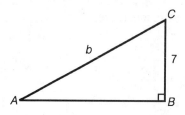

Now compute b:

$$\sin A = \frac{7}{b}, \text{ so } b = \frac{7}{\sin A} = \frac{7}{\frac{2}{5}} = \frac{35}{2}$$

23. We draw a triangle:

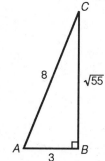

Using the Pythagorean theorem, we have:

$$3^2 + b^2 = 8^2$$
$$9 + b^2 = 64$$
$$b^2 = 55$$
$$b = \sqrt{55}$$

From the triangle, we have:

$$\sin A = \frac{\sqrt{55}}{8} \text{ and } \cot A = \frac{3}{\sqrt{55}} = \frac{3\sqrt{55}}{55}$$

25. We draw the triangle:

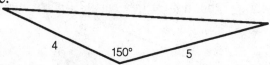

Now compute the area:

$$\tfrac{1}{2}ab\sin\theta = \tfrac{1}{2}(5)(4)\sin 150° = 10 \bullet \tfrac{1}{2} = 5 \text{ square units}$$

27. We draw a triangle:

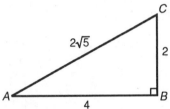

Using the Pythagorean theorem, we have:
$$4^2 + 2^2 = c^2$$
$$16 + 4 = c^2$$
$$c^2 = 20$$
$$c = 2\sqrt{5}$$
From the triangle, we have:
$$\sin A = \frac{2}{2\sqrt{5}} = \frac{1}{\sqrt{5}} = \frac{\sqrt{5}}{5} \text{ and } \cos B = 0$$
So:
$$\sin^2 A + \cos^2 B = \left(\frac{\sqrt{5}}{5}\right)^2 + (0)^2 = \frac{5}{25} + 0 = \frac{1}{5}$$

29. We draw a triangle:

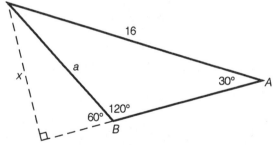

We first find x:
$$\sin 30° = \frac{x}{16}, \text{ so } x = 16\sin 30° = \frac{1}{2}(16) = 8$$
Now, using the smaller triangle:
$$\sin 60° = \frac{8}{a}, \text{ so } a = \frac{8}{\sin 60°} = \frac{8}{\frac{\sqrt{3}}{2}} = \frac{16}{\sqrt{3}} = \frac{16\sqrt{3}}{3}$$

31. We draw a triangle:

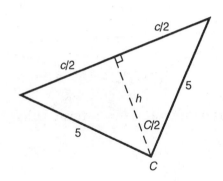

Notice that the bisector drawn for C bisects the opposite side (of length c) into two pieces of length $\frac{c}{2}$. This is true since $a = b = 5$. Using one of the smaller triangles, we have:

$$\sin\left(\tfrac{1}{2}C\right) = \frac{\frac{c}{2}}{5}$$
$$\frac{9}{10} = \frac{c}{10}$$
$$9 = c$$

To find the height h of the triangle, we use the Pythagorean theorem:

$$\left(\tfrac{c}{2}\right)^2 + h^2 = 5^2$$
$$\left(\tfrac{9}{2}\right)^2 + h^2 = 25$$
$$\tfrac{81}{4} + h^2 = 25$$
$$h^2 = \tfrac{19}{4}$$
$$h = \tfrac{\sqrt{19}}{2}$$

So the area of $\triangle ABC$ is given by:

$$\tfrac{1}{2}(\text{base})(\text{height}) = \tfrac{1}{2}(9)\left(\tfrac{\sqrt{19}}{2}\right) = \tfrac{9\sqrt{19}}{4}$$

33. Letting z be the side opposite the $90° - \beta$ angle (adjacent to β), we have:

$$\cot\alpha = \frac{x+z}{y} \text{ and } \cot\beta = \frac{z}{y}$$

Thus:

$$\cot\alpha - \cot\beta = \frac{x+z}{y} - \frac{z}{y}$$
$$\cot\alpha - \cot\beta = \frac{x}{y}$$
$$y(\cot\alpha - \cot\beta) = x$$
$$y = \frac{x}{\cot\alpha - \cot\beta}$$

35. Since $\sin^2\theta + \cos^2\theta = 1$, we have:

$$\left(\frac{2p^2q^2}{p^4+q^4}\right)^2 + \cos^2\theta = 1$$

$$\frac{4p^4q^4}{\left(p^4+q^4\right)^2} + \cos^2\theta = 1$$

$$\cos^2\theta = \frac{\left(p^4+q^4\right)^2 - 4p^4q^4}{\left(p^4+q^4\right)^2}$$

$$\cos^2\theta = \frac{p^8 - 2p^4q^4 + q^8}{\left(p^4+q^4\right)^2}$$

$$\cos^2\theta = \frac{\left(p^4-q^4\right)^2}{\left(p^4+q^4\right)^2}$$

$$\cos\theta = \frac{p^4-q^4}{p^4+q^4}$$

Now find $\tan\theta$:

$$\tan\theta = \frac{\sin\theta}{\cos\theta} = \frac{2p^2q^2}{p^4-q^4}$$

37. Using the identities for $\sec A$ and $\csc A$, we have:

$$\frac{\sin A + \cos A}{\sec A + \csc A} = \frac{\sin A + \cos A}{\frac{1}{\cos A} + \frac{1}{\sin A}} = \frac{\sin A \cos A(\sin A + \cos A)}{\sin A + \cos A} = \sin A \cos A$$

39. Using the identities for $\sec A$, $\tan A$ and $\cot A$, we have:

$$\frac{\sin A \sec A}{\tan A + \cot A} = \frac{\frac{\sin A}{\cos A}}{\frac{\sin A}{\cos A} + \frac{\cos A}{\sin A}} = \frac{\sin^2 A}{\sin^2 A + \cos^2 A} = \sin^2 A$$

41. Using the identities for $\tan A$ and $\cot A$, we have:

$$\frac{\cos A}{1-\tan A} + \frac{\sin A}{1-\cot A} = \frac{\cos A}{1-\frac{\sin A}{\cos A}} + \frac{\sin A}{1-\frac{\cos A}{\sin A}}$$

$$= \frac{\cos^2 A}{\cos A - \sin A} + \frac{\sin^2 A}{\sin A - \cos A}$$

$$= \frac{\cos^2 A}{\cos A - \sin A} - \frac{\sin^2 A}{\cos A - \sin A}$$

$$= \frac{\cos^2 A - \sin^2 A}{\cos A - \sin A}$$

$$= \frac{(\cos A + \sin A)(\cos A - \sin A)}{\cos A - \sin A}$$

$$= \cos A + \sin A$$

43. Using the identities for $\sec A$ and $\csc A$, we have:

$$(\sec A + \csc A)^{-1}\left[(\sec A)^{-1} + (\csc A)^{-1}\right] = \frac{1}{\sec A + \csc A} \cdot \left[\frac{1}{\sec A} + \frac{1}{\csc A}\right]$$

$$= \frac{1}{\sec A + \csc A} \cdot \frac{\csc A + \sec A}{\sec A \csc A}$$

$$= \frac{1}{\sec A \csc A}$$

$$= \frac{1}{\frac{1}{\cos A} \cdot \frac{1}{\sin A}}$$

$$= \sin A \cos A$$

45. Simplifying the complex fraction, we have:

$$\frac{\frac{\sin A + \cos A}{\sin A - \cos A} - \frac{\sin A - \cos A}{\sin A + \cos A}}{\frac{\sin A + \cos A}{\sin A - \cos A} + \frac{\sin A - \cos A}{\sin A + \cos A}} = \frac{(\sin A + \cos A)^2 - (\sin A - \cos A)^2}{(\sin A + \cos A)^2 + (\sin A - \cos A)^2}$$

$$= \frac{1 + 2\sin A \cos A - 1 + 2\sin A \cos A}{1 + 2\sin A \cos A + 1 - 2\sin A \cos A}$$

$$= \frac{4\sin A \cos A}{2}$$

$$= 2\sin A \cos A$$

47. The x-coordinate of P is approximately 0.9848, so $\cos 10° \approx 0.9848$.

49. The y-coordinate of P is approximately 0.1736, so $\sin 10° \approx 0.1736$.

51. We draw a triangle:

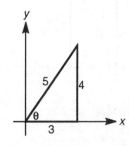

Using the Pythagorean theorem, we have:

$$3^2 + b^2 = 5^2$$
$$9 + b^2 = 25$$
$$b^2 = 16$$
$$b = 4$$

Since $0° < \theta < 90°$, both $\sin \theta$ and $\tan \theta$ are positive, so:

$$\sin \theta = \tfrac{4}{5} \text{ and } \tan \theta = \tfrac{4}{3}$$

53. We draw a triangle:

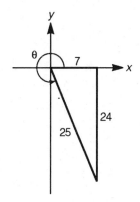

Using the Pythagorean theorem, we have:
$$7^2 + b^2 = 25^2$$
$$49 + b^2 = 625$$
$$b^2 = 576$$
$$b = 24$$

Since $270° < \theta < 360°$, $\tan \theta$ is negative, so:
$$\tan \theta = -\tfrac{24}{7}$$

55. We draw a triangle:

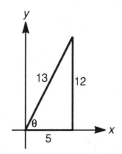

Using the Pythagorean theorem, we have:
$$a^2 + 12^2 = 13^2$$
$$a^2 + 144 = 169$$
$$a^2 = 25$$
$$a = 5$$

Since $0° < \theta < 90°$, $\cot \theta$ is positive, so:
$$\cot \theta = \tfrac{5}{12}$$

57. We draw a triangle:

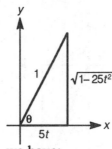

Using the Pythagorean theorem, we have:

$$(5t)^2 + b^2 = 1^2$$
$$25t^2 + b^2 = 1$$
$$b^2 = 1 - 25t^2$$
$$b = \sqrt{1 - 25t^2}$$

Since $0° < \theta < 90°$, $\tan(90° - \theta)$ is positive, so:

$$\tan(90° - \theta) = \cot\theta = \frac{5t}{\sqrt{1 - 25t^2}} = \frac{5t\sqrt{1 - 25t^2}}{1 - 25t^2}$$

59. We draw a triangle:

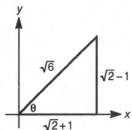

Using the Pythagorean theorem, we have:

$$\left(\sqrt{2} + 1\right)^2 + \left(\sqrt{2} - 1\right)^2 = c^2$$
$$2 + 2\sqrt{2} + 1 + 2 - 2\sqrt{2} + 1 = c^2$$
$$6 = c^2$$
$$\sqrt{6} = c$$

Since $0° < \theta < 90°$, $\sin\theta$ is positive, so:

$$\sin\theta = \frac{\sqrt{2} - 1}{\sqrt{6}} = \frac{2\sqrt{3} - \sqrt{6}}{6}$$

61. Using the identities for $\tan\theta$ and $\cot\theta$, we have:

$$\tan\theta + \cot\theta = 2$$
$$\frac{\sin\theta}{\cos\theta} + \frac{\cos\theta}{\sin\theta} = 2$$
$$\frac{\sin^2\theta + \cos^2\theta}{\sin\theta\cos\theta} = 2$$
$$1 = 2\sin\theta\cos\theta$$

Now $(\sin\theta + \cos\theta)^2 = \sin^2\theta + 2\sin\theta\cos\theta + \cos^2\theta = 1 + 1 = 2$, so $\sin\theta + \cos\theta = \sqrt{2}$ since $0° < \theta < 90°$.

63. We draw the figure:

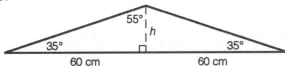

Note that the altitude of the triangle must bisect the upper vertex angle, which is $110°$ $(180° - 35° - 35°)$. Now we find the height h:

$$\tan 35° = \frac{h}{60}, \text{ so } h = 60\tan 35°$$

Thus the area is given by:

$$\tfrac{1}{2}(\text{base})(\text{height}) = \tfrac{1}{2}(120)(60\tan 35°) = 3600\tan 35° \approx 2521 \text{ cm}^2$$

65. We draw the figure:

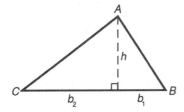

Now:

$$\cot B = \frac{b_1}{h}, \text{ so } b_1 = h\cot B$$

$$\cot C = \frac{b_2}{h}, \text{ so } b_2 = h\cot C$$

So we have:

$$b_1 + b_2 = a$$
$$h\cot B + h\cot C = a$$
$$h(\cot B + \cot C) = a$$
$$h = \frac{a}{\cot B + \cot C}$$

67. We draw the figure:

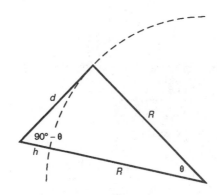

We have:
$$d^2 + R^2 = (R+h)^2$$
$$d^2 = h^2 + 2Rh$$
$$d = \sqrt{2Rh + h^2}$$

So:
$$\cot\theta = \frac{R}{d} = \frac{R}{\sqrt{2Rh + h^2}}$$

69. Using the identities $\sin(90° - \theta) = \cos\theta$ and $\tan(90° - \theta) = \cot\theta$, we have:

$$\sin^2(90° - \theta)\csc\theta - \tan^2(90° - \theta)\sin\theta = \cos^2\theta \bullet \frac{1}{\sin\theta} - \cot^2\theta\sin\theta$$
$$= \frac{\cos^2\theta}{\sin\theta} - \frac{\cos^2\theta}{\sin^2\theta} \bullet \sin\theta$$
$$= \frac{\cos^2\theta}{\sin\theta} - \frac{\cos^2\theta}{\sin\theta}$$
$$= 0$$

71. Working from the right-hand side and using the identity for $\csc A$, we have:

$$\cdot \csc A - \sin A = \frac{1}{\sin A} - \sin A$$
$$= \frac{1 - \sin^2 A}{\sin A}$$
$$= \frac{\cos^2 A}{\sin A}$$
$$= \frac{\cos A}{\sin A} \bullet \cos A$$
$$= \cos A \cot A$$

73. Using the identity for $\cot A$, we have:

$$\frac{\cot A - 1}{\cot A + 1} = \frac{\frac{\cos A}{\sin A} - 1}{\frac{\cos A}{\sin A} + 1} = \frac{\cos A - \sin A}{\cos A + \sin A}$$

75. Using the identity $\sin^2\theta = 1 - \cos^2\theta$, we have:

$$\cos^2\theta - \sin^2\theta = \cos^2\theta - (1 - \cos^2\theta) = 2\cos^2\theta - 1$$

77. Working from the right-hand side and using the identity for $\tan A$, we have:

$$\frac{1 - \tan A}{1 + \tan A} = \frac{1 - \frac{\sin A}{\cos A}}{1 + \frac{\sin A}{\cos A}} = \frac{\cos A - \sin A}{\cos A + \sin A}$$

79. Using the identities for $\sec A$ and $\tan A$, we have:

$$\frac{1-\sin A}{\cos A} - \frac{1}{\sec A + \tan A} = \frac{1-\sin A}{\cos A} - \frac{1}{\frac{1}{\cos A} + \frac{\sin A}{\cos A}}$$

$$= \frac{1-\sin A}{\cos A} - \frac{\cos A}{1+\sin A} \bullet \frac{1-\sin A}{1-\sin A}$$

$$= \frac{1-\sin A}{\cos A} - \frac{\cos A(1-\sin A)}{1-\sin^2 A}$$

$$= \frac{1-\sin A}{\cos A} - \frac{\cos A(1-\sin A)}{\cos^2 A}$$

$$= \frac{1-\sin A}{\cos A} - \frac{1-\sin A}{\cos A}$$

$$= 0$$

81. Simplifying the left-hand side, we have:

$$\frac{\sin A}{1+\cos A} + \frac{1+\cos A}{\sin A} = \frac{\sin^2 A + (1+\cos A)^2}{\sin A(1+\cos A)}$$

$$= \frac{\sin^2 A + 1 + 2\cos A + \cos^2 A}{\sin A(1+\cos A)}$$

$$= \frac{2+2\cos A}{\sin A(1+\cos A)}$$

$$= \frac{2(1+\cos A)}{\sin A(1+\cos A)}$$

$$= \frac{2}{\sin A}$$

$$= 2\csc A$$

83. Simplifying the left-hand side, we have:

$$\frac{1}{1-\cos A} + \frac{1}{1+\cos A} = \frac{1+\cos A + 1 - \cos A}{(1-\cos A)(1+\cos A)} = \frac{2}{1-\cos^2 A} = \frac{2}{\sin^2 A}$$

Simplifying the right-hand side using the identity for $\cot A$, we have:

$$2 + 2\cot^2 A = 2 + 2 \bullet \frac{\cos^2 A}{\sin^2 A} = \frac{2\sin^2 A + 2\cos^2 A}{\sin^2 A} = \frac{2(\sin^2 A + \cos^2 A)}{\sin^2 A} = \frac{2}{\sin^2 A}$$

Since both sides of the equality simplify to the same quantity, the original equality is an identity.

85. First simplifying two of the fractions using the identities for $\sec A$ and $\csc A$, we have:

$$\frac{1}{1+\sec^2 A} = \frac{1}{1+\frac{1}{\cos^2 A}} = \frac{\cos^2 A}{\cos^2 A + 1}$$

$$\frac{1}{1+\csc^2 A} = \frac{1}{1+\frac{1}{\sin^2 A}} = \frac{\sin^2 A}{\sin^2 A + 1}$$

Replacing the original fractions with these simplified fractions, the left-hand side becomes:

$$\frac{1}{1+\sin^2 A}+\frac{1}{1+\cos^2 A}+\frac{\cos^2 A}{1+\cos^2 A}+\frac{\sin^2 A}{1+\sin^2 A}$$
$$=\frac{1+\sin^2 A}{1+\sin^2 A}+\frac{1+\cos^2 A}{1+\cos^2 A}$$
$$=2$$

87. Using the identities for $\sec A$ and $\csc A$, we have:

$$\frac{\sec A-\csc A}{\sec A+\csc A}=\frac{\frac{1}{\cos A}-\frac{1}{\sin A}}{\frac{1}{\cos A}+\frac{1}{\sin A}}=\frac{\sin A-\cos A}{\sin A+\cos A}$$

89. Working from the right-hand side and using the identity for $\tan A$, we have:

$$\frac{\tan A}{1+\tan^2 A}=\frac{\frac{\sin A}{\cos A}}{1+\frac{\sin^2 A}{\cos^2 A}}=\frac{\sin A\cos A}{\cos^2 A+\sin^2 A}=\sin A\cos A$$

91. (a) Since $OP=1$, then $PN=\sin\theta$.

(b) Since $OP=1$, then $ON=\cos\theta$.

(c) Since $\cos\theta=\dfrac{PN}{PT}$ (note $\theta=\angle TPN$), then $\cos\theta=\dfrac{\sin\theta}{PT}$. Hence $PT=\tan\theta$.

(d) Here $\cos\theta=\dfrac{1}{OT}$ (using ΔOPT), so $OT=\dfrac{1}{\cos\theta}=\sec\theta$.

(e) We simplify $NA=1-ON=1-\cos\theta$.

(f) Since $NT=OT-ON$, we have:
$$NT=\sec\theta-\cos\theta$$
$$=\frac{1}{\cos\theta}-\cos\theta$$
$$=\frac{1-\cos^2\theta}{\cos\theta}$$
$$=\frac{\sin^2\theta}{\cos\theta}$$
$$=\sin\theta\bullet\frac{\sin\theta}{\cos\theta}$$
$$=\sin\theta\tan\theta$$

93. We re-draw the figure (note the labels):

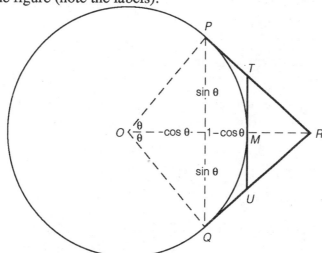

Using $\triangle QOR$:

$$\cos\theta = \frac{1}{1+MR}$$

$$1+MR = \frac{1}{\cos\theta}$$

$$MR = \frac{1}{\cos\theta} - 1$$

Since $\triangle RTU$ and $\triangle RPQ$ are similar, then the corresponding sides are proportional:

$$\frac{TU}{PQ} = \frac{MR}{1-\cos\theta + MR}$$

But $PQ = 2\sin\theta$ and $MR = \dfrac{1}{\cos\theta} - 1$, so we have:

$$\frac{TU}{2\sin\theta} = \frac{\frac{1}{\cos\theta}-1}{1-\cos\theta + \frac{1}{\cos\theta}-1} = \frac{1-\cos\theta}{1-\cos^2\theta} = \frac{1-\cos\theta}{\sin^2\theta}$$

$$TU = \frac{2(1-\cos\theta)}{\sin\theta}$$

Thus the area is given by:

$$A = \frac{1}{2}(TU)(MR)$$

$$= \frac{1}{2} \cdot \frac{2(1-\cos\theta)}{\sin\theta} \cdot \left(\frac{1}{\cos\theta} - 1\right)$$

$$= \frac{1-\cos\theta}{\sin\theta} \cdot \frac{1-\cos\theta}{\cos\theta}$$

$$= \frac{(1-\cos\theta)^2}{\sin\theta\cos\theta}$$

95. We must find expressions (in terms of θ) for AO and OC. Note that $\angle OAC = \theta$, since they are both complementary to the same angle ($\angle AOP$). Also $OP = 1$ (a unit circle), then using $\triangle APO$:

$$\sin\theta = \frac{OP}{AO} = \frac{1}{AO}, \text{ so } AO = \frac{1}{\sin\theta}$$

Using $\triangle OPC$:

$$\cos\theta = \frac{OP}{OC} = \frac{1}{OC}, \text{ so } OC = \frac{1}{\cos\theta}$$

So the area is $\dfrac{1}{2}(OC)(AO) = \dfrac{1}{2\sin\theta\cos\theta} = \dfrac{1}{2}\sec\theta\csc\theta$.

97. (a) Using $\triangle ODC$, $OD = \cos\theta$ and $DC = \sin\theta$ (recall that $OC = 1$).

Thus, using $\triangle ADC$, $\tan\dfrac{\theta}{2} = \dfrac{CD}{AD} = \dfrac{\sin\theta}{1+\cos\theta}$, since $AO = 1$.

(b) We compute $\tan 15°$ using the formula:

$$\tan 15° = \frac{\sin 30°}{1+\cos 30°}$$

$$= \frac{\frac{1}{2}}{1+\frac{\sqrt{3}}{2}}$$

$$= \frac{1}{2+\sqrt{3}} \cdot \frac{2-\sqrt{3}}{2-\sqrt{3}}$$

$$= \frac{2-\sqrt{3}}{4-3}$$

$$= 2-\sqrt{3}$$

We compute $\tan 22.5°$ using the formula:

$$\tan 22.5° = \frac{\sin 45°}{1+\cos 45°}$$

$$= \frac{\frac{\sqrt{2}}{2}}{1+\frac{\sqrt{2}}{2}}$$

$$= \frac{\sqrt{2}}{2+\sqrt{2}} \cdot \frac{2-\sqrt{2}}{2-\sqrt{2}}$$

$$= \frac{2\sqrt{2}-2}{4-2}$$

$$= \sqrt{2}-1$$

99. We draw the figure:

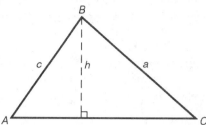

Using the left-side right triangle, we have:

$\sin A = \dfrac{h}{c}$, so $h = c \sin A$

Using the right-side triangle, we have:

$\sin C = \dfrac{h}{a}$, so $h = a \sin C$

Thus $h = c \sin A = a \sin C$.

101. (a) Since S is the area of the triangle, it can be written in the following forms:

$$S = \tfrac{1}{2}ab\sin C = \tfrac{1}{2}bc\sin A = \tfrac{1}{2}ac\sin B$$

From Exercise 100 we have $r = \dfrac{2S}{a+b+c}$, so:

$$\frac{1}{r} = \frac{a+b+c}{2S}$$
$$= \frac{a}{2S} + \frac{b}{2S} + \frac{c}{2S}$$
$$= \frac{a}{ab\sin C} + \frac{b}{bc\sin A} + \frac{c}{ac\sin B}$$
$$= \frac{1}{b\sin C} + \frac{1}{c\sin A} + \frac{1}{a\sin B}$$

(b) Since the altitudes are $b\sin C$, $c\sin A$, and $a\sin B$ respectively, the result is verified.

Chapter Two Test

1. (a) Since the terminal side of $30°$ lies in the first quadrant, $\tan 30° > 0$, so $\tan 30° = \frac{\sqrt{3}}{3}$.

(b) Since the terminal side of $45°$ lies in the first quadrant, $\sec 45° > 0$, so $\sec 45° = \sqrt{2}$.

(c) Since $\sin^2 \theta + \cos^2 \theta = 1$ for all angles θ, then $\sin^2 25° + \cos^2 25° = 1$.

(d) Since $\sin 53° = \cos(90° - 53°) = \cos 37°$, then:
$$\sin 53° - \cos 37° = \cos 37° - \cos 37° = 0$$

2. (a) Since $\cos\theta$ is negative, then $\sec\theta$ is negative.
    (b) Since $\sin\theta$ is negative, then $\csc\theta$ is negative.
    (c) Since both $\sin\theta$ and $\cos\theta$ are negative, then $\cot\theta$ is positive.

3. (a) Since $-270°$ corresponds to the point $(0,1)$ on the unit circle, then $\sin(-270°) = 1$.
    (b) Since $180°$ corresponds to the point $(-1,0)$ on the unit circle, then $\cos 180° = -1$.
    (c) Since $720°$ corresponds to the point $(1,0)$ on the unit circle, then:
    $$\tan 720° = \frac{y}{x} = \frac{0}{1} = 0$$

4. Factoring as a trinomial, we have:
    $$2\cot^2\theta + 11\cot\theta + 12 = (2\cot\theta + 3)(\cot\theta + 4)$$

5. Using the formula for area, we have:
    $$\tfrac{1}{2}ab\sin\theta = \tfrac{1}{2}(8)(9)\sin 150° = 36 \bullet \tfrac{1}{2} = 18 \text{ cm}^2$$

6. (a) Since the terminal side of $-225°$ lies in quadrant 2, $\sin(-225°) > 0$. Using a reference angle of $45°$, we have:
    $$\sin(-225°) = \sin 45° = \tfrac{\sqrt{2}}{2}$$

    (b) Since the terminal side of $330°$ lies in quadrant 4, $\tan 330° < 0$. Using a reference angle of $30°$, we have:
    $$\tan 330° = -\tan 30° = -\tfrac{\sqrt{3}}{3}$$

    (c) Since the terminal side of $120°$ lies in quadrant 2, $\sec 120° < 0$. Using a reference angle of $60°$, we have:
    $$\sec 120° = -\sec 60° = -2$$

7. We draw a triangle:

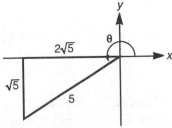

    Using the Pythagorean theorem, we have:
    $$a^2 + \left(\sqrt{5}\right)^2 = 5^2$$
    $$a^2 + 5 = 25$$
    $$a^2 = 20$$
    $$a = 2\sqrt{5}$$
    Since $180° < \theta < 270°$, $\cos\theta < 0$ and $\tan\theta > 0$, so:
    $$\cos\theta = -\tfrac{2\sqrt{5}}{5} \quad\text{and}\quad \tan\theta = \tfrac{\sqrt{5}}{2\sqrt{5}} = \tfrac{1}{2}$$

8. We draw a triangle:

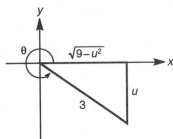

Using the Pythagorean theorem, we have:
$$a^2 + u^2 = 3^2$$
$$a^2 + u^2 = 9$$
$$a^2 = 9 - u^2$$
$$a = \sqrt{9 - u^2}$$

Since $270° < \theta < 360°$, $\cos\theta > 0$ and $\cot\theta < 0$, so:
$$\cos\theta = \frac{\sqrt{9 - u^2}}{3} \quad \text{and} \quad \cot\theta = -\frac{\sqrt{9 - u^2}}{u}$$

9. Multiplying the numerator and denominator by $\cos\theta$, we have:
$$\frac{\frac{\cos\theta + 1}{\cos\theta} + 1}{\frac{\cos\theta - 1}{\cos\theta} - 1} = \frac{\cos\theta + 1 + \cos\theta}{\cos\theta - 1 - \cos\theta} = \frac{2\cos\theta + 1}{-1} = -2\cos\theta - 1$$

10. The central angle of a triangle drawn from the center to two adjacent vertices is given by $9\theta = 360°$, so $\theta = 40°$. The area of this triangle is therefore:
$$\tfrac{1}{2}ab\sin\theta = \tfrac{1}{2}(3)(3)\sin 40° = \tfrac{9}{2}\sin 40°$$

Since there are nine such triangles which form the polygon, its area is:
$$9\left(\tfrac{9}{2}\sin 40°\right) = \tfrac{81}{2}\sin 40° \approx 26.033 \text{ m}^2$$

11. Using the identities for $\cot\theta$, $\tan\theta$, $\csc\theta$ and $\sec\theta$, we have:
$$\frac{\cot^2\theta}{\csc^2\theta} + \frac{\tan^2\theta}{\sec^2\theta} = \frac{\frac{\cos^2\theta}{\sin^2\theta}}{\frac{1}{\sin^2\theta}} + \frac{\frac{\sin^2\theta}{\cos^2\theta}}{\frac{1}{\cos^2\theta}} = \cos^2\theta + \sin^2\theta = 1$$

12. We will find BD and BC. Using the smaller right triangle, we have:
$$\tan 25° = \frac{BD}{50}, \text{ so } BD = 50\tan 25°$$

Using the larger right triangle, we have:
$$\tan 55° = \frac{BC}{50}, \text{ so } BC = 50\tan 55°$$

Since $CD = BC - BD$, then:
$$CD = 50\tan 55° - 50\tan 25° = 50(\tan 55° - \tan 25°)$$

13. (a) Since 85° has a much larger y-coordinate than 5°, sin 85° is larger.

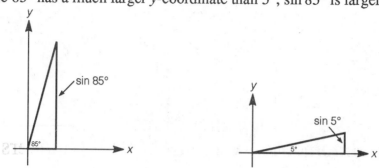

(b) Since the x-coordinate is much larger at 5° than the y-coordinate, cos 5° is larger.

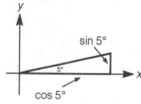

(c) Since the terminal side for 175° lies in quadrant 2, tan 175° < 0. Since the terminal side for 185° lies in quadrant 3, tan 185° > 0. Thus tan 185° is larger. Note that a sketch isn't necessary to make this comparison.

14. (a) To find AD, we use $\triangle ADC$:
$$\cos 20° = \frac{AD}{2.75}, \text{ so } AD = 2.75\cos 20°$$
To find CD, we use $\triangle ADC$:
$$\sin 20° = \frac{CD}{2.75}, \text{ so } CD = 2.75\sin 20°$$
To find DB, note that $DB = AB - AD$, so:
$$DB = 3.25 - 2.75\cos 20°$$

(b) Using the Pythagorean theorem in $\triangle CDB$, we have:
$$(CB)^2 = (DB)^2 + (CD)^2$$
$$(CB)^2 = (3.25 - 2.75\cos 20°)^2 + (2.75\sin 20°)^2$$
$$(CB)^2 = 10.5625 - 17.875\cos 20° + 7.5625\cos^2 20° + 7.5625\sin^2 20°$$
$$(CB)^2 = 10.5625 - 17.875\cos 20° + 7.5625\left(\cos^2 20° + \sin^2 20°\right)$$
$$(CB)^2 = 10.5625 - 17.875\cos 20° + 7.5625$$
$$(CB)^2 = 18.125 - 17.875\cos 20°$$
$$CB = \sqrt{18.125 - 17.875\cos 20°} \approx 1.15 \text{ cm}$$

Chapter Three
Trigonometric Functions of Real Numbers

3.1 Radian Measure

1. (a) Multiplying by the conversion factor $\frac{\pi}{180°}$, we have:
$$45° \bullet \frac{\pi}{180°} = \frac{\pi}{4} \approx 0.79 \text{ radians}$$

(b) Multiplying by the conversion factor $\frac{\pi}{180°}$, we have:
$$90° \bullet \frac{\pi}{180°} = \frac{\pi}{2} \approx 1.57 \text{ radians}$$

(c) Multiplying by the conversion factor $\frac{\pi}{180°}$, we have:
$$135° \bullet \frac{\pi}{180°} = \frac{3\pi}{4} \approx 2.36 \text{ radians}$$

3. (a) Multiplying by the conversion factor $\frac{\pi}{180°}$, we have:
$$0° \bullet \frac{\pi}{180°} = 0 \text{ radians}$$

(b) Multiplying by the conversion factor $\frac{\pi}{180°}$, we have:
$$360° \bullet \frac{\pi}{180°} = 2\pi \approx 6.28 \text{ radians}$$

(c) Multiplying by the conversion factor $\frac{\pi}{180°}$, we have:
$$450° \bullet \frac{\pi}{180°} = \frac{5\pi}{2} \approx 7.85 \text{ radians}$$

5. (a) Multiplying by the conversion factor $\frac{180°}{\pi}$, we have:

$$\frac{\pi}{12} \bullet \frac{180°}{\pi} = 15°$$

(b) Multiplying by the conversion factor $\frac{180°}{\pi}$, we have:

$$\frac{\pi}{6} \bullet \frac{180°}{\pi} = 30°$$

(c) Multiplying by the conversion factor $\frac{180°}{\pi}$, we have:

$$\frac{\pi}{4} \bullet \frac{180°}{\pi} = 45°$$

7. (a) Multiplying by the conversion factor $\frac{180°}{\pi}$, we have:

$$\frac{\pi}{3} \bullet \frac{180°}{\pi} = 60°$$

(b) Multiplying by the conversion factor $\frac{180°}{\pi}$, we have:

$$\frac{5\pi}{3} \bullet \frac{180°}{\pi} = 300°$$

(c) Multiplying by the conversion factor $\frac{180°}{\pi}$, we have:

$$4\pi \bullet \frac{180°}{\pi} = 720°$$

9. (a) Multiplying by the conversion factor $\frac{180°}{\pi}$, we have:

$$2 \bullet \frac{180°}{\pi} = \frac{360°}{\pi} \approx 114.59°$$

(b) Multiplying by the conversion factor $\frac{180°}{\pi}$, we have:

$$3 \bullet \frac{180°}{\pi} = \frac{540°}{\pi} \approx 171.89°$$

(c) Multiplying by the conversion factor $\frac{180°}{\pi}$, we have:

$$\pi^2 \bullet \frac{180°}{\pi} = 180\pi° \approx 565.49°$$

11. Since a right angle has radian measure of $\frac{\pi}{2}$, which is larger than $\frac{3}{2}$ (since π is larger than 3), then this angle is smaller than a right angle.

13. Multiplying by the conversion factor $\frac{\pi}{180°}$, we have:

$$30° = 30° \bullet \frac{\pi}{180°} = \frac{\pi}{6} \text{ radians}$$

$$45° = 45° \bullet \frac{\pi}{180°} = \frac{\pi}{4} \text{ radians}$$

$$60° = 60° \bullet \frac{\pi}{180°} = \frac{\pi}{3} \text{ radians}$$

$$120° = 120° \bullet \frac{\pi}{180°} = \frac{2\pi}{3} \text{ radians}$$

$$135° = 135° \bullet \frac{\pi}{180°} = \frac{3\pi}{4} \text{ radians}$$

$$150° = 150° \bullet \frac{\pi}{180°} = \frac{5\pi}{6} \text{ radians}$$

15. Using the formula $\theta = \frac{s}{r}$, where $s = 5$ cm and $r = 2$ cm, we have:

$$\theta = \frac{s}{r} = \frac{5 \text{ cm}}{2 \text{ cm}} = 2.5 \text{ radians}$$

17. Using the formula $\theta = \frac{s}{r}$ where $s = 200$ cm $= 2$ m and $r = 1$ m, we have:

$$\theta = \frac{s}{r} = \frac{2 \text{ m}}{1 \text{ m}} = 2 \text{ radians}$$

19. Using the formula $s = r\theta$ where $r = 3$ ft and $\theta = \frac{4\pi}{3}$, we have:

$$s = r\theta = 3 \bullet \frac{4\pi}{3} = 4\pi \text{ ft}$$

21. We must first convert $45°$ to radian measure:

$$45° = 45° \bullet \frac{\pi}{180°} = \frac{\pi}{4} \text{ radians}$$

Now using the formula $s = r\theta$ where $r = 2$ cm and $\theta = \frac{\pi}{4}$, we have:

$$s = r\theta = 2 \bullet \frac{\pi}{4} = \frac{\pi}{2} \text{ cm}$$

23. Since $s = r\theta$, and $s = 1$ cm while $r = 6$ cm, we have:

$$1 = 6\theta$$
$$\theta = \frac{1}{6}$$

The radian measure of θ is $\frac{1}{6}$.

25. (a) The area is given by:

$$A = \frac{1}{2}r^2\theta = \frac{1}{2}(6)^2\left(\frac{2\pi}{3}\right) = 12\pi \text{ cm}^2 \approx 37.70 \text{ cm}^2$$

(b) We first convert θ to radian measure:

$$\theta = 80° \bullet \frac{\pi}{180°} = \frac{4\pi}{9} \text{ radians}$$

The area is given by:

$$A = \frac{1}{2}r^2\theta = \frac{1}{2}(5)^2\left(\frac{4\pi}{9}\right) = \frac{50\pi}{9} \text{ m}^2 \approx 17.45 \text{ m}^2$$

(c) The area is given by:

$$A = \frac{1}{2}r^2\theta = \frac{1}{2}(24)^2\left(\frac{\pi}{20}\right) = \frac{72\pi}{5} \text{ m}^2 \approx 45.24 \text{ m}^2$$

(d) We first convert θ to radian measure:

$$\theta = 144° \bullet \frac{\pi}{180°} = \frac{4\pi}{5} \text{ radians}$$

The area is given by:

$$A = \frac{1}{2}r^2\theta = \frac{1}{2}(1.8)^2\left(\frac{4\pi}{5}\right) = \frac{12.96\pi}{10} = 1.296\pi \text{ cm}^2 \approx 4.07 \text{ cm}^2$$

27. We have $r = 1$ cm and $A = \frac{\pi}{5}$ cm^2. Substituting into $A = \frac{1}{2}r^2\theta$, we have:

$$\frac{\pi}{5} = \frac{1}{2}(1)^2\theta$$
$$\frac{\pi}{5} = \frac{\theta}{2}$$
$$\theta = \frac{2\pi}{5} \text{ radians}$$

29. (a) Each revolution of the wheel is 2π radians, so in 6 revolutions there are $\theta = 6(2\pi) = 12\pi$ radians. Consequently, we have:

$$\omega = \frac{\theta}{t} = \frac{12\pi \text{ radians}}{1 \sec} = 12\pi \frac{\text{radians}}{\sec}$$

(b) Using the formula $s = r\theta$, where $r = 12$ cm and $\theta = 12\pi$ radians, we have $s = (12 \text{ cm})(12\pi \text{ radians}) = 144\pi$ cm. The linear speed, therefore, is:

$$v = \frac{d}{t} = \frac{144\pi \text{ cm}}{1\sec} = 144\pi \frac{\text{cm}}{\sec}$$

(c) Using the formula $s = r\theta$, where $r = 6$ cm and $\theta = 12\pi$ radians, we have $s = (6 \text{ cm})(12\pi \text{ radians}) = 72\pi$ cm. Thus, we have:

$$v = \frac{d}{t} = \frac{72\pi \text{ cm}}{1\sec} = 72\pi \frac{\text{cm}}{\sec}$$

31. (a) In 1 second, the wheel has rotated $1080° \bullet \frac{\pi \text{ radians}}{180°} = 6\pi$ radians. Consequently, we have:

$$\omega = \frac{\theta}{t} = \frac{6\pi \text{ radians}}{1\sec} = 6\pi \frac{\text{radians}}{\sec}$$

(b) Using $\theta = 6\pi$ radians and $r = 25$ cm, then:

$$s = r\theta = (25 \text{ cm})(6\pi \text{ radians}) = 150\pi \text{ cm}$$

Thus, we have:

$$v = \frac{150\pi \text{ cm}}{1\sec} = 150\pi \frac{\text{cm}}{\sec}$$

(c) Using $\theta = 6\pi$ radians and $r = \frac{25}{2}$ cm, then:

$$s = r\theta = \left(\frac{25}{2} \text{ cm}\right)(6\pi \text{ radians}) = 75\pi \text{ cm}$$

Thus, we have:

$$v = \frac{75\pi \text{ cm}}{1\sec} = 75\pi \frac{\text{cm}}{\sec}$$

33. (a) In 1 minute, the wheel has rotated 500 rev. Since each revolution is equal to 2π radians, then we have:

$$\theta = (500)(2\pi) = 1000\pi \text{ radians}$$

Consequently, we have:

$$\omega = \frac{\theta}{t} = \frac{1000\pi \text{ radians}}{60 \sec} = \frac{50\pi}{3} \frac{\text{radians}}{\sec}$$

(b) Using $\theta = 1000\pi$ radians and $r = 45$ cm, then:
$$s = r\theta = (45 \text{ cm})(1000\pi \text{ radians}) = 45000\pi \text{ cm}$$
Thus, we have:
$$v = \frac{45000\pi \text{ cm}}{60 \text{ sec}} = 750\pi \frac{\text{cm}}{\text{sec}}$$

(c) Using $\theta = 1000\pi$ radians and $r = \frac{45}{2}$ cm, then:
$$s = r\theta = \left(\frac{45}{2} \text{ cm}\right)(1000\pi \text{ radians}) = 22500\pi \text{ cm}$$
Thus, we have:
$$v = \frac{22500\pi \text{ cm}}{60 \text{ sec}} = 375\pi \frac{\text{cm}}{\text{sec}}$$

35. (a) Each revolution is 2π radians, and 24 hr = 86400 sec, so:
$$\omega = \frac{\theta}{t} = \frac{2\pi \text{ radians}}{86400 \text{ sec}} \approx 0.000073 \frac{\text{radians}}{\text{sec}}$$

(b) Using $\theta = 2\pi$ radians and $r = 3960$ mi, then:
$$s = r\theta = (3960 \text{ mi})(2\pi \text{ radians}) = 7920\pi \text{ mi}$$
Thus, we have:
$$v = \frac{s}{t} = \frac{7920\pi \text{ mi}}{24 \text{ hr}} = 330\pi \frac{\text{mi}}{\text{hr}} \approx 1040 \text{ mph}$$

37. Since the triangle is equilateral with sides 1 ft, then the circle formed will have a radius of 1 ft and arc length of 1 ft, so:
$$\theta = \frac{s}{r} = \frac{1}{1} = 1 \text{ radian}$$
Converting to degrees, the angle at A will have a measure of:
$$1 \bullet \frac{180°}{\pi} = \frac{180°}{\pi} \approx 57.30°$$

39. (a) The angular speed of the larger wheel is given by:
$$\omega = 100 \frac{\text{rev}}{\text{min}} \bullet 2\pi \frac{\text{rad}}{\text{rev}} = 200\pi \frac{\text{rad}}{\text{min}}$$

(b) Since $v = R\omega$ and $R = 10$ cm, we have:
$$v = R\omega = 10 \text{ cm} \bullet 200\pi \frac{\text{rad}}{\text{min}} = 2000\pi \frac{\text{cm}}{\text{min}}$$

(c) Since the linear speed on the smaller wheel must also be $2000\pi \frac{\text{cm}}{\text{min}}$, and the radius of the smaller wheel is $r = 6$ cm, then its angular speed is:
$$\omega = \frac{v}{r} = \frac{2000\pi \text{ cm/min}}{6 \text{ cm}} = \frac{1000}{3}\pi \frac{\text{radians}}{\text{min}}$$

(d) In rpm, the angular speed of the smaller wheel is:
$$\frac{1 \text{ rev}}{2\pi \text{ rad}} \bullet \frac{1000}{3}\pi \frac{\text{rad}}{\text{min}} = \frac{500}{3} \text{ rpm}$$

41. We first find the linear speed in terms of r. Since $v = R\omega$, we have:

$$v = R\omega = 2r \bullet 6\,\tfrac{\text{rad}}{\text{sec}} = 12r$$

Now find the angular speed of the smaller wheel:

$$\omega = \tfrac{v}{r} = \tfrac{12r}{r} = 12 \text{ radians per second}$$

43. We first convert θ to radian measure:

$$\theta = 71°23' = 71\tfrac{23°}{60} \bullet \tfrac{\pi}{180°} = \tfrac{4283\pi}{10800}$$

Now compute the arc length:

$$s = r\theta = 3960 \text{ mi} \bullet \tfrac{4283\pi}{10800} = \tfrac{47113\pi}{30} \text{ mi} \approx 4930 \text{ mi}$$

45. We first convert θ to radian measure:

$$\theta = 21°19' = 21\tfrac{19°}{60} \bullet \tfrac{\pi}{180°} = \tfrac{1279\pi}{10800}$$

Now compute the arc length:

$$s = r\theta = 3960 \text{ mi} \bullet \tfrac{1279\pi}{10800} = \tfrac{14069\pi}{30} \text{ mi} \approx 1470 \text{ mi}$$

47. We first convert θ to radian measure:

$$\theta = 38°54' = 38\tfrac{54°}{60} \bullet \tfrac{\pi}{180°} = \tfrac{2334\pi}{10800}$$

Now compute the arc length:

$$s = r\theta = 3960 \text{ mi} \bullet \tfrac{2334\pi}{10800} = \tfrac{8558\pi}{10} \text{ mi} \approx 2690 \text{ mi}$$

49. (a) We first find the arc length $s = r(2\theta) = 2r\theta$. Since the perimeter is 12 cm, we have:

$$r + r + 2r\theta = 12$$
$$2r + 2r\theta = 12$$
$$r + r\theta = 6$$
$$r(1+\theta) = 6$$
$$r = \tfrac{6}{1+\theta}$$

(b) The area is given by:

$$A = \tfrac{1}{2}r^2(2\theta) = \tfrac{1}{2}\left(\tfrac{6}{1+\theta}\right)^2(2\theta) = \tfrac{36\theta}{(1+\theta)^2}$$

3.2 Trigonometric Functions of Real Numbers

1. (a) The reference angle is $\tfrac{\pi}{4}$.

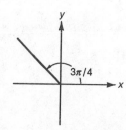

(b)   The reference angle is $\frac{\pi}{6}$.

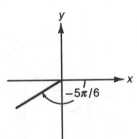

(c)   The reference angle is $\frac{\pi}{3}$.

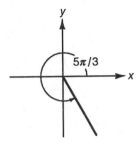

3.   (a)   The reference angle is $\frac{\pi}{4}$.

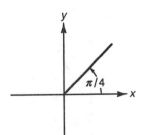

(b)   The reference angle is $\frac{\pi}{4}$.

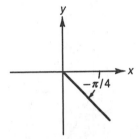

(c)   The reference angle is $\frac{\pi}{4}$.

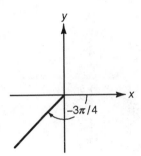

5.   (a)   The reference angle is $\frac{\pi}{3}$.

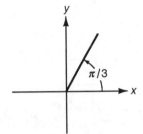

(b)   The reference angle is $\frac{\pi}{3}$.

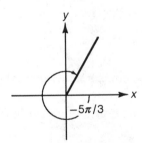

(c)   The reference angle is $\frac{\pi}{3}$.

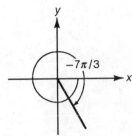

7. The reference number is $\pi - 2$.

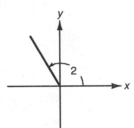

9. The reference number is $4 - \pi$.

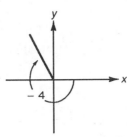

11. (a)   Since $\cos\frac{\pi}{3} = \cos 60° = \frac{1}{2}$, the correct answer is (B).

(b)   Since $\sin\frac{\pi}{3} = \sin 60° = \frac{\sqrt{3}}{2}$, the correct answer is (A).

(c)   Since $\tan\frac{\pi}{3} = \tan 60° = \sqrt{3}$, the correct answer is (D).

(d)   Since $\cos\frac{\pi}{6} = \cos 30° = \frac{\sqrt{3}}{2}$, the correct answer is (A).

(e)   Since $\sin\frac{\pi}{6} = \sin 30° = \frac{1}{2}$, the correct answer is (B).

(f)   Since $\tan\frac{\pi}{6} = \tan 30° = \frac{\sqrt{3}}{3}$, the correct answer is (C).

13. (a)   The reference angle for $\frac{5\pi}{3}$ is $\frac{\pi}{3}$, and $\cos\frac{\pi}{3} = \frac{1}{2}$. The terminal side of $\frac{5\pi}{3}$ lies in the fourth quadrant where the x-coordinates are positive, so:
$$\cos\frac{5\pi}{3} = \cos\frac{\pi}{3} = \frac{1}{2}$$

(b)   The reference angle for $-\frac{5\pi}{3}$ is $\frac{\pi}{3}$, and $\cos\frac{\pi}{3} = \frac{1}{2}$. The terminal side of $-\frac{5\pi}{3}$ lies in the first quadrant where the x-coordinates are positive, so:
$$\cos\left(-\frac{5\pi}{3}\right) = \cos\frac{\pi}{3} = \frac{1}{2}$$

(c)   The reference angle for $\frac{5\pi}{3}$ is $\frac{\pi}{3}$, and $\sin\frac{\pi}{3} = \frac{\sqrt{3}}{2}$. The terminal side of $\frac{5\pi}{3}$ lies in the fourth quadrant where the y-coordinates are negative, so:
$$\sin\frac{5\pi}{3} = -\sin\frac{\pi}{3} = -\frac{\sqrt{3}}{2}$$

(d)   The reference angle for $-\frac{5\pi}{3}$ is $\frac{\pi}{3}$, and $\sin\frac{\pi}{3} = \frac{\sqrt{3}}{2}$. The terminal side of $-\frac{5\pi}{3}$ lies in the first quadrant where the y-coordinates are positive, so:

$$\sin\left(-\frac{5\pi}{3}\right) = \sin\frac{\pi}{3} = \frac{\sqrt{3}}{2}$$

15. (a)   The reference angle for $\frac{7\pi}{6}$ is $\frac{\pi}{6}$, and $\cos\frac{\pi}{6} = \frac{\sqrt{3}}{2}$. The terminal side of $\frac{7\pi}{6}$ lies in the third quadrant where the x-coordinates are negative, so:

$$\cos\frac{7\pi}{6} = -\cos\frac{\pi}{6} = -\frac{\sqrt{3}}{2}$$

(b)   The reference angle for $-\frac{7\pi}{6}$ is $\frac{\pi}{6}$, and $\cos\frac{\pi}{6} = \frac{\sqrt{3}}{2}$. The terminal side of $-\frac{7\pi}{6}$ lies in the second quadrant where the x-coordinates are negative, so:

$$\cos\left(-\frac{7\pi}{6}\right) = -\cos\frac{\pi}{6} = -\frac{\sqrt{3}}{2}$$

(c)   The reference angle for $\frac{7\pi}{6}$ is $\frac{\pi}{6}$, and $\sin\frac{\pi}{6} = \frac{1}{2}$. The terminal side of $\frac{7\pi}{6}$ lies in the third quadrant where the y-coordinates are negative, so:

$$\sin\frac{7\pi}{6} = -\sin\frac{\pi}{6} = -\frac{1}{2}$$

(d)   The reference angle for $-\frac{7\pi}{6}$ is $\frac{\pi}{6}$, and $\sin\frac{\pi}{6} = \frac{1}{2}$. The terminal side of $-\frac{7\pi}{6}$ lies in the second quadrant where the y-coordinates are positive, so:

$$\sin\left(-\frac{7\pi}{6}\right) = \sin\frac{\pi}{6} = \frac{1}{2}$$

17. (a)   The reference angle for $\frac{4\pi}{3}$ is $\frac{\pi}{3}$, and $\cos\frac{\pi}{3} = \frac{1}{2}$. The terminal side of $\frac{4\pi}{3}$ lies in the third quadrant where the x-coordinates are negative, so:

$$\cos\frac{4\pi}{3} = -\cos\frac{\pi}{3} = -\frac{1}{2}$$

(b)   The reference angle for $-\frac{4\pi}{3}$ is $\frac{\pi}{3}$, and $\cos\frac{\pi}{3} = \frac{1}{2}$. The terminal side of $-\frac{4\pi}{3}$ lies in the second quadrant where the x-coordinates are negative, so:

$$\cos\left(-\frac{4\pi}{3}\right) = -\cos\frac{\pi}{3} = -\frac{1}{2}$$

(c)   The reference angle for $\frac{4\pi}{3}$ is $\frac{\pi}{3}$, and $\sin\frac{\pi}{3} = \frac{\sqrt{3}}{2}$. The terminal side of $\frac{4\pi}{3}$ lies in the third quadrant where the y-coordinates are negative, so:

$$\sin\frac{4\pi}{3} = -\sin\frac{\pi}{3} = -\frac{\sqrt{3}}{2}$$

(d)   The reference angle for $-\frac{4\pi}{3}$ is $\frac{\pi}{3}$, and $\sin\frac{\pi}{3} = \frac{\sqrt{3}}{2}$. The terminal side of $-\frac{4\pi}{3}$ lies in the second quadrant where the y-coordinates are positive, so:

$$\sin\left(-\frac{4\pi}{3}\right) = \sin\frac{\pi}{3} = \frac{\sqrt{3}}{2}$$

19. (a)   The reference angle for $\frac{5\pi}{4}$ is $\frac{\pi}{4}$, and $\cos\frac{\pi}{4} = \frac{\sqrt{2}}{2}$. The terminal side of $\frac{5\pi}{4}$ lies in the third quadrant where the x-coordinates are negative, so:

$$\cos\frac{5\pi}{4} = -\cos\frac{\pi}{4} = -\frac{\sqrt{2}}{2}$$

(b)   The reference angle for $-\frac{5\pi}{4}$ is $\frac{\pi}{4}$, and $\cos\frac{\pi}{4} = \frac{\sqrt{2}}{2}$. The terminal side of $-\frac{5\pi}{4}$ lies in the second quadrant where the x-coordinates are negative, so:

$$\cos\left(-\frac{5\pi}{4}\right) = -\cos\frac{\pi}{4} = -\frac{\sqrt{2}}{2}$$

(c)   The reference angle for $\frac{5\pi}{4}$ is $\frac{\pi}{4}$, and $\sin\frac{\pi}{4} = \frac{\sqrt{2}}{2}$. The terminal side of $\frac{5\pi}{4}$ lies in the third quadrant where the y-coordinates are negative, so:

$$\sin\frac{5\pi}{4} = -\sin\frac{\pi}{4} = -\frac{\sqrt{2}}{2}$$

(d)   The reference angle for $-\frac{5\pi}{4}$ is $\frac{\pi}{4}$, and $\sin\frac{\pi}{4} = \frac{\sqrt{2}}{2}$. The terminal side of $-\frac{5\pi}{4}$ lies in the second quadrant where the y-coordinates are positive, so:

$$\sin\left(-\frac{5\pi}{4}\right) = \sin\frac{\pi}{4} = \frac{\sqrt{2}}{2}$$

21. (a)   The reference angle for $\frac{4\pi}{3}$ is $\frac{\pi}{3}$, and $\sec\frac{\pi}{3} = 2$. The terminal side of $\frac{4\pi}{3}$ lies in the third quadrant where the x-coordinates are negative, so:

$$\sec\frac{4\pi}{3} = -\sec\frac{\pi}{3} = -2$$

(b)   The reference angle for $-\frac{4\pi}{3}$ is $\frac{\pi}{3}$, and $\csc\frac{\pi}{3} = \frac{2\sqrt{3}}{3}$. The terminal side of $-\frac{4\pi}{3}$ lies in the second quadrant where the y-coordinates are positive, so:

$$\csc\left(-\frac{4\pi}{3}\right) = \csc\frac{\pi}{3} = \frac{2\sqrt{3}}{3}$$

(c)   The reference angle for $\frac{4\pi}{3}$ is $\frac{\pi}{3}$, and $\tan\frac{\pi}{3} = \sqrt{3}$. The terminal side of $\frac{4\pi}{3}$ lies in the third quadrant where both the x- and y-coordinates are negative, so:

$$\tan\frac{4\pi}{3} = \tan\frac{\pi}{3} = \sqrt{3}$$

(d)   The reference angle for $-\frac{4\pi}{3}$ is $\frac{\pi}{3}$, and $\cot\frac{\pi}{3} = \frac{\sqrt{3}}{3}$. The terminal side of $-\frac{4\pi}{3}$ lies in the second quadrant where the x-coordinates are negative while the y-coordinates are positive, so:

$$\cot\left(-\frac{4\pi}{3}\right) = -\cot\frac{\pi}{3} = -\frac{\sqrt{3}}{3}$$

23. (a)   The reference angle for $\frac{17\pi}{6}$ is $\frac{\pi}{6}$, and $\sec\frac{\pi}{6} = \frac{2\sqrt{3}}{3}$. The terminal side of $\frac{17\pi}{6}$ lies in the second quadrant where the x-coordinates are negative, so:

$$\sec\frac{17\pi}{6} = -\sec\frac{\pi}{6} = -\frac{2\sqrt{3}}{3}$$

(b)   The reference angle for $-\frac{17\pi}{6}$ is $\frac{\pi}{6}$, and $\csc\frac{\pi}{6} = 2$. The terminal side of $-\frac{17\pi}{6}$ lies in the third quadrant where the y-coordinates are negative, so:

$$\csc\left(-\frac{17\pi}{6}\right) = -\csc\frac{\pi}{6} = -2$$

(c)   The reference angle for $\frac{17\pi}{6}$ is $\frac{\pi}{6}$, and $\tan\frac{\pi}{6} = \frac{\sqrt{3}}{3}$. The terminal side of $\frac{17\pi}{6}$ lies in the second quadrant where the x-coordinates are negative while the y-coordinates are positive, so:

$$\tan\frac{17\pi}{6} = -\tan\frac{\pi}{6} = -\frac{\sqrt{3}}{3}$$

(d)   The reference angle for $-\frac{17\pi}{6}$ is $\frac{\pi}{6}$, and $\cot\frac{\pi}{6} = \sqrt{3}$. The terminal side of $-\frac{17\pi}{6}$ lies in the third quadrant where both the x- and y-coordinates are negative, so:

$$\cot\left(-\frac{17\pi}{6}\right) = \cot\frac{\pi}{6} = \sqrt{3}$$

25. (a)   We need to choose positive radian values such that the x-coordinate of the point on the unit circle is 0. Four such values are $t = \frac{\pi}{2}, \frac{3\pi}{2}, \frac{5\pi}{2}$, and $\frac{7\pi}{2}$ (other answers are possible).

(b)   We need to choose negative radian values such that the x-coordinate of the point on the unit circle is 0. Four such values are $t = -\frac{\pi}{2}, -\frac{3\pi}{2}, -\frac{5\pi}{2}$, and $-\frac{7\pi}{2}$ (other answers are possible).

27. From the figure we have $0.5 < \cos 1 < 0.6$ and $0.8 < \sin 1 < 0.9$. Using a calculator, we have $\cos 1 \approx 0.54$ and $\sin 1 \approx 0.84$.

29. From the figure we have $0.5 < \cos(-1) < 0.6$ and $-0.9 < \sin(-1) < -0.8$. Using a calculator, we have $\cos(-1) \approx 0.54$ and $\sin(-1) \approx -0.84$.

31. From the figure we have $-0.7 < \cos 4 < -0.6$ and $-0.8 < \sin 4 < -0.7$. Using a calculator, we have $\cos 4 \approx -0.65$ and $\sin 4 \approx -0.76$.

33. From the figure, we have $-0.7 < \cos(-4) < -0.6$ and $0.7 < \sin(-4) < 0.8$. Using a calculator, we have $\cos(-4) \approx -0.65$ and $\sin(-4) \approx 0.76$.

35. Since $\sin(1 + 2\pi) = \sin 1$, we can use the figure to obtain $0.8 < \sin(1 + 2\pi) < 0.9$. Using a calculator, we have $\sin(1 + 2\pi) \approx 0.84$.

37. (a) Using a calculator to two decimal place accuracy, we have:

$$\sin 2.06 \approx 0.88 \qquad \cos 2.06 \approx -0.47 \qquad \tan 2.06 \approx -1.88$$
$$\sec 2.06 \approx -2.13 \qquad \csc 2.06 \approx 1.13 \qquad \cot 2.06 \approx -0.53$$

(b) Using a calculator to two decimal place accuracy, we have:

$$\sin(-2.06) \approx -0.88 \qquad \cos(-2.06) \approx -0.47 \qquad \tan(-2.06) \approx 1.88$$
$$\sec(-2.06) \approx -2.13 \qquad \csc(-2.06) \approx -1.13 \qquad \cot(-2.06) \approx 0.53$$

39. (a) Using a calculator to two decimal place accuracy, we have:

$$\sin\tfrac{\pi}{6} \approx 0.50 \qquad \cos\tfrac{\pi}{6} \approx 0.87 \qquad \tan\tfrac{\pi}{6} \approx 1.73$$
$$\sec\tfrac{\pi}{6} \approx 1.15 \qquad \csc\tfrac{\pi}{6} \approx 2.00 \qquad \cot\tfrac{\pi}{6} \approx 0.58$$

(b) Since $\tfrac{\pi}{6} + 2\pi$ will intersect the unit circle at the same location as in (a), all six trigonometric functions will have the same values as in (a):

$$\sin\left(\tfrac{\pi}{6} + 2\pi\right) \approx 0.50 \qquad \cos\left(\tfrac{\pi}{6} + 2\pi\right) \approx 0.87 \qquad \tan\left(\tfrac{\pi}{6} + 2\pi\right) \approx 1.73$$
$$\sec\left(\tfrac{\pi}{6} + 2\pi\right) \approx 1.15 \qquad \csc\left(\tfrac{\pi}{6} + 2\pi\right) \approx 2.00 \qquad \cot\left(\tfrac{\pi}{6} + 2\pi\right) \approx 0.58$$

41. Since the shaded region is the area of the circular sector less the area of the triangle, we first find the area of the sector and the triangle:

$$A_{\text{sector}} = \tfrac{1}{2}r^2\theta = \tfrac{1}{2}(7)^2\left(\tfrac{2\pi}{5}\right) = \tfrac{49\pi}{5} \approx 30.788 \text{ cm}^2$$
$$A_{\text{triangle}} = \tfrac{1}{2}ab\sin\theta = \tfrac{1}{2}(7)(7)\sin\tfrac{2\pi}{5} = \tfrac{49}{2}\sin\tfrac{2\pi}{5} \approx 23.301 \text{ cm}^2$$

So the area of the shaded region is given by:

$$A = \tfrac{49\pi}{5} - \tfrac{49}{2}\sin\tfrac{2\pi}{5} \approx 30.788 - 23.301 \approx 7.49 \text{ cm}^2$$

43. We first find the area of the sector and the area of the triangle:

$$A_{\text{sector}} = \tfrac{1}{2}r^2\theta = \tfrac{1}{2}(\sqrt{2})^2\theta = \theta$$
$$A_{\text{triangle}} = \tfrac{1}{2}ab\sin(180° - \theta) = \tfrac{1}{2}(\sqrt{2})(\sqrt{2})\sin\theta = \sin\theta$$

So the area of the shaded region is given by:

$$A(\theta) = A_{\text{sector}} + A_{\text{triangle}} = \theta + \sin\theta$$

45. We complete the table:

θ	$\theta - \frac{\theta^3}{6}$	$\sin\theta$
0.1	0.099833	0.099833
0.2	0.198667	0.198669
0.3	0.295500	0.295520

47. (a) Since the x-coordinate is $\frac{1}{2}$ and A is a point on the unit circle, then $x^2 + y^2 = 1$ and so:

$$\left(\tfrac{1}{2}\right)^2 + y^2 = 1$$
$$\tfrac{1}{4} + y^2 = 1$$
$$y^2 = \tfrac{3}{4}$$
$$y = \tfrac{\sqrt{3}}{2}$$

(b) By symmetry, the y-coordinate of B is $\frac{1}{2}$, and by a solution similar to part (a) the x-coordinate must be $\frac{\sqrt{3}}{2}$. For C, the x-coordinate is $\frac{1}{2}$ and since the y-coordinate of A is $\frac{\sqrt{3}}{2}$, then the y-coordinate for C must be $1 - \frac{\sqrt{3}}{2}$. For D, the y-coordinate is $\frac{1}{2}$ and since the x-coordinate of B is $\frac{\sqrt{3}}{2}$, then the x-coordinate for D must be $1 - \frac{\sqrt{3}}{2}$. Thus the coordinates of B, C and D are:

$$B\left(\tfrac{\sqrt{3}}{2}, \tfrac{1}{2}\right), \quad C\left(\tfrac{1}{2}, 1 - \tfrac{\sqrt{3}}{2}\right), \quad D\left(1 - \tfrac{\sqrt{3}}{2}, \tfrac{1}{2}\right)$$

(c) Using the coordinates for A and B as found in parts (a) and (b), then AB is given by (using the distance formula):

$$AB = \sqrt{\left(\tfrac{\sqrt{3}}{2} - \tfrac{1}{2}\right)^2 + \left(\tfrac{1}{2} - \tfrac{\sqrt{3}}{2}\right)^2} = \sqrt{\tfrac{(\sqrt{3}-1)^2}{4} + \tfrac{(\sqrt{3}-1)^2}{4}} = \tfrac{\sqrt{3}-1}{2} \bullet \sqrt{2} = \tfrac{\sqrt{3}-1}{\sqrt{2}}$$

Since the area of the square is $(AB)^2$, then:

$$\text{Area of square} = \left(\tfrac{\sqrt{3}-1}{\sqrt{2}}\right)^2 = \tfrac{3 - 2\sqrt{3} + 1}{2} = \tfrac{4 - 2\sqrt{3}}{2} = 2 - \sqrt{3}$$

(d) Since the coordinates of D are $D\left(1 - \tfrac{\sqrt{3}}{2}, \tfrac{1}{2}\right)$, then $DQ = \frac{1}{2}$ and

$$QP = 1 - \left(1 - \tfrac{\sqrt{3}}{2}\right) = \tfrac{\sqrt{3}}{2}. \text{ Thus:}$$

$$\tan \angle DPQ = \frac{DQ}{QP} = \frac{\tfrac{1}{2}}{\tfrac{\sqrt{3}}{2}} = \frac{1}{\sqrt{3}} = \frac{\sqrt{3}}{3}$$

Thus $\angle DPQ = \frac{\pi}{6}$.

(e)   Since the three angles must sum to $\frac{\pi}{2}$, we have:

$$\tfrac{\pi}{6} + \angle DPA + \tfrac{\pi}{6} = \tfrac{\pi}{2}$$
$$\angle DPA + \tfrac{\pi}{3} = \tfrac{\pi}{2}$$
$$\angle DPA = \tfrac{\pi}{6}$$

(f)   The area of the triangle and the sector are given by:

$$A_{\text{triangle}} = \tfrac{1}{2} ab \sin \theta = \tfrac{1}{2}(1)(1)\sin \tfrac{\pi}{6} = \tfrac{1}{2} \cdot \tfrac{1}{2} = \tfrac{1}{4}$$
$$A_{\text{sector}} = \tfrac{1}{2} r^2 \theta = \tfrac{1}{2}(1)^2 \cdot \tfrac{\pi}{6} = \tfrac{\pi}{12}$$

So the area of the shaded segment is given by:

$$\text{Area of shaded segment} = A_{\text{sector}} - A_{\text{triangle}} = \tfrac{\pi}{12} - \tfrac{1}{4}$$

(g)   Since the shaded area consists of the square plus four sectors, we have:

$$A_{\text{total}} = 2 - \sqrt{3} + 4\left(\tfrac{\pi}{12} - \tfrac{1}{4}\right) = 2 - \sqrt{3} + \tfrac{\pi}{3} - 1 = 1 - \sqrt{3} + \tfrac{\pi}{3} = \tfrac{3 + \pi - 3\sqrt{3}}{3}$$

3.3 Some Fundamental Identities

1.   (a)   Checking the identity $\sin^2 t + \cos^2 t = 1$, we have:

$$\sin^2 \tfrac{\pi}{3} + \cos^2 \tfrac{\pi}{3} = \left(\tfrac{\sqrt{3}}{2}\right)^2 + \left(\tfrac{1}{2}\right)^2 = \tfrac{3}{4} + \tfrac{1}{4} = 1$$

(b)   Checking the identity $\sin^2 t + \cos^2 t = 1$, we have:

$$\sin^2 \tfrac{5\pi}{4} + \cos^2 \tfrac{5\pi}{4} = \left(-\tfrac{\sqrt{2}}{2}\right)^2 + \left(-\tfrac{\sqrt{2}}{2}\right)^2 = \tfrac{2}{4} + \tfrac{2}{4} = 1$$

(c)   Checking the identity $\sin^2 t + \cos^2 t = 1$ and using a calculator, we have:

$$\sin^2(-53) + \cos^2(-53) \approx (-0.3959)^2 + (-0.9183)^2 \approx 0.1568 + 0.8432 = 1$$

3.   (a)   Checking the identity $\cot^2 t + 1 = \csc^2 t$, we have:

$$\cot^2\left(-\tfrac{\pi}{6}\right) + 1 = \left(-\sqrt{3}\right)^2 + 1 = 3 + 1 = 4$$
$$\csc^2\left(-\tfrac{\pi}{6}\right) = (-2)^2 = 4$$

(b)   Checking the identity $\cot^2 t + 1 = \csc^2 t$, we have:

$$\cot^2\left(\tfrac{7\pi}{4}\right) + 1 = (-1)^2 + 1 = 1 + 1 = 2$$
$$\csc^2\left(\tfrac{7\pi}{4}\right) = \left(-\sqrt{2}\right)^2 = 2$$

(c)   Checking the identity $\cot^2 t + 1 = \csc^2 t$ and using a calculator, we have:

$$\cot^2(0.12) + 1 \approx 68.7787 + 1 = 69.7787$$
$$\csc^2(0.12) \approx 69.7787$$

5. (a) Checking the identity $\sin(-t) = -\sin t$, we have:
$$\sin\left(-\tfrac{3\pi}{2}\right) = 1$$
$$-\sin\tfrac{3\pi}{2} = -(-1) = 1$$

   (b) Checking the identity $\sin(-t) = -\sin t$, we have:
$$\sin\tfrac{5\pi}{6} = \tfrac{1}{2}$$
$$-\sin\left(-\tfrac{5\pi}{6}\right) = -\left(-\tfrac{1}{2}\right) = \tfrac{1}{2}$$

   (c) Checking the identity $\sin(-t) = -\sin t$, and using a calculator, we have:
$$\sin(-13.24) \approx -0.6238$$
$$-\sin 13.24 \approx -(0.6238) = -0.6238$$

7. (a) Checking the identity $\sin(t + 2\pi) = \sin t$, we have:
$$\sin\left(\tfrac{5\pi}{3} + 2\pi\right) = \sin\left(\tfrac{11\pi}{3}\right) = -\tfrac{\sqrt{3}}{2}$$
$$\sin\tfrac{5\pi}{3} = -\tfrac{\sqrt{3}}{2}$$

   (b) Checking the identity $\sin(t + 2\pi) = \sin t$, we have:
$$\sin\left(-\tfrac{3\pi}{2} + 2\pi\right) = \sin\tfrac{\pi}{2} = 1$$
$$\sin\left(-\tfrac{3\pi}{2}\right) = 1$$

   (c) Checking the identity $\sin(t + 2\pi) = \sin t$ and using a calculator, we have:
$$\sin(\sqrt{19} + 2\pi) \approx \sin(10.6421) \approx -0.9382$$
$$\sin(\sqrt{19}) \approx \sin(4.3589) \approx -0.9382$$

9. Evaluating each side of the "equality" when $t = \tfrac{\pi}{6}$, we have:
$$\cos\left(2 \bullet \tfrac{\pi}{6}\right) = \cos\tfrac{\pi}{3} = \tfrac{1}{2}$$
$$2\cos\tfrac{\pi}{6} = 2 \bullet \tfrac{\sqrt{3}}{2} = \sqrt{3}$$
Since these results are unequal, then $\cos 2t = 2\cos t$ is not an identity.

11. Using $\sin^2 t + \cos^2 t = 1$, we have:
$$\cos t = \pm\sqrt{1 - \sin^2 t} = \pm\sqrt{1 - \tfrac{9}{25}} = \pm\tfrac{4}{5}$$

Since $\pi < t < \tfrac{3\pi}{2}$, then $\cos t = -\tfrac{4}{5}$ and thus:
$$\tan t = \frac{\sin t}{\cos t} = \frac{-\tfrac{3}{5}}{-\tfrac{4}{5}} = \frac{3}{4}$$

13. Since $\frac{\pi}{2} < t < \pi$, $\cos t < 0$ and thus:
$$\cos t = -\sqrt{1 - \sin^2 t} = -\sqrt{1 - \frac{3}{16}} = -\sqrt{\frac{13}{16}} = -\frac{\sqrt{13}}{4}$$

Thus:
$$\tan t = \frac{\sin t}{\cos t} = \frac{\frac{\sqrt{3}}{4}}{-\frac{\sqrt{13}}{4}} = -\frac{\sqrt{3}}{\sqrt{13}} = -\frac{\sqrt{39}}{13}$$

15. Using the identity $1 + \tan^2 \alpha = \sec^2 \alpha$, we have:
$$1 + \left(\frac{12}{5}\right)^2 = \sec^2 \alpha$$
$$1 + \frac{144}{25} = \sec^2 \alpha$$
$$\frac{169}{25} = \sec^2 \alpha$$
$$\pm \frac{13}{5} = \sec \alpha$$

Since $\cos \alpha > 0$, $\sec \alpha > 0$ and thus we pick $\sec \alpha = \frac{13}{5}$. Thus $\cos \alpha = \frac{5}{13}$, and we use $\sin^2 \alpha + \cos^2 \alpha = 1$ to get:
$$\sin^2 \alpha = 1 - \left(\frac{5}{13}\right)^2 = \frac{144}{169}, \text{ thus } \sin \alpha = \pm \frac{12}{13}$$

We pick the positive value since, if both the tangent and cosine are positive, then so is the sine. Thus $\sin \alpha = \frac{12}{13}$.

17. For $0 < \theta < \frac{\pi}{2}$, we have:
$$\sqrt{9 - x^2} = \sqrt{9 - (3\sin \theta)^2} = \sqrt{9(1 - \sin^2 \theta)} = \sqrt{9\cos^2 \theta} = 3\cos \theta$$

We choose the positive root since $\cos \theta > 0$ for $0 < \theta < \frac{\pi}{2}$.

19. Since $0 < \theta < \frac{\pi}{2}$, then $\tan \theta > 0$. Thus we have:
$$\frac{1}{\left(u^2 - 25\right)^{3/2}} = \frac{1}{\left(25\sec^2 \theta - 25\right)^{3/2}} = \frac{1}{125\tan^3 \theta} = \frac{\cot^3 \theta}{125}$$

21. Since $\sec \theta > 0$, we have:
$$\frac{1}{\sqrt{u^2 + 7}} = \frac{1}{\sqrt{7\tan^2 \theta + 7}} = \frac{1}{\sqrt{7} \sec \theta} = \frac{\cos \theta}{\sqrt{7}} = \frac{\sqrt{7} \cos \theta}{7}$$

23. (a) Since $\sin(-t) = -\sin t$, we have:
$$\sin(-t) = -\sin t = -\frac{2}{3}$$

(b) Since $\sin(-\phi) = -\sin\phi$, we have:
$$\sin(-\phi) = -\sin\phi = -\left(-\tfrac{1}{4}\right) = \tfrac{1}{4}$$

(c) Since $\cos(-\alpha) = \cos\alpha$, we have:
$$\cos(-\alpha) = \cos\alpha = \tfrac{1}{5}$$

(d) Since $\cos(-s) = \cos s$, we have:
$$\cos(-s) = \cos s = -\tfrac{1}{5}$$

25. (a) Since $\cos t = -\tfrac{1}{3}$ and $\tfrac{\pi}{2} < t < \pi$, we have $\sin t > 0$ and thus:
$$\sin t = \sqrt{1 - \tfrac{1}{9}} = \sqrt{\tfrac{8}{9}} = \tfrac{2\sqrt{2}}{3}$$
Therefore we have the values:
$$\sin(-t) = -\sin t = -\tfrac{2\sqrt{2}}{3}$$
$$\cos(-t) = \cos t = -\tfrac{1}{3}$$
Thus:
$$\sin(-t) + \cos(-t) = -\tfrac{2\sqrt{2}}{3} - \tfrac{1}{3} = -\tfrac{1+2\sqrt{2}}{3}$$

(b) Note that $\sin^2(-t) + \cos^2(-t) = 1$, regardless of the value of t.

27. (a) Using the identity $\cos(t + 2\pi k) = \cos t$, we have:
$$\cos\left(\tfrac{\pi}{4} + 2\pi\right) = \cos\tfrac{\pi}{4} = \tfrac{\sqrt{2}}{2}$$

(b) Using the identity $\sin(t + 2\pi k) = \sin t$, we have:
$$\sin\left(\tfrac{\pi}{3} + 2\pi\right) = \sin\tfrac{\pi}{3} = \tfrac{\sqrt{3}}{2}$$

(c) Using the identity $\sin(t + 2\pi k) = \sin t$, we have:
$$\sin\left(\tfrac{\pi}{2} - 6\pi\right) = \sin\tfrac{\pi}{2} = 1$$

29. Using the Pythagorean identities, we have:
$$\frac{\sin^2 t + \cos^2 t}{\tan^2 t + 1} = \frac{1}{\sec^2 t} = \cos^2 t$$

31. Using the Pythagorean identities, we have:
$$\frac{\sec^2\theta - \tan^2\theta}{1 + \cot^2\theta} = \frac{\tan^2\theta + 1 - \tan^2\theta}{\csc^2\theta} = \frac{1}{\csc^2\theta} = \sin^2\theta$$

33. Working from the right-hand side and using the identity for $\cot t$, we have:

$$\sin t + \cot t \cos t = \sin t + \frac{\cos t}{\sin t}(\cos t) = \frac{\sin^2 t}{\sin t} + \frac{\cos^2 t}{\sin t} = \frac{\sin^2 t + \cos^2 t}{\sin t} = \frac{1}{\sin t} = \csc t$$

35. Combining fractions on the left-hand side, we have:

$$\frac{1}{1+\sec s} + \frac{1}{1-\sec s} = \frac{(1-\sec s)+(1+\sec s)}{(1-\sec s)(1+\sec s)} = \frac{2}{1-\sec^2 s} = \frac{-2}{\tan^2 s} = -2\cot^2 s$$

37. Denoting $\sin \theta$ by S and $\cos \theta$ by C, we have:

$$\frac{\cot \theta}{1-\tan \theta} + \frac{\tan \theta}{1-\cot \theta} = \frac{\frac{C}{S}}{1-\frac{S}{C}} + \frac{\frac{S}{C}}{1-\frac{C}{S}}$$

$$= \frac{C^2}{S(C-S)} - \frac{S^2}{C(C-S)}$$

$$= \frac{C^3 - S^3}{SC(C-S)}$$

$$= \frac{(C-S)(C^2 + SC + S^2)}{SC(C-S)}$$

$$= \frac{1+SC}{SC}$$

$$= \frac{1}{SC}+1$$

Now we replace C with $\cos \theta$ and S with $\sin \theta$:

$$\frac{1}{SC}+1 = \frac{\cos^2 \theta + \sin^2 \theta}{\sin \theta \cos \theta}+1 = \frac{\cos \theta}{\sin \theta} + \frac{\sin \theta}{\cos \theta}+1 = \cot \theta + \tan \theta +1$$

39. The expression on the left-hand side becomes:

$$\tan \theta - (\tan \theta \cot \theta)(\cot \theta) + \cot \theta - (\cot \theta \tan \theta)(\tan \theta)$$

This multiplies out to become:

$$\tan \theta - \cot \theta + \cot \theta - \tan \theta = 0$$

41. If $\sec t = \frac{13}{5}$, then $\cos t = \frac{5}{13}$. Given $\frac{3\pi}{2} < t < 2\pi$, then $\sin t < 0$ (fourth quadrant), thus:

$$\sin t = -\sqrt{1-\cos^2 t} = -\sqrt{1-\frac{25}{169}} = -\sqrt{\frac{144}{169}} = -\frac{12}{13}$$

So we have:

$$\frac{2\sin t - 3\cos t}{4\sin t - 9\cos t} = \frac{2\left(-\frac{12}{13}\right)-3\left(\frac{5}{13}\right)}{4\left(-\frac{12}{13}\right)-9\left(\frac{5}{13}\right)} = \frac{-24-15}{-48-45} = \frac{-39}{-93} = \frac{13}{31}$$

43. Assuming that the radius of the circle is 1, the coordinates of the point labeled (x, y) are $(\cos t, \sin t)$, and the coordinates of the point labeled $(-x, -y)$ are $\big(\cos(t + \pi), \sin(t + \pi)\big)$. So we have $y = \sin t$ and $-y = \sin(t + \pi)$, from which it follows that $\sin(t + \pi) = -\sin t$. Similarly, $-x = \cos(t + \pi)$ and $x = \cos t$, from which it follows that $\cos(t + \pi) = -\cos t$. Since $t - \pi$ results in the same intersection point with the unit circle as $t + \pi$, identities (ii) and (iv) follow in a similar manner.

45. The substitutions are $x = \frac{\sqrt{2}}{2}(X - Y)$ and $y = \frac{\sqrt{2}}{2}(X + Y)$. Thus we have:
$$x^2 = \tfrac{1}{2}\big(X^2 - 2XY + Y^2\big) \text{ and } y^2 = \tfrac{1}{2}\big(X^2 + 2XY + Y^2\big)$$
Squaring again, we obtain:
$$x^4 = \tfrac{1}{4}\big(X^4 - 4X^3Y + 6X^2Y^2 - 4XY^3 + Y^4\big)$$
$$y^4 = \tfrac{1}{4}\big(X^4 + 4X^3Y + 6X^2Y^2 + 4XY^3 + Y^4\big)$$

We also find $x^2y^2 = \tfrac{1}{4}\big(X^2 - Y^2\big)^2 = \tfrac{1}{4}\big(X^4 - 2X^2Y^2 + Y^4\big)$.

Now we use the expressions that have been found for $x^2, y^2, x^4,$ and y^4 to substitute in the expression $x^4 + 6x^2y^2 + y^4$ to obtain:
$$\tfrac{X^4}{4} - X^3Y + \tfrac{3}{2}X^2Y^2 - XY^3 + \tfrac{Y^4}{4} + \tfrac{3}{2}X^4 - 3X^2Y^2 + \tfrac{3}{2}Y^4 + \tfrac{X^4}{4} + X^3Y + \tfrac{3}{2}X^2Y^2 + XY^3 + \tfrac{Y^4}{4}$$
which is $2X^4 + 2Y^4$. In light of this result, the equation $x^4 + 6x^2y^2 + y^4 = 32$ is equivalent to $2X^4 + 2Y^4 = 32$, or $X^4 + Y^4 = 16$, as required.

47. (a) When $t = \frac{\pi}{6}$, we have:
$$2\sin^2 \tfrac{\pi}{6} - \sin \tfrac{\pi}{6} = 2\big(\tfrac{1}{2}\big)^2 - \tfrac{1}{2} = \tfrac{1}{2} - \tfrac{1}{2} = 0$$
$$2\sin \tfrac{\pi}{6}\cos \tfrac{\pi}{6} - \cos \tfrac{\pi}{6} = 2 \bullet \tfrac{1}{2} \bullet \tfrac{\sqrt{3}}{2} - \tfrac{\sqrt{3}}{2} = \tfrac{\sqrt{3}}{2} - \tfrac{\sqrt{3}}{2} = 0$$

(b) When $t = \frac{\pi}{4}$, we have:
$$2\sin^2 \tfrac{\pi}{4} - \sin \tfrac{\pi}{4} = 2\big(\tfrac{\sqrt{2}}{2}\big)^2 - \tfrac{\sqrt{2}}{2} = 1 - \tfrac{\sqrt{2}}{2} = \tfrac{2-\sqrt{2}}{2}$$
$$2\sin \tfrac{\pi}{4}\cos \tfrac{\pi}{4} - \cos \tfrac{\pi}{4} = 2 \bullet \tfrac{\sqrt{2}}{2} \bullet \tfrac{\sqrt{2}}{2} - \tfrac{\sqrt{2}}{2} = 1 - \tfrac{\sqrt{2}}{2} = \tfrac{2-\sqrt{2}}{2}$$

(c) No. Using the value $t = 0$, we have:
$$2\sin^2 0 - \sin 0 = 2(0)^2 - 0 = 0 - 0 = 0$$
$$2\sin 0\cos 0 - \cos 0 = 2(0)(1) - 1 = 0 - 1 = -1$$
Since the two sides of the equation are not equal, the given equation is not an identity.

3.4 Graphs of the Sine and Cosine Functions

1. A cycle is completed every 2 units, so the period is 2. The curve has high and low points of 1 and −1, respectively, so the amplitude is 1.

3. A cycle is completed every 4 units, so the period is 4. The curve has high and low points of 6 and −6, respectively, so the amplitude is 6.

5. A cycle is completed every 4 units, so the period is 4. The curve has high and low points of 6 and 2, respectively, so the amplitude is $\frac{6-2}{2} = 2$.

7. A cycle is completed every 6 units, so the period is 6. The curve has high and low points of −3 and −6, respectively, so the amplitude is $\frac{-3-(-6)}{2} = \frac{3}{2}$.

9. The coordinates of point C are $\left(-\frac{7\pi}{2}, 1\right) \approx (-10.996, 1)$.

11. The coordinates of point G are $\left(\frac{5\pi}{2}, 1\right) \approx (7.854, 1)$.

13. The coordinates of point B are $(-4\pi, 0) \approx (-12.566, 0)$.

15. The coordinates of point D are $(-3\pi, 0) \approx (-9.425, 0)$.

17. The coordinates of point E are $(-\pi, 0) \approx (-3.142, 0)$.

19. Referring to the graph of $y = \sin x$, we see that $y = \sin x$ is increasing on the interval $\frac{3\pi}{2} < x < 2\pi$.

21. Referring to the graph of $y = \sin x$, we see that $y = \sin x$ is decreasing on the interval $\frac{5\pi}{2} < x < \frac{7\pi}{2}$.

23. We complete the table:

x	$\sin x$	$x - \sin x$
0.453	0.43766	0.01534
0.253	0.25031	0.00269
0.0253	0.02530	0.00000
0.00253	0.00253	0.00000

25. The coordinates of point J are $\left(\frac{9\pi}{2},0\right) \approx (14.137,0)$.

27. The coordinates of point A are $(-4\pi,1) \approx (-12.566,1)$.

29. The coordinates of point E are $\left(\frac{\pi}{2},0\right) \approx (1.571,0)$.

31. The coordinates of point I are $(4\pi,1) \approx (12.566,1)$.

33. The coordinates of point B are $\left(-\frac{5\pi}{2},0\right) \approx (-7.854,0)$.

35. Referring to the graph of $y = \cos x$, we see that $y = \cos x$ is decreasing on the interval $0 < x < \pi$.

37. Referring to the graph of $y = \cos x$, we see that $y = \cos x$ is increasing on the interval $-\frac{\pi}{2} < x < 0$.

39. (a) Since C lies on the unit circle, its coordinates are $C(\cos \theta, \sin \theta)$.

    (b) Since $\angle AOB = \theta = \angle COD$ and $AO = CO$, the two triangles are congruent.

    (c) Since $AB = CD$ and $OB = OD$, the coordinates of A are $A(-\sin \theta, \cos \theta)$.

    (d) Matching up x- and y-coordinates, we have:
    $$\cos\left(\theta + \tfrac{\pi}{2}\right) = -\sin \theta \text{ and } \sin\left(\theta + \tfrac{\pi}{2}\right) = \cos \theta$$

3.5　Graphs of $y = A \sin(Bx - C)$ and $y = A \cos(Bx - C)$

1. (a) The amplitude is 2, the period is 2π, and the x-intercepts are 0, π, 2π. The function is increasing on the intervals $\left(0,\frac{\pi}{2}\right)$ and $\left(\frac{3\pi}{2},2\pi\right)$.

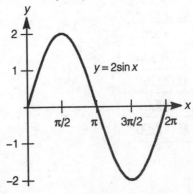

(b)   The amplitude is 1 and the period is $\frac{2\pi}{2} = \pi$. The x-intercepts are $0, \frac{\pi}{2}, \pi$, and the

function is increasing on the interval $\left(\frac{\pi}{4}, \frac{3\pi}{4}\right)$. Notice that the graph is a reflection of

$y = \sin 2x$ across the x-axis.

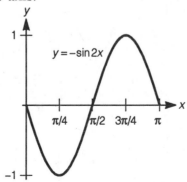

3.   (a)   The amplitude is 1 and the period is $\frac{2\pi}{2} = \pi$. The x-intercepts are $\frac{\pi}{4}, \frac{3\pi}{4}$, and the

function is increasing on the interval $\left(\frac{\pi}{2}, \pi\right)$.

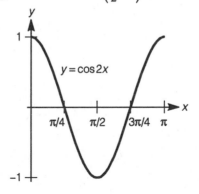

(b)   The amplitude is 2 and the period is $\frac{2\pi}{2} = \pi$. The x-intercepts are $\frac{\pi}{4}, \frac{3\pi}{4}$, and the

function is increasing on the interval $\left(\frac{\pi}{2}, \pi\right)$.

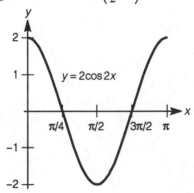

5. (a) The amplitude is 3 and the period is $\frac{2\pi}{\pi/2} = 4$. The x-intercepts are 0, 2, 4, and the function is increasing on the intervals $(0, 1)$ and $(3, 4)$.

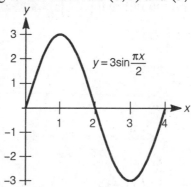

(b) The amplitude is 3 and the period is $\frac{2\pi}{\pi/2} = 4$. The x-intercepts are 0, 2, 4, and the function is increasing on the interval $(1, 3)$. Notice that the graph is a reflection of $y = 3\sin\frac{\pi x}{2}$ across the x-axis.

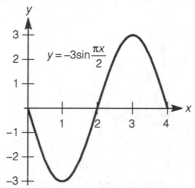

7. (a) The amplitude is 1 and the period is $\frac{2\pi}{2\pi} = 1$. The x-intercepts are $\frac{1}{4}, \frac{3}{4}$, and the function is increasing on the interval $\left(\frac{1}{2}, 1\right)$.

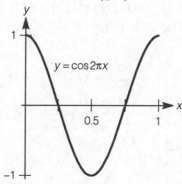

(b)   The amplitude is 4 and the period is $\frac{2\pi}{2\pi} = 1$. The x-intercepts are $\frac{1}{4}, \frac{3}{4}$, and the function is increasing on the interval $\left(0, \frac{1}{2}\right)$. Notice that the graph is a reflection of $y = 4\cos 2\pi x$ across the x-axis.

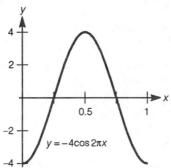

9.   The amplitude is 1 and the period is $\frac{2\pi}{2} = \pi$. The x-intercept is $\frac{3\pi}{4}$, and the function is increasing on the intervals $\left(0, \frac{\pi}{4}\right)$ and $\left(\frac{3\pi}{4}, \pi\right)$. Notice that the graph is a displacement of $y = \sin 2x$ up 1 unit.

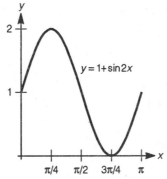

11.   The amplitude is 1 and the period is $\frac{2\pi}{\pi/3} = 6$. The x-intercepts are 0, 6, and the function is increasing on the interval $(0, 3)$. Notice that the graph is a reflection of $y = \cos\frac{\pi x}{3}$ across the x-axis, then a displacement up 1 unit.

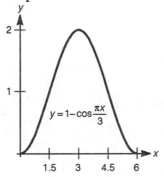

13. The amplitude is 1, the period is 2π, and the phase shift is $\frac{\pi}{6}$. The x-intercepts are

$\frac{\pi}{6}, \frac{7\pi}{6}, \frac{13\pi}{6}$, the high point is $\left(\frac{2\pi}{3}, 1\right)$ and the low point is $\left(\frac{5\pi}{3}, -1\right)$.

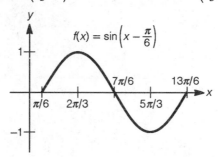

15. The amplitude is 1, the period is 2π, and the phase shift is $-\frac{\pi}{4}$. The x-intercepts are

$\frac{\pi}{4}, \frac{5\pi}{4}$, the high point is $\left(\frac{3\pi}{4}, 1\right)$, and the low points are $\left(-\frac{\pi}{4}, -1\right)$ and $\left(\frac{7\pi}{4}, -1\right)$. Notice that

the graph is a reflection of $y = \cos\left(x + \frac{\pi}{4}\right)$ across the x-axis.

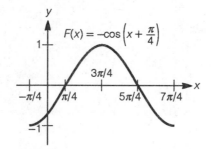

17. The amplitude is 1, the period is $\frac{2\pi}{2} = \pi$, and the phase shift is $\frac{\pi/2}{2} = \frac{\pi}{4}$. The x-intercepts

are $\frac{\pi}{4}, \frac{3\pi}{4}, \frac{5\pi}{4}$, the high point is $\left(\frac{\pi}{2}, 1\right)$, and the low point is $(\pi, -1)$.

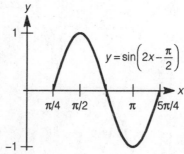

19. The amplitude is 1, the period is $\frac{2\pi}{2} = \pi$, and the phase shift is $\frac{\pi}{2}$. The x-intercepts are $\frac{3\pi}{4}, \frac{5\pi}{4}$, the high points are $\left(\frac{\pi}{2}, 1\right)$ and $\left(\frac{3\pi}{2}, 1\right)$, and the low point is $(\pi, -1)$.

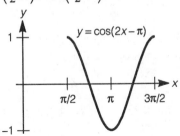

21. The amplitude is 3, the period is $\frac{2\pi}{1/2} = 4\pi$, and the phase shift is $\frac{-\pi/6}{1/2} = -\frac{\pi}{3}$. The x-intercepts are $-\frac{\pi}{3}, \frac{5\pi}{3}, \frac{11\pi}{3}$, the high point is $\left(\frac{2\pi}{3}, 3\right)$, and the low point is $\left(\frac{8\pi}{3}, -3\right)$.

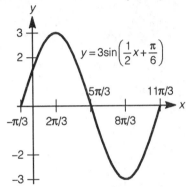

23. The amplitude is 4, the period is $\frac{2\pi}{3}$, and the phase shift is $\frac{\pi/4}{3} = \frac{\pi}{12}$. The x-intercepts are $\frac{\pi}{4}, \frac{7\pi}{12}$, the high points are $\left(\frac{\pi}{12}, 4\right)$ and $\left(\frac{3\pi}{4}, 4\right)$, and the low point is $\left(\frac{5\pi}{12}, -4\right)$.

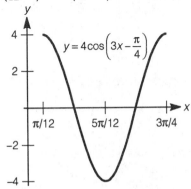

25. The amplitude is $\frac{1}{2}$, the period is $\frac{2\pi}{\pi/2} = 4$, and the phase shift is $\frac{\pi^2}{\pi/2} = 2\pi$. The

x-intercepts are 2π, $2\pi + 2$, $2\pi + 4$, the high point is $\left(2\pi + 1, \frac{1}{2}\right)$, and the low point is

$\left(2\pi + 3, -\frac{1}{2}\right)$.

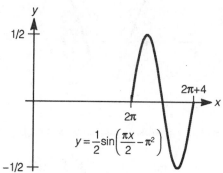

27. The amplitude is 1, the period is $\frac{2\pi}{2} = \pi$, and the phase shift is $\frac{\pi/3}{2} = \frac{\pi}{6}$. The x-intercepts

are $\frac{\pi}{6}$ and $\frac{7\pi}{6}$, the high point is $\left(\frac{2\pi}{3}, 2\right)$, and the low points are $\left(\frac{\pi}{6}, 0\right)$ and $\left(\frac{7\pi}{6}, 0\right)$. Notice

that the graph is a reflection of $y = \cos\left(2x - \frac{\pi}{3}\right)$ across the x-axis, then a displacement up
1 unit.

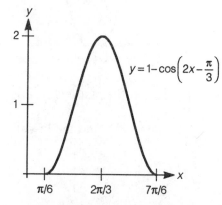

29. This is a sine function where the amplitude is 2, so $A = 2$. Since the period is 4π, we
have:

$$\frac{2\pi}{B} = 4\pi$$
$$2\pi = 4\pi B$$
$$\frac{1}{2} = B$$

31. This is a sine function (reflected across the x-axis) where the amplitude is 3, so $A = -3$. Since the period is 2, we have:
$$\frac{2\pi}{B} = 2$$
$$2\pi = 2B$$
$$\pi = B$$

33. This is a cosine function (reflected across the x-axis) where the amplitude is 4, so $A = -4$. Since the period is 10π, we have:
$$\frac{2\pi}{B} = 10\pi$$
$$2\pi = 10\pi B$$
$$\tfrac{1}{5} = B$$

35. We graph $y = \frac{1}{2} - \frac{1}{2}\cos 2x$, which has an amplitude of $\frac{1}{2}$ and a period of $\frac{2\pi}{2} = \pi$. The graph will be a reflection of $y = \frac{1}{2}\cos 2x$ across the x-axis, then a displacement up $\frac{1}{2}$ unit.

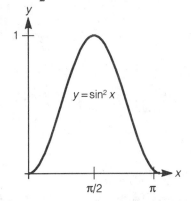

37. We graph $y = \frac{1}{2}\sin 2x$, which has an amplitude of $\frac{1}{2}$ and a period of $\frac{2\pi}{2} = \pi$.

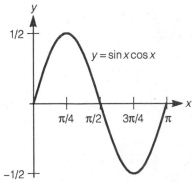

39. We know that the curve $y = \cos x$ begins a period when $x = 0$ and completes that period when $x = 2\pi$. Thus $y = A \cos Bx$ will begin a period when $Bx = 0$ and complete that period when $Bx = 2\pi$, which yields $x = 0$ and $x = \frac{2\pi}{B}$. Thus the period of $y = A \cos Bx$ is $\frac{2\pi}{B}$.

<u>Section 3.5 TI-81 Graphing Calculator Exercises</u>

1.   (a)   Using the indicated settings, the graph should appear as:

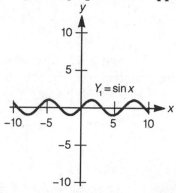

(b)   Using the indicated settings, the graph should appear as:

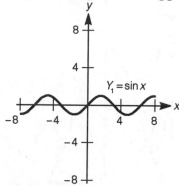

(c)   The x-coordinate for the intercept between 0 and 4 is π. The x-coordinate for the intercept between 4 and 8 is 2π.

(d)   The x-coordinate for the intercept between 0 and 4 is $\frac{\pi}{2}$. Using the same RANGE settings as in (b), the graph should appear as:

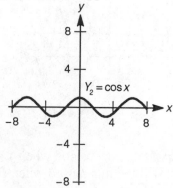

(e)   Using ZOOM-7, the graph appears as:

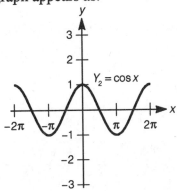

(f)   The graph of $Y_1 = \sin x$ would have to be shifted $\frac{\pi}{2}$ to the left to coincide with the graph of $Y_2 = \cos x$. Using the same settings as in (e), the graphs of both functions appear as:

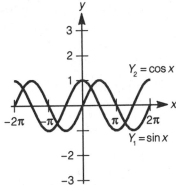

3.   (a)   Using ZOOM-7, the graphs of all three functions appear as:

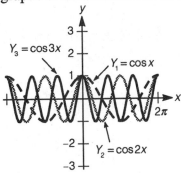

(b)   The amplitude for all three functions is 1. The period of $Y_1 = \cos x$ is 2π, the period

of $Y_2 = \cos 2x$ is $\frac{2\pi}{2} = \pi$, and the period of $Y_3 = \cos 3x$ is $\frac{2\pi}{3}$. Using the indicated

RANGE settings, the graphs of all three functions appear as:

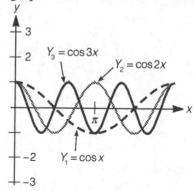

5.   (a)   The amplitude is 2.5, the period is $\frac{2\pi}{3\pi} = \frac{2}{3}$, and the phase shift is $-\frac{4}{3\pi}$.

(b)   We use $X_{min} = -\frac{4}{3\pi} \approx -0.4244$, $X_{max} = -\frac{4}{3\pi} + \frac{4}{3} \approx 0.9089$, $Y_{min} = -2.5$ and

$Y_{max} = 2.5$ to obtain the graph:

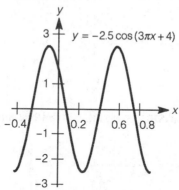

(c)   The highest points occur at $\left(\frac{1}{3} - \frac{4}{3\pi}, 2.5\right) \approx (-0.09, 2.5)$ and

$\left(1 - \frac{4}{3\pi}, 2.5\right) \approx (0.58, 2.5)$, and the lowest points occur at

$\left(-\frac{4}{3\pi}, -2.5\right) \approx (-0.42, -2.5)$, $\left(\frac{2}{3} - \frac{4}{3\pi}, -2.5\right) \approx (0.24, -2.5)$, and

$\left(\frac{4}{3} - \frac{4}{3\pi}, -2.\dot{5}\right) \approx (0.91, -2.5)$.

(d)   Tracing the curve verifies these results.

7. (a) The amplitude is 1, the period is $\frac{2\pi}{0.5} = 4\pi$, and the phase shift is $\frac{-0.75}{0.5} = -1.5$.

    (b) We use $X_{min} = -1.5$, $X_{max} = -1.5 + 8\pi \approx 23.63$, $Y_{min} = -1$ and $Y_{max} = 1$ to obtain the graph:

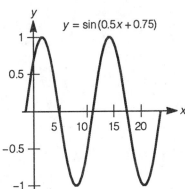

$y = \sin(0.5x + 0.75)$

    (c) The highest points occur at $(-1.5 + \pi, 1) \approx (1.64, 1)$ and $(-1.5 + 5\pi, 1) \approx (14.2, 1)$, and the lowest points occur at $(-1.5 + 3\pi, -1) \approx (7.92, -1)$ and $(-1.5 + 7\pi, -1) \approx (20.49, -1)$.

    (d) Tracing the curve verifies these results.

9. (a) The amplitude is 0.02, the period is $\frac{2\pi}{100\pi} = 0.02$, and the phase shift is

    $-\frac{4\pi}{100\pi} = -0.04$.

    (b) We use $X_{min} = -0.04$, $X_{max} = 0$, $Y_{min} = -0.02$ and $Y_{max} = 0.02$ to obtain the graph:

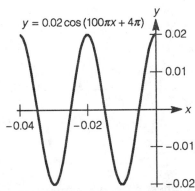

$y = 0.02\cos(100\pi x + 4\pi)$

    (c) The highest points occur at $(-0.04, 0.02)$, $(-0.02, 0.02)$ and $(0, 0.02)$, and the lowest points occur at $(-0.03, -0.02)$ and $(-0.01, -0.02)$.

    (d) Tracing the curve verifies these results.

11. **(b)**   The graph of $Y_1 = \sin x$ with the indicated settings appears as:

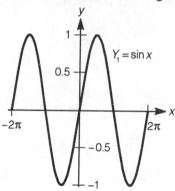

The graph of $Y_2 = \sin(\sin x)$ with ZOOM-7 settings appears as:

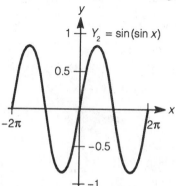

Although similar, the two graphs are not identical. The second graph appears to have a smaller amplitude as illustrated in part (c).

(c)   We graph the two functions together:

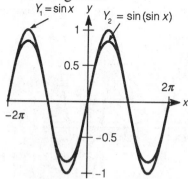

(d)   The coordinates of the highest point are approximately $(1.58, 0.84)$. The y-coordinate gives the amplitude.

(e) Since $\sin x$ is largest when $x = \frac{\pi}{2}$ (then $\sin x = 1$), and $\sin x$ is increasing for

$0 < x < \frac{\pi}{2}$, then $\sin (\sin x)$ should have its largest value when $x = \frac{\pi}{2}$. Its value then

is $\sin\left(\sin \frac{\pi}{2}\right) = \sin 1$.

(f) Since $\sin 1 \approx 0.84$, our results from (d) and (e) are confirmed.

3.6 Graphs of the Tangent and the Reciprocal Functions

1. (a) The x-intercept is $-\frac{\pi}{4}$, the y-intercept is 1, and the asymptotes are $x = -\frac{3\pi}{4}$ and

$x = \frac{\pi}{4}$. Notice that this graph is $y = \tan x$ displaced $\frac{\pi}{4}$ units to the left.

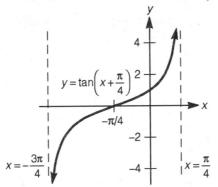

(b) The x-intercept is $-\frac{\pi}{4}$, the y-intercept is -1, and the asymptotes are $x = -\frac{3\pi}{4}$ and

$x = \frac{\pi}{4}$. Notice that this graph is $y = \tan\left(x + \frac{\pi}{4}\right)$ reflected across the x-axis.

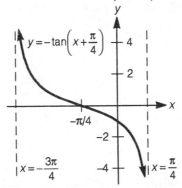

3. (a) The x- and y-intercepts are both 0, and the asymptotes are $x = -\frac{3\pi}{2}$ and $x = \frac{3\pi}{2}$.

Notice that the period is $\frac{\pi}{1/3} = 3\pi$.

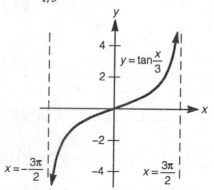

(b) The x- and y-intercepts are both 0, and the asymptotes are $x = -\frac{3\pi}{2}$ and $x = \frac{3\pi}{2}$.

Notice that this graph is $y = \tan\frac{x}{3}$ reflected across the x-axis.

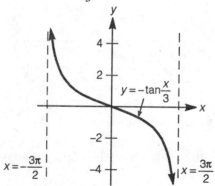

5. The x- and y-intercepts are both 0, and the asymptotes are $x = -1$ and $x = 1$. Notice that the period is $\frac{\pi}{\pi/2} = 2$.

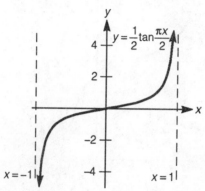

7.   The x-intercept is 1, there is no y-intercept, and the asymptotes are $x = 0$ and $x = 2$.

Notice that the period is $\frac{\pi}{\pi/2} = 2$.

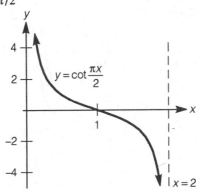

9.   The x-intercept is $\frac{3\pi}{4}$, the y-intercept is 1, and the asymptotes are $x = \frac{\pi}{4}$ and $x = \frac{5\pi}{4}$.

Notice that this graph is $y = \cot x$ displaced $\frac{\pi}{4}$ units to the right, then reflected across the x-axis.

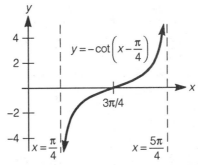

11.   The x-intercept is $\frac{\pi}{4}$, there is no y-intercept, and the asymptotes are $x = 0$ and $x = \frac{\pi}{2}$.

Notice that the period is $\frac{\pi}{2}$.

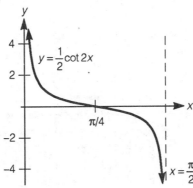

13. There is no x-intercept, the y-intercept is $-\sqrt{2}$, and the asymptotes are $x = -\frac{3\pi}{4}$, $x = \frac{\pi}{4}$, and $x = \frac{5\pi}{4}$. Notice that this graph is $y = \csc x$ displaced $\frac{\pi}{4}$ units to the right.

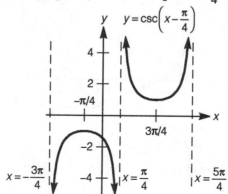

15. There are no x- or y-intercepts, and the asymptotes are $x = -2\pi$, $x = 0$ and $x = 2\pi$. Notice that the period is $\frac{2\pi}{1/2} = 4\pi$, and that this graph is $y = \csc \frac{x}{2}$ reflected across the x-axis.

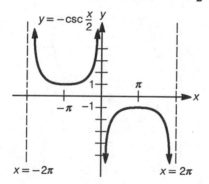

17. There are no x- or y-intercepts, and the asymptotes are $x = -1$, $x = 0$, and $x = 1$. Notice that the period is $\frac{2\pi}{\pi} = 2$.

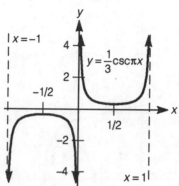

19. There is no x-intercept, the y-intercept is -1, and the asymptotes are $x = -\frac{\pi}{2}$, $x = \frac{\pi}{2}$, and $x = \frac{3\pi}{2}$. Notice that this graph is $y = \sec x$ reflected across the x-axis.

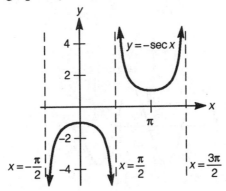

21. There is no x-intercept, the y-intercept is -1, and the asymptotes are $x = \frac{\pi}{2}$, $x = \frac{3\pi}{2}$, and $x = \frac{5\pi}{2}$. Notice that this graph is $y = \sec x$ displaced π units to the right.

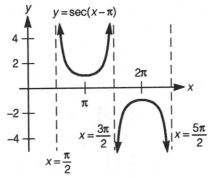

23. There is no x-intercept, the y-intercept is 3, and the asymptotes are $x = -1$, $x = 1$, and $x = 3$. Notice that the period is $\frac{2\pi}{\pi/2} = 4$.

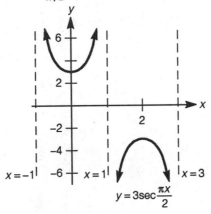

25. (a)   The x-intercepts are $-\frac{11}{18}, -\frac{5}{18}, \frac{1}{18}, \frac{7}{18}$ and $\frac{13}{18}$, and the y-intercept is -1, and there

are no asymptotes. Notice that the period is $\frac{2\pi}{3\pi} = \frac{2}{3}$ and the phase shift is $\frac{\pi/6}{3\pi} = \frac{1}{18}$.

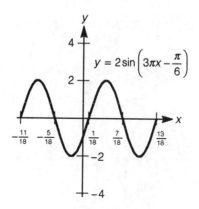

(b)   There are no x-intercepts, the y-intercept is -4, and the asymptotes are $x = -\frac{11}{18}$,

$x = -\frac{5}{18}$, $x = \frac{1}{18}$, $x = \frac{7}{18}$ and $x = \frac{13}{18}$. Notice that the period is $\frac{2\pi}{3\pi} = \frac{2}{3}$ and the phase

shift is $\frac{\pi/6}{3\pi} = \frac{1}{18}$.

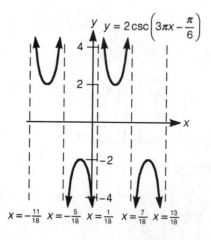

27. (a)   The x-intercepts are $-\frac{5}{8}$, $-\frac{1}{8}$, $\frac{3}{8}$, and $\frac{7}{8}$, and the y-intercept is $-\frac{3\sqrt{2}}{2}$, and there are no

asymptotes. Notice that the period is $\frac{2\pi}{2\pi} = 1$, the phase shift is $\frac{\pi/4}{2\pi} = \frac{1}{8}$, and that this

graph is $y = 3\cos\left(2\pi x - \frac{\pi}{4}\right)$ reflected across the x-axis.

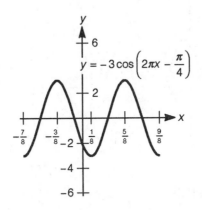

(b)   There are no x-intercepts, the y-intercept is $-3\sqrt{2}$, and the asymptotes are $x = -\frac{5}{8}$,

$x = -\frac{1}{8}$, $x = \frac{3}{8}$ and $x = \frac{7}{8}$. Notice that the period is $\frac{2\pi}{2\pi} = 1$, the phase shift is $\frac{\pi/4}{2\pi} = \frac{1}{8}$,

and that this graph is $y = 3\sec\left(2\pi x - \frac{\pi}{4}\right)$ reflected across the x-axis.

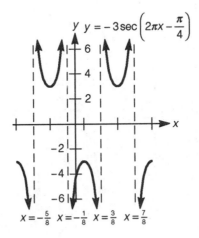

29.   (a)   We first find the composition:
$$(f \circ h)(x) = f\left(\pi x - \tfrac{\pi}{6}\right) = \sin\left(\pi x - \tfrac{\pi}{6}\right)$$

The period is $\frac{2\pi}{\pi} = 2$, the amplitude is 1, and the phase shift is $\frac{\pi/6}{\pi} = \frac{1}{6}$.

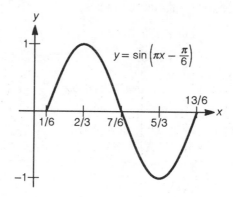

(b)   We first find the composition:
$$(g \circ h)(x) = g\left(\pi x - \tfrac{\pi}{6}\right) = \csc\left(\pi x - \tfrac{\pi}{6}\right)$$

The period is $\frac{2\pi}{\pi} = 2$, the phase shift is $\frac{\pi/6}{\pi} = \frac{1}{6}$, and the asymptotes are $x = \frac{1}{6}$, $x = \frac{7}{6}$ and $x = \frac{11}{6}$.

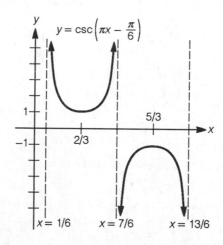

31.   (a)   We first find the composition:
$$(f \circ H)(x) = f\left(\pi x + \tfrac{\pi}{4}\right) = \sin\left(\pi x + \tfrac{\pi}{4}\right)$$

The period is $\frac{2\pi}{\pi} = 2$, the amplitude is 1, and the phase shift is $\frac{-\pi/4}{\pi} = -\frac{1}{4}$.

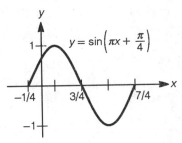

(b) We first find the composition:

$$(g \circ H)(x) = g\left(\pi x + \frac{\pi}{4}\right) = \csc\left(\pi x + \frac{\pi}{4}\right)$$

The period is $\frac{2\pi}{\pi} = 2$, the phase shift is $\frac{-\pi/4}{\pi} = -\frac{1}{4}$, and the asymptotes are $x = -\frac{1}{4}$, $x = \frac{3}{4}$ and $x = \frac{7}{4}$.

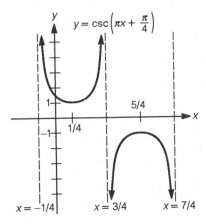

33. We first find the composition:

$$(A \circ T)(x) = A(\tan x) = |\tan x|$$

Now draw the graph, noting that values of x where $\tan x < 0$ will be reflected across the x-axis.

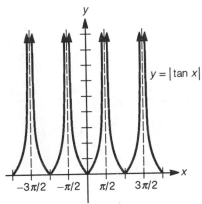

35. We first find the composition:

$$(A \circ f)(x) = A(\csc x) = |\csc x|$$

Now draw the graph, noting that values of x where $\csc x < 0$ will be reflected across the x-axis.

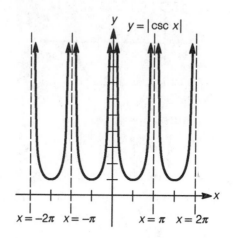

37. (a) Since P and Q are both points on the unit circle, the coordinates are $P(\cos s, \sin s)$

and $Q\left(\cos\left(s - \frac{\pi}{2}\right), \sin\left(s - \frac{\pi}{2}\right)\right)$.

(b) Since $\triangle OAP$ is congruent to $\triangle OBQ$ (labeling the third vertex B), then $OA = OB$ and $AP = BQ$. Because the y-coordinate at Q is negative, we have concluded what was required.

(c) Restating part (b), we have shown that:

$$\cos\left(s - \frac{\pi}{2}\right) = \sin s \quad \text{and} \quad \sin\left(s - \frac{\pi}{2}\right) = -\cos s$$

(d) We compute $\cot s$:

$$\cot s = \frac{\cos s}{\sin s} = \frac{-\sin\left(s - \frac{\pi}{2}\right)}{\cos\left(s - \frac{\pi}{2}\right)} = -\tan\left(s - \frac{\pi}{2}\right)$$

<u>Section 3.6 TI-81 Graphing Calculator Exercises</u>

1.   Using the indicated settings, the graph appears as:

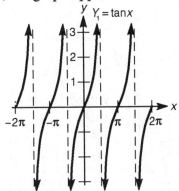

3.   Using the same settings as in Exercise 1, notice that the two graphs are identical,

verifying the identity $\cot x = -\tan\left(x - \frac{\pi}{2}\right)$.

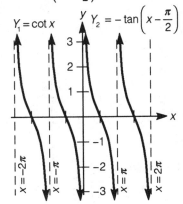

5.   (a)   Since the period is $\frac{\pi}{1/4} = 4\pi$, we adjust $X_{min} = -2\pi \approx -6.28$ and $X_{max} = 6\pi \approx 18.85$
        to obtain the graph:

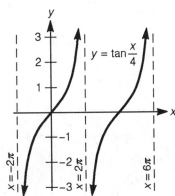

(b)   Since the period is $\frac{\pi}{4}$, we adjust $X_{min} = -\frac{\pi}{8} \approx -0.39$ and $X_{max} = \frac{3\pi}{8} \approx 1.18$ to obtain the graph:

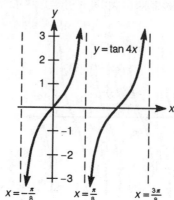

7.   (a)   Since the period is $\frac{\pi}{\pi} = 1$, we use the settings $X_{min} = -0.5$ and $X_{max} = 1.5$ to obtain the graph:

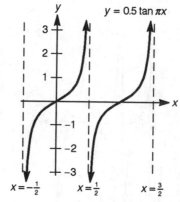

(b)   The period is the same as in (a), and the phase shift is $\frac{-\pi/3}{\pi} = -\frac{1}{3}$. We use the settings $X_{min} = -\frac{1}{2} - \frac{1}{3} = -\frac{5}{6} \approx -0.83$ and $X_{max} = \frac{3}{2} - \frac{1}{3} = \frac{7}{6} \approx 1.17$ to obtain the graph:

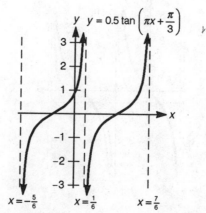

(c)   The period is the same as in (a), and the phase shift is $-\frac{1}{\pi}$. We use the settings

$X_{min} = -\frac{1}{2} - \frac{1}{\pi} \approx -0.82$ and $X_{max} = \frac{3}{2} - \frac{1}{\pi} \approx 1.18$ to obtain the graph:

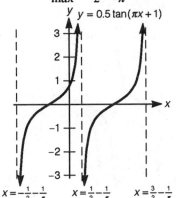

9.   (a)   Since the period is $\frac{\pi}{1/2} = 2\pi$, we use the settings $X_{min} = -\pi \approx -3.14$ and $X_{max} = 3\pi \approx 9.42$ to obtain the graph:

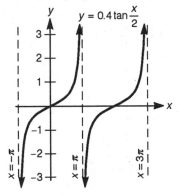

(b)   Since the period is $\frac{\pi}{1/3} = 3\pi$, we use the settings $X_{min} = \frac{-3\pi}{2} \approx -4.71$ and

$X_{max} = \frac{9\pi}{2} \approx 14.14$ to obtain the graph:

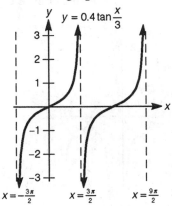

(c)   Since the period is $\frac{\pi}{1/5} = 5\pi$, we use the settings $X_{min} = \frac{-5\pi}{2} \approx -7.85$ and

$X_{max} = \frac{15\pi}{2} \approx 23.56$ to obtain the graph:

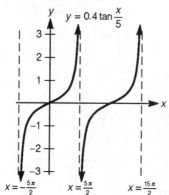

11.   It appears that $\sin x = \csc x$ at odd multiples of $\frac{\pi}{2}$, such as $-\frac{3\pi}{2}, -\frac{\pi}{2}, \frac{\pi}{2}$, and $\frac{3\pi}{2}$. Since $\sin x$ and $\csc x$ have the same sign, there are no points in which $\sin x = -\csc x$. Using the indicated settings, the graphs appear as:

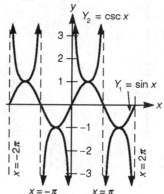

13.   Since the period is $\frac{2\pi}{1/2} = 4\pi$, we use the settings $X_{min} = -4\pi \approx -12.57$, $X_{max} = 4\pi \approx 12.57$, $Y_{min} = -2$ and $Y_{max} = 2$ to obtain the graph:

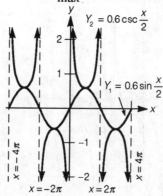

15. Since the period is $\frac{2\pi}{2} = \pi$ and the phase shift is $\frac{\pi/6}{2} = \frac{\pi}{12}$, we use the settings

$X_{min} = -\frac{\pi}{4} + \frac{\pi}{12} = -\frac{\pi}{6} \approx -0.52$, $X_{max} = \frac{7\pi}{4} + \frac{\pi}{12} = \frac{11\pi}{6} \approx 5.76$, $Y_{min} = -6$ and $Y_{max} = 6$ to obtain the graph:

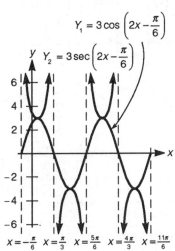

$$Y_1 = 3\cos\left(2x - \frac{\pi}{6}\right)$$

$$Y_2 = 3\sec\left(2x - \frac{\pi}{6}\right)$$

17. The two graphs are identical. This demonstrates that $\tan^2 x = \sec^2 x - 1$ is a trigonometric identity. Using the ZOOM-7 setting we obtain the graph:

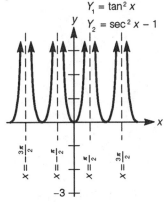

$Y_1 = \tan^2 x$

$Y_2 = \sec^2 x - 1$

19. (a) Using the ZOOM-7 setting we obtain the graph:

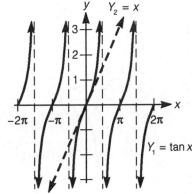

$Y_2 = x$

$Y_1 = \tan x$

(b) Using the settings $X_{min} = -1.57$ and $X_{max} = 1.57$, the graph appears as:

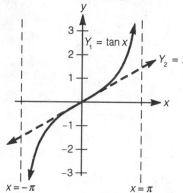

We complete the table:

x	0.000123	0.01	0.05	0.1
$\tan x$	0.000123	0.010000	0.050042	0.100335
x	0.2	0.3	0.4	0.5
$\tan x$	0.202710	0.309336	0.422793	0.546302

(c) The graphs of $Y_1 = \tan x$ and $Y_3 = x + \frac{x^3}{3}$ appear as:

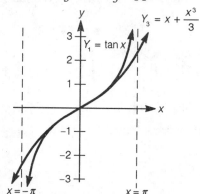

Notice how similar the two curves are near $x = 0$.

(d) We complete the table:

x	0.000123	0.01	0.05	0.1
$\tan x$	0.000123	0.010000	0.050042	0.100335
$x + \frac{x^3}{3}$	0.000123	0.010000	0.050042	0.100333
x	0.2	0.3	0.4	0.5
$\tan x$	0.202710	0.309336	0.422793	0.546302
$x + \frac{x^3}{3}$	0.202667	0.309	0.421333	0.541667

Note that the values of $x + \frac{x^3}{3}$ are much closer to $\tan x$ than are those for x.

21. (a)   The graphs of $Y_1 = \tan x$ and $Y_2 = x$ appear as:

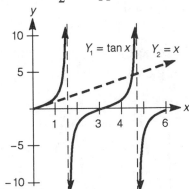

The two graphs appear to intersect at approximately $x \approx 4.5$.

(b)   The intersection point appears to occur at approximately $x \approx 4.493$.

3.7 Simple Harmonic Motion

1.   (a)   Evaluating $s = 4\cos\frac{\pi t}{2}$ at the given values of t, we have:

$t = 0$: $s = 4\cos 0 = 4$ cm

$t = 0.5$: $s = 4\cos\frac{\pi}{4} = 2\sqrt{2}$ cm ≈ 2.83 cm

$t = 1$: $s = 4\cos\frac{\pi}{2} = 0$ cm

$t = 2$: $s = 4\cos\pi = -4$ cm

(b)   The amplitude is 4 cm, the period is $\frac{2\pi}{\pi/2} = 4$ sec, and the frequency is $\frac{1}{4}$ cycles/sec. We sketch the graph over the interval $0 \le t \le 8$:

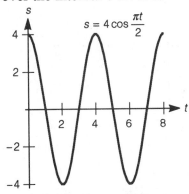

(c)   The mass is farthest from the origin at high and low points, which occur at $t = 0$, $t = 2$, $t = 4$, $t = 6$ and $t = 8$ sec.

(d)   The mass is passing through the origin when $s = 0$, which occurs at $t = 1$, $t = 3$, $t = 5$ and $t = 7$ sec.

(e)   The mass is moving to the right when the s-coordinate is increasing, which occurs during the intervals $2 < t < 4$ and $6 < t < 8$.

3.   (a)   The amplitude is 3 feet, the period is $\frac{2\pi}{\pi/3} = 6$ sec, and the frequency is $\frac{1}{6}$ cycles/sec. We sketch the graph over the interval $0 \le t \le 12$:

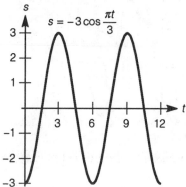

(b)   The mass is moving upward when the s-coordinate is increasing, which occurs during the intervals $0 < t < 3$ and $6 < t < 9$.

(c)   The mass is moving downward when the s-coordinate is decreasing, which occurs during the intervals $3 < t < 6$ and $9 < t < 12$.

(d)   We graph the velocity function for $0 \le t \le 12$, noting its period is $\frac{2\pi}{\pi/3} = 6$ sec and its amplitude is π feet/sec.

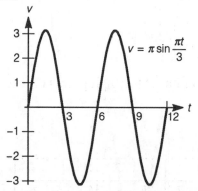

(e)   We note that $v = 0$ when $t = 0$, $t = 3$, $t = 6$, $t = 9$, and $t = 12$ sec. At these times the mass (s-coordinate) is at -3, 3, -3, 3 and -3 feet, respectively.

(f)   The velocity is a maximum at the high points of this graph, which occur at $t = 1.5$ and $t = 7.5$ sec. At these times the mass is at 0 feet.

(g)   The velocity is a minimum at the low points of this graph, which occur at $t = 4.5$ and $t = 10.5$ sec. At these times the mass is at 0 feet.

(h)   We graph the velocity function and position function on the same set of axes for $0 \leq t \leq 12$:

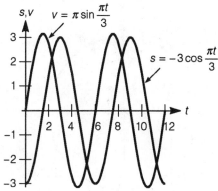

5.   (a)   The amplitude is 170 volts and the period is $\frac{2\pi}{120\pi} = \frac{1}{60}$ sec, so the frequency is 60 cycles/sec.

(b)   We graph the voltage for $0 \leq t \leq \frac{1}{30}$ sec:

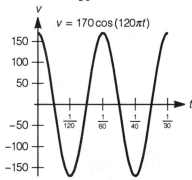

(c)   The voltage is a maximum at the high points of this graph, which occur at $t = 0$, $t = \frac{1}{60}$ and $t = \frac{1}{30}$ sec.

7.   (a)   We complete the table:

t(sec)	0	1	2	3	4	5	6	7
θ(radians)	0	$\frac{\pi}{3}$	$\frac{2\pi}{3}$	π	$\frac{4\pi}{3}$	$\frac{5\pi}{3}$	2π	$\frac{7\pi}{3}$

(b)   Since the x-coordinate of the point Q is $\cos\theta$, the corresponding x-coordinates are $\cos 0 = 1$, $\cos\frac{\pi}{3} = \frac{1}{2}$, $\cos\frac{2\pi}{3} = -\frac{1}{2}$, $\cos\pi = -1$, $\cos\frac{4\pi}{3} = -\frac{1}{2}$, $\cos\frac{5\pi}{3} = \frac{1}{2}$, $\cos 2\pi = 1$, and $\cos\frac{7\pi}{3} = \frac{1}{2}$.

(c)   We sketch P and Q when $t = 1$ second, thus $\theta = \frac{\pi}{3}$ radians:

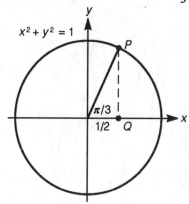

(d)   We sketch P and Q when $t = 2$ seconds, thus $\theta = \frac{2\pi}{3}$ radians:

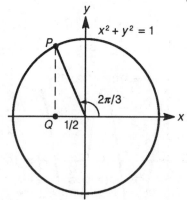

We sketch P and Q when $t = 3$ seconds, thus $\theta = \pi$ radians. Note that P and Q coincide at the same point, since P lies on the x-axis:

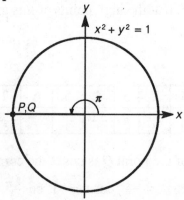

We sketch P and Q when $t = 4$ seconds, thus $\theta = \frac{4\pi}{3}$ radians:

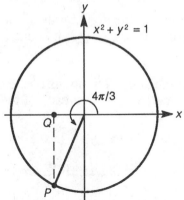

(e)   The x-coordinate of Q is the same as the x-coordinate of P. But P is a point on the unit circle and θ is the radian measure to that point, thus the x-coordinate is $\cos \theta$.

(f)   The period of this function is $\frac{2\pi}{\pi/3} = 6$, the amplitude is 1, and the frequency is $\frac{1}{6}$. We graph the function for two complete cycles, or $0 \le t \le 12$:

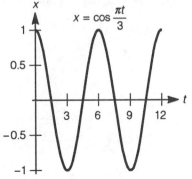

(g)   The period is $\frac{2\pi}{\pi/3} = 6$ and the amplitude is $\frac{\pi}{3}$. We graph the velocity for two complete cycles, or $0 \le t \le 12$:

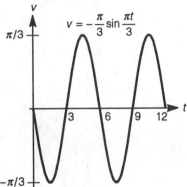

(h)   The velocity is 0 when $t = 0, 3, 6, 9$ and 12 sec. The corresponding x-coordinates are $1, -1, 1, -1$ and 1, respectively.

(i)   The velocity is a maximum at the high points of the curve, which occur at $t = 4.5$ and $t = 10.5$ sec. The x-coordinate of Q is 0 at each of these times, so Q is located at the origin.

(j)   The velocity is a minimum at the low points of the curve, which occur at $t = 1.5$ and $t = 7.5$ sec. The x-coordinate of Q is 0 at each of these times, so Q is located at the origin.

Chapter Three Review Exercises

1.   (a)   Since the terminal side of $\frac{5\pi}{3}$ lies in the fourth quadrant, then $\sin \frac{5\pi}{3} < 0$. Using a reference angle of $\frac{\pi}{3}$, we have:

$$\sin \frac{5\pi}{3} = -\sin \frac{\pi}{3} = -\frac{\sqrt{3}}{2}$$

(b)   Since the terminal side of $\frac{11\pi}{6}$ lies in the fourth quadrant, then $\cot \frac{11\pi}{6} < 0$. Using a reference angle of $\frac{\pi}{6}$, we have:

$$\cot \frac{11\pi}{6} = -\cot \frac{\pi}{6} = -\frac{\cos \frac{\pi}{6}}{\sin \frac{\pi}{6}} = -\frac{\frac{\sqrt{3}}{2}}{\frac{1}{2}} = -\sqrt{3}$$

3.   Using the identities $\cos(-t) = \cos t$ and $\sin(-t) = -\sin t$, we have:
$$\cos t - \cos(-t) + \sin t - \sin(-t) = \cos t - \cos t + \sin t - (-\sin t) = 2\sin t$$

5.   The period is $\frac{2\pi}{2\pi} = 1$ and the phase shift is $\frac{3}{2\pi}$. The asymptotes will occur at

$$x = -\frac{1}{4} + \frac{3}{2\pi}, \; x = \frac{1}{4} + \frac{3}{2\pi} \text{ and } x = \frac{3}{4} + \frac{3}{2\pi}.$$

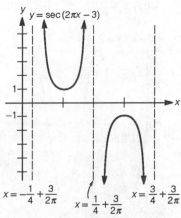

$y = \sec(2\pi x - 3)$

$x = -\frac{1}{4} + \frac{3}{2\pi}$ $x = \frac{1}{4} + \frac{3}{2\pi}$ $x = \frac{3}{4} + \frac{3}{2\pi}$

7. The amplitude is 1, the period is $\frac{2\pi}{2} = \pi$, and the phase shift is $\frac{\pi}{2}$. Notice that this graph is $y = \sin(2x - \pi)$ reflected across the x-axis.

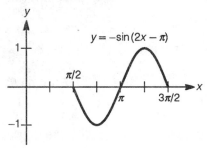

9. Since the terminal side of π radians intersects the unit circle at the point $(-1, 0)$, then $\cos \pi = -1$.

11. Since the terminal side of $\frac{2\pi}{3}$ radians lies in the second quadrant, $\csc \frac{2\pi}{3} > 0$. Using a reference angle of $\frac{\pi}{3}$ radians, we have:

$$\csc \frac{2\pi}{3} = \csc \frac{\pi}{3} = \frac{1}{\sin \frac{\pi}{3}} = \frac{2}{\sqrt{3}} = \frac{2\sqrt{3}}{3}$$

13. Since the terminal side of $\frac{11\pi}{6}$ radians lies in the fourth quadrant, $\tan \frac{11\pi}{6} < 0$. Using a reference angle of $\frac{\pi}{6}$ radians, we have:

$$\tan \frac{11\pi}{6} = -\tan \frac{\pi}{6} = -\frac{\sin \frac{\pi}{6}}{\cos \frac{\pi}{6}} = -\frac{\frac{1}{2}}{\frac{\sqrt{3}}{2}} = -\frac{1}{\sqrt{3}} = -\frac{\sqrt{3}}{3}$$

15. Since the terminal side of $\frac{\pi}{6}$ radians lies in the first quadrant, $\sin \frac{\pi}{6} > 0$. Thus $\sin \frac{\pi}{6} = \frac{1}{2}$.

17. Since the terminal side of $\frac{5\pi}{4}$ radians lies in the third quadrant, $\cot \frac{5\pi}{4} > 0$. Using a reference angle of $\frac{\pi}{4}$ radians, we have:

$$\cot \frac{5\pi}{4} = \cot \frac{\pi}{4} = \frac{\cos \frac{\pi}{4}}{\sin \frac{\pi}{4}} = \frac{\frac{\sqrt{2}}{2}}{\frac{\sqrt{2}}{2}} = 1$$

19. Since the terminal side of $-\frac{5\pi}{6}$ radians lies in the third quadrant, $\csc\left(-\frac{5\pi}{6}\right) < 0$. Using a reference angle of $\frac{\pi}{6}$ radians, we have:

$$\csc\left(-\frac{5\pi}{6}\right) = -\csc \frac{\pi}{6} = -\frac{1}{\sin \frac{\pi}{6}} = -\frac{1}{\frac{1}{2}} = -2$$

21. Using a calculator, we find $\sin 1 \approx 0.841$.

23. Since $\frac{3\pi}{2}$ radians intersects the unit circle at the point $(0,-1)$, then $\sin\frac{3\pi}{2} = -1$. A calculator also verifies this value.

25. Using a calculator, we find $\sin(\sin 0.0123) \approx 0.0123$.

27. Since $\sin 1776 \approx -0.842$ and $\cos 1776 \approx -0.540$, we have:
$$\sin^2 1776 + \cos^2 1776 \approx (-0.842)^2 + (-0.540)^2 = 1$$

29. Since $\cos(0.25) \approx 0.969$ and $\sin(0.25) \approx 0.247$, then:
$$\cos^2(0.25) - \sin^2(0.25) \approx (0.969)^2 - (0.247)^2 \approx 0.878$$
Since $\cos(0.5) \approx 0.878$, the equality is verified.

31. Since $0 < \theta < \frac{\pi}{2}$, then $\cos\theta > 0$ and thus:
$$\sqrt{25 - x^2} = \sqrt{25 - 25\sin^2\theta} = 5\sqrt{1 - \sin^2\theta} = 5\sqrt{\cos^2\theta} = 5\cos\theta$$

33. Since $0 < \theta < \frac{\pi}{2}$, $\tan\theta > 0$ and thus:
$$\left(x^2 - 100\right)^{1/2} = \left(100\sec^2\theta - 100\right)^{1/2} = 10\left(\sec^2\theta - 1\right)^{1/2} = 10\left(\tan^2\theta\right)^{1/2} = 10\tan\theta$$

35. Since $0 < \theta < \frac{\pi}{2}$, $\sec\theta > 0$ and thus:
$$\left(x^2 + 5\right)^{-1/2} = \left(5\tan^2\theta + 5\right)^{-1/2} = \frac{\sqrt{5}}{5}\left(\sec^2\theta\right)^{-1/2} = \frac{\sqrt{5}}{5}\cos\theta$$

37. Since $\sin\theta$ is negative, then:
$$\sin\theta = -\sqrt{1 - \cos^2\theta} = -\sqrt{1 - \left(\frac{8}{17}\right)^2} = -\sqrt{1 - \frac{64}{289}} = -\sqrt{\frac{225}{289}} = -\frac{15}{17}$$
Thus:
$$\tan\theta = \frac{\sin\theta}{\cos\theta} = \frac{-\frac{15}{17}}{\frac{8}{17}} = -\frac{15}{8}$$

39. Since $\angle BPA = \theta$, then $\angle APC = \pi - \theta$, and the area of the sector formed by $\angle APC$ is:
$$\tfrac{1}{2}r^2\theta = \tfrac{1}{2}\left(\sqrt{2}\right)^2(\pi - \theta) = \pi - \theta$$
Now using the area formula for $\triangle APC$:
$$\text{Area}_{\triangle APC} = \tfrac{1}{2}ab\sin(\pi - \theta) = \tfrac{1}{2}\left(\sqrt{2}\right)\left(\sqrt{2}\right)\sin\theta = \sin\theta$$
Thus, the area of the shaded region is given by:
$$A(\theta) = \pi - \theta - \sin\theta$$

41. Using the hint, we have two congruent shaded regions each with $r = 1$ cm and $\theta = \frac{\pi}{2}$, and thus the total area is:

$$2 \bullet \tfrac{1}{2}\left(\tfrac{\pi}{2} - 1\right) \text{ cm}^2 = \tfrac{\pi - 2}{2} \text{ cm}^2$$

43. Recognizing the left-hand side as the factoring for a difference of squares, we have:

$$\begin{aligned}
(\cos x + \sin x - 1)(\cos x + \sin x + 1) &= (\cos x + \sin x)^2 - 1 \\
&= \cos^2 x + 2\sin x \cos x + \sin^2 x - 1 \\
&= \cos^2 x + \sin^2 x + 2\sin x \cos x - 1 \\
&= 1 + 2\sin x \cos x - 1 \\
&= 2\sin x \cos x
\end{aligned}$$

45. Recognizing the left-hand side as the factoring for a difference of squares, we have:

$$\begin{aligned}
\left(9 - 4\sin x - \cos^2 x\right)&\left(9 + 4\sin x - \cos^2 x\right) \\
&= \left(9 - \cos^2 x\right)^2 - (4\sin x)^2 \\
&= 81 - 18\cos^2 x + \cos^4 x - 16\sin^2 x \\
&= 81 - 18\left(1 - \sin^2 x\right) + \left(1 - \sin^2 x\right)^2 - 16\sin^2 x \\
&= 81 - 18 + 18\sin^2 x + 1 - 2\sin^2 x + \sin^4 x - 16\sin^2 x \\
&= \sin^4 x + 64
\end{aligned}$$

47. Multiplying out the left-hand side and using the identity $\sin^2 x + \cos^2 x = 1$, we have:

$$\begin{aligned}
(a\sin x - b\cos x)^2 &+ (a\cos x + b\sin x)^2 \\
&= a^2 \sin^2 x - 2ab\cos x \sin x + b^2 \cos^2 x + a^2 \cos^2 x + 2ab\cos x \sin x + b^2 \sin^2 x \\
&= \left(a^2 + b^2\right)\sin^2 x + \left(a^2 + b^2\right)\cos^2 x \\
&= \left(a^2 + b^2\right)\left(\sin^2 x + \cos^2 x\right) \\
&= a^2 + b^2
\end{aligned}$$

49. This is a sine function where the amplitude is 4, so $A = 4$. Since the period is 2π, we have:

$$\frac{2\pi}{B} = 2\pi$$
$$2\pi = 2\pi B$$
$$1 = B$$

So the equation is $y = 4\sin x$.

51. This is a cosine function (reflected across the x-axis) where the amplitude is 2, so $A = -2$.

Since the period is $\frac{\pi}{2}$, we have:

$$\frac{2\pi}{B} = \frac{\pi}{2}$$
$$4\pi = \pi B$$
$$4 = B$$

So the equation is $y = -2\cos x$.

53. The x-intercepts are $\frac{\pi}{8}$ and $\frac{3\pi}{8}$, the high point is $\left(\frac{\pi}{4},3\right)$, and the low points are $(0,-3)$ and

$\left(\frac{\pi}{2},-3\right)$. Notice that the period is $\frac{2\pi}{4}=\frac{\pi}{2}$, and that this graph is $y=3\cos 4x$ reflected across the x-axis.

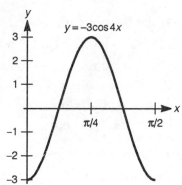

55. The x-intercepts are $\frac{1}{2},\frac{5}{2}$ and $\frac{9}{2}$, the high point is $\left(\frac{3}{2},2\right)$, and the low point is $\left(\frac{7}{2},-2\right)$.

Notice that the period is $\frac{2\pi}{\pi/2}=4$, and the phase shift is $\frac{\pi/4}{\pi/2}=\frac{1}{2}$.

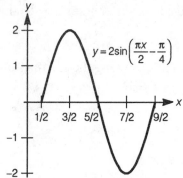

57. The x-intercepts are $\frac{5}{2}$ and $\frac{11}{2}$, the high points are $(1,3)$ and $(7,3)$, and the low point is

$(4,-3)$. Notice that the period is $\frac{2\pi}{\pi/6}=6$, and the phase shift is $\frac{\pi/3}{\pi/3}=1$.

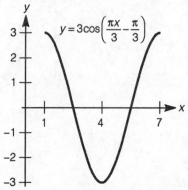

59. (a) Notice that the period is $\frac{\pi}{\pi/4} = 4$, and the asymptotes occur at $x = 2$ and $x = -2$.

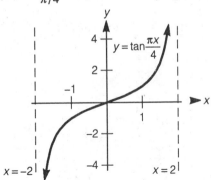

(b) Notice that the period is $\frac{\pi}{\pi/4} = 4$, and the asymptotes occur at $x = 0$ and $x = 4$.

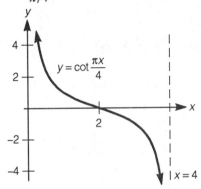

61. (a) Notice that the period is $\frac{2\pi}{1/4} = 8\pi$, and the asymptotes occur at $x = -2\pi$, $x = 2\pi$ and $x = 6\pi$.

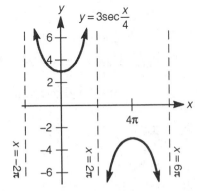

(b) Notice that the period is $\frac{2\pi}{1/4} = 8\pi$, and the asymptotes occur at $x = -4\pi$, $x = 0$ and $x = 4\pi$.

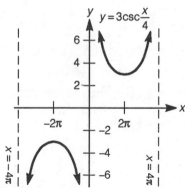

63. (a) The amplitude is 2.5 cm, the period is $\frac{2\pi}{\pi/8} = 16$ sec, and the frequency is $\frac{1}{16}$ cycles/sec. We sketch the graph over the interval $0 \le t \le 32$:

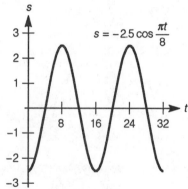

(b) The mass is farthest from the origin at high and low points, which occur at $t = 0$, $t = 8$, $t = 16$, $t = 24$ and $t = 32$ sec.

(c) The mass is passing through the equilibrium position when $s = 0$, which occurs at $t = 4$, $t = 12$, $t = 20$ and $t = 28$ sec.

65. (a) The angular speed of the smaller wheel is given by:
$$\omega = 600 \, \tfrac{\text{rev}}{\text{min}} \bullet 2\pi \, \tfrac{\text{rad}}{\text{rev}} = 1200\pi \, \tfrac{\text{rad}}{\text{min}}$$

(b) Since $v = r\omega$ and $r = 15$ cm, we have:
$$v = r\omega = 15 \text{ cm} \bullet 1200\pi \, \tfrac{\text{rad}}{\text{min}} = 18000\pi \, \tfrac{\text{cm}}{\text{min}}$$

(c) Since the linear speed on the larger wheel must also be $18000\pi \, \tfrac{\text{cm}}{\text{min}}$, and the radius of the larger wheel is $R = 25$ cm, then its angular speed is:
$$\omega = \frac{v}{r} = \frac{18000\pi \text{ cm/min}}{25 \text{ cm}} = 720\pi \, \tfrac{\text{radians}}{\text{min}}$$

(d)   In rpm, the angular speed of the larger wheel is:

$$\frac{1 \text{ rev}}{2\pi \text{ rad}} \cdot 720\pi \frac{\text{rad}}{\text{min}} = 360 \text{ rpm}$$

Chapter Three Test

1.   (a)   Since the terminal side of $\frac{4\pi}{3}$ radians lies in the third quadrant, $\cos\frac{4\pi}{3} < 0$. Using a

reference angle of $\frac{\pi}{3}$ radians, we have:

$$\cos\frac{4\pi}{3} = -\cos\frac{\pi}{3} = -\frac{1}{2}$$

(b)   Since the terminal side of $-\frac{5\pi}{6}$ radians lies in the third quadrant, $\csc\left(-\frac{5\pi}{6}\right) < 0$.

Using a reference angle of $\frac{\pi}{6}$ radians, we have:

$$\csc\left(-\frac{5\pi}{6}\right) = -\csc\frac{\pi}{6} = -\frac{1}{\sin\frac{\pi}{6}} = -\frac{1}{\frac{1}{2}} = -2$$

(c)   Since $\sin^2 t + \cos^2 t = 1$ for all values of t, then:

$$\sin^2\frac{3\pi}{4} + \cos^2\frac{3\pi}{4} = 1$$

2.   The amplitude is 2 and the period is $\frac{2\pi}{4\pi} = \frac{1}{2}$. We graph the function for $0 \le x \le 1$, noting
that the graph is $y = 2\sin(4\pi x)$ reflected across the x-axis.

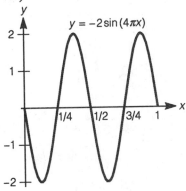

3.   Substituting $t = 4\sin u$ and noting that $\sin u > 0$ when $0 < u < \frac{\pi}{2}$, we have:

$$\frac{1}{\sqrt{16 - t^2}} = \frac{1}{\sqrt{16 - 16\cos^2 u}} = \frac{1}{\sqrt{16(1 - \cos^2 u)}} = \frac{1}{\sqrt{16\sin^2 u}} = \frac{1}{4\sin u} = \frac{1}{4}\csc u$$

4. Noting that the period is $\frac{2\pi}{4\pi} = \frac{1}{2}$, and the phase shift is $\frac{1}{4\pi}$, then the asymptotes are
$x = -\frac{1}{8} + \frac{1}{4\pi}$, $x = \frac{1}{8} + \frac{1}{4\pi}$, and $x = \frac{3}{8} + \frac{1}{4\pi}$.

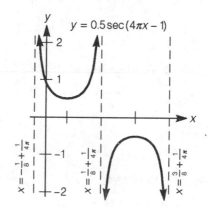

5. The amplitude is 1, the period is $\frac{2\pi}{3}$, and the phase shift is $\frac{\pi/4}{3} = \frac{\pi}{12}$. Notice that this
graph is $y = \sin\left(3x - \frac{\pi}{4}\right)$ reflected across the x-axis.

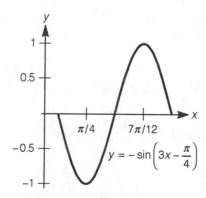

6. Given $\csc t = -\frac{3}{2}$, then $\sin t = -\frac{2}{3}$. Since $\frac{3\pi}{2} < t < 2\pi$, then $\cos t > 0$ and thus:
$$\cos t = \sqrt{1 - \sin^2 t} = \sqrt{1 - \left(-\frac{2}{3}\right)^2} = \sqrt{1 - \frac{4}{9}} = \sqrt{\frac{5}{9}} = \frac{\sqrt{5}}{3}$$

Therefore:
$$\cot t = \frac{\cos t}{\sin t} = \frac{\frac{\sqrt{5}}{3}}{-\frac{2}{3}} = -\frac{\sqrt{5}}{2}$$

7. Notice that the period is $\frac{\pi}{\pi/4} = 4$, and that the asymptote occurs at $x = 2$.

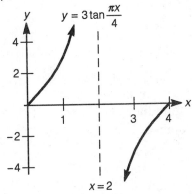

8. Sketching 2° in standard position, we have:

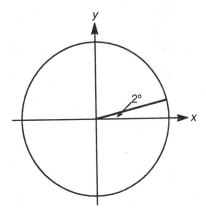

Sketching 2 radians in standard position, we have:

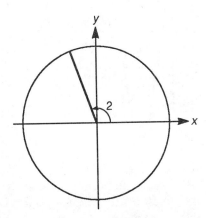

Since the y-coordinate at 2 radians is larger than the y-coordinate at 2°, then $\sin 2$ is larger.

9.   (a)   Multiplying by the conversion factor $\frac{\pi}{180°}$, we have:
$$175° \bullet \frac{\pi}{180°} = \frac{35}{36}\pi \text{ radians}$$

(b)   Multiplying by the conversion factor $\frac{180°}{\pi}$, we have:
$$5 \bullet \frac{180°}{\pi} = \frac{900°}{\pi}$$

10.   First converting $\theta = \angle BAC$ to radians, we have:
$$\theta = 75° \bullet \frac{\pi}{180°} = \frac{5\pi}{12} \text{ radians}$$

Thus the arc length is given by:
$$S = r\theta = 5\,\text{cm} \bullet \frac{5\pi}{12} = \frac{25\pi}{12} \text{ cm}$$

11.   The area of the sector is given by:
$$A = \tfrac{1}{2}r^2\theta = \tfrac{1}{2}(5)^2 \bullet \frac{5\pi}{12} = \frac{125\pi}{24} \text{ cm}^2$$

12.   Converting the angular speed to radians, we have:
$$\omega = 25\,\tfrac{\text{rev}}{\text{sec}} \bullet 2\pi\,\tfrac{\text{rad}}{\text{rev}} = 50\pi\,\tfrac{\text{rad}}{\text{sec}}$$

13.   Using $r = 5$ cm, we have:
$$v = r\omega = 5\,\text{cm} \bullet 50\pi\,\tfrac{\text{rad}}{\text{sec}} = 250\pi\,\tfrac{\text{cm}}{\text{sec}}$$

14.   (a)   The period is $\frac{2\pi}{\pi/3} = 6$ sec, and the amplitude is 10 cm. We graph the function over the interval $0 \le t \le 12$:

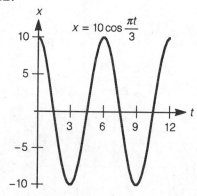

(b)   The point is passing through the origin when $x = 0$, which occurs when $t = 1.5$, $t = 4.5$, $t = 7.5$ and $t = 10.5$ sec. The point is farthest from the origin at the high and low points of this graph, which occur when $t = 0$, $t = 3$, $t = 6$, $t = 9$ and $t = 12$ sec.

15. We will simplify each side of the equation, using the identities $\sin(-\theta) = -\sin\theta$ and $\cos(-\theta) = \cos\theta$:

$$\frac{\cot\theta}{1+\tan(-\theta)} + \frac{\tan\theta}{1+\cot(-\theta)} = \frac{\frac{\cos\theta}{\sin\theta}}{1+\frac{\sin(-\theta)}{\cos(-\theta)}} + \frac{\frac{\sin\theta}{\cos\theta}}{1+\frac{\cos(-\theta)}{\sin(-\theta)}}$$

$$= \frac{\frac{\cos\theta}{\sin\theta}}{1-\frac{\sin\theta}{\cos\theta}} + \frac{\frac{\sin\theta}{\cos\theta}}{1-\frac{\cos\theta}{\sin\theta}}$$

$$= \frac{\cos^2\theta}{\sin\theta\cos\theta - \sin^2\theta} + \frac{\sin^2\theta}{\sin\theta\cos\theta - \cos^2\theta}$$

$$= \frac{\cos^2\theta}{\sin\theta(\cos\theta - \sin\theta)} + \frac{\sin^2\theta}{\cos\theta(\sin\theta - \cos\theta)}$$

$$= \frac{-\cos^3\theta}{\sin\theta\cos\theta(\sin\theta - \cos\theta)} + \frac{\sin^3\theta}{\sin\theta\cos\theta(\sin\theta - \cos\theta)}$$

$$= \frac{\sin^3\theta - \cos^3\theta}{\sin\theta\cos\theta(\sin\theta - \cos\theta)}$$

$$= \frac{(\sin\theta - \cos\theta)(\sin^2\theta + \sin\theta\cos\theta + \cos^2\theta)}{\sin\theta\cos\theta(\sin\theta - \cos\theta)}$$

$$= \frac{1+\sin\theta\cos\theta}{\sin\theta\cos\theta}$$

$$\cot\theta + \tan\theta + 1 = \frac{\cos\theta}{\sin\theta} + \frac{\sin\theta}{\cos\theta} + 1 = \frac{\cos^2\theta + \sin^2\theta + \sin\theta\cos\theta}{\sin\theta\cos\theta} = \frac{1+\sin\theta\cos\theta}{\sin\theta\cos\theta}$$

Since both sides of the equation simplify to the same quantity, the equation is an identity.

Chapter Four
Analytic Trigonometry

4.1 The Addition Formulas for Sine and Cosine

1. Using the identity $\sin(s+t) = \sin s \cos t + \cos s \sin t$ where $s = \theta$ and $t = 2\theta$, we have:
$$\sin\theta\cos 2\theta + \cos\theta\sin 2\theta = \sin(\theta + 2\theta) = \sin 3\theta$$

3. Using the identity $\sin(s-t) = \sin s \cos t - \cos s \sin t$ where $s = 3\theta$ and $t = \theta$, we have:
$$\sin 3\theta\cos\theta - \cos 3\theta\sin\theta = \sin(3\theta - \theta) = \sin 2\theta$$

5. Using the identity $\cos(s+t) = \cos s \cos t - \sin s \sin t$ where $s = 2u$ and $t = 3u$, we have:
$$\cos 2u\cos 3u - \sin 2u\sin 3u = \cos(2u + 3u) = \cos 5u$$

7. Using the identity $\cos(s-t) = \cos s \cos t + \sin s \sin t$ where $s = \frac{2\pi}{9}$ and $t = \frac{\pi}{18}$, we have:
$$\cos\frac{2\pi}{9}\cos\frac{\pi}{18} + \sin\frac{2\pi}{9}\sin\frac{\pi}{18} = \cos\left(\frac{2\pi}{9} - \frac{\pi}{18}\right) = \cos\frac{3\pi}{18} = \cos\frac{\pi}{6} = \frac{\sqrt{3}}{2}$$

9. Using the identity $\sin(s-t) = \sin s \cos t - \cos s \sin t$ where $s = A + B$ and $t = A$, we have:
$$\sin(A+B)\cos A - \cos(A+B)\sin A = \sin(A + B - A) = \sin B$$

11. Using the identity $\sin(s-t) = \sin s \cos t - \cos s \sin t$ with $s = \theta$ and $t = \frac{3\pi}{2}$, we have:
$$\sin\left(\theta - \frac{3\pi}{2}\right) = \sin\theta\cos\frac{3\pi}{2} - \cos\theta\sin\frac{3\pi}{2} = \sin\theta\bullet 0 - \cos\theta\bullet(-1) = \cos\theta$$

13. Using the identity $\cos(s+t) = \cos s \cos t - \sin s \sin t$ with $s = \theta$ and $t = \pi$, we have:
$$\cos(\theta + \pi) = \cos\theta\cos\pi - \sin\theta\sin\pi = \cos\theta\bullet(-1) - \sin\theta\bullet 0 = -\cos\theta$$

15. Using the identity $\sin(s+t) = \sin s \cos t + \cos s \sin t$ with $s = t$ and $t = 2\pi$, we have:
$$\sin(t+2\pi) = \sin t \cos 2\pi + \cos t \sin 2\pi = \sin t \bullet 1 + \cos t \bullet 0 = \sin t$$
Notice that this verifies the formula $\sin(t+2\pi) = \sin t$.

17. Using the identity $\cos(s+t) = \cos s \cos t - \sin s \sin t$, we have:
$$\begin{aligned}
\cos 75° &= \cos(45°+30°)\\
&= \cos 45° \cos 30° - \sin 45° \sin 30°\\
&= \frac{\sqrt{2}}{2} \bullet \frac{\sqrt{3}}{2} - \frac{\sqrt{2}}{2} \bullet \frac{1}{2}\\
&= \frac{\sqrt{6}-\sqrt{2}}{4}
\end{aligned}$$

19. Using the identity $\sin(s+t) = \sin s \cos t + \cos s \sin t$, we have:
$$\sin\tfrac{7\pi}{12} = \sin\left(\tfrac{\pi}{3}+\tfrac{\pi}{4}\right) = \sin\tfrac{\pi}{3}\cos\tfrac{\pi}{4} + \sin\tfrac{\pi}{4}\cos\tfrac{\pi}{3} = \tfrac{\sqrt{3}}{2}\bullet\tfrac{\sqrt{2}}{2}+\tfrac{\sqrt{2}}{2}\bullet\tfrac{1}{2} = \tfrac{\sqrt{6}+\sqrt{2}}{4}$$

21. Using the identities $\sin(s+t) = \sin s \cos t + \cos s \sin t$ and
$\sin(s-t) = \sin s \cos t - \cos s \sin t$, we have:
$$\begin{aligned}
\sin\left(\tfrac{\pi}{4}+s\right) - \sin\left(\tfrac{\pi}{4}-s\right) &= \left(\sin\tfrac{\pi}{4}\cos s + \cos\tfrac{\pi}{4}\sin s\right) - \left(\sin\tfrac{\pi}{4}\cos s - \cos\tfrac{\pi}{4}\sin s\right)\\
&= \tfrac{\sqrt{2}}{2}\cos s + \tfrac{\sqrt{2}}{2}\sin s - \tfrac{\sqrt{2}}{2}\cos s + \tfrac{\sqrt{2}}{2}\sin s\\
&= \sqrt{2}\sin s
\end{aligned}$$

23. Using the identities $\cos(s+t) = \cos s \cos t - \sin s \sin t$ and
$\cos(s-t) = \cos s \cos t + \sin s \sin t$, we have:
$$\begin{aligned}
\cos\left(\tfrac{\pi}{3}-\theta\right) - \cos\left(\tfrac{\pi}{3}+\theta\right) &= \left(\cos\tfrac{\pi}{3}\cos\theta + \sin\tfrac{\pi}{3}\sin\theta\right) - \left(\cos\tfrac{\pi}{3}\cos\theta - \sin\tfrac{\pi}{3}\sin\theta\right)\\
&= \tfrac{1}{2}\cos\theta + \tfrac{\sqrt{3}}{2}\sin\theta - \tfrac{1}{2}\cos\theta + \tfrac{\sqrt{3}}{2}\sin\theta\\
&= \sqrt{3}\sin\theta
\end{aligned}$$

25. We must first compute $\cos\alpha$ and $\sin\beta$. Since $\tfrac{\pi}{2} < \alpha < \pi$, then $\cos\alpha < 0$ and thus:
$$\cos\alpha = -\sqrt{1-\sin^2\alpha} = -\sqrt{1-\left(\tfrac{12}{13}\right)^2} = -\sqrt{1-\tfrac{144}{169}} = -\sqrt{\tfrac{25}{169}} = -\tfrac{5}{13}$$

Since $\pi < \beta < \tfrac{3\pi}{2}$, then $\sin\beta < 0$ and thus:
$$\sin\beta = -\sqrt{1-\cos^2\beta} = -\sqrt{1-\left(-\tfrac{3}{5}\right)^2} = -\sqrt{1-\tfrac{9}{25}} = -\sqrt{\tfrac{16}{25}} = -\tfrac{4}{5}$$

(a) Using the identity for $\sin(\alpha+\beta)$, we have:
$$\begin{aligned}
\sin(\alpha+\beta) &= \sin\alpha\cos\beta + \cos\alpha\sin\beta\\
&= \left(\tfrac{12}{13}\right)\bullet\left(-\tfrac{3}{5}\right) + \left(-\tfrac{5}{13}\right)\bullet\left(-\tfrac{4}{5}\right)\\
&= -\tfrac{36}{65} + \tfrac{20}{65}\\
&= -\tfrac{16}{65}
\end{aligned}$$

(b)   Using the identity for $\cos(\alpha + \beta)$, we have:
$$\cos(\alpha + \beta) = \cos\alpha\cos\beta - \sin\alpha\sin\beta$$
$$= \left(-\tfrac{5}{13}\right)\bullet\left(-\tfrac{3}{5}\right) - \left(\tfrac{12}{13}\right)\bullet\left(-\tfrac{4}{5}\right)$$
$$= \tfrac{15}{65} + \tfrac{48}{65}$$
$$= \tfrac{63}{65}$$

27.   We use the value for $\cos\alpha$ computed in Exercise 26. Since $-2\pi < \theta < -\tfrac{3\pi}{2}$, then $\sin\theta > 0$ and thus:
$$\sin\theta = \sqrt{1-\cos^2\theta} = \sqrt{1-\left(\tfrac{7}{25}\right)^2} = \sqrt{1-\tfrac{49}{625}} = \sqrt{\tfrac{576}{625}} = \tfrac{24}{25}$$

(a)   Using the identity for $\sin(\theta - \beta)$, we have:
$$\sin(\theta - \beta) = \sin\theta\cos\beta - \cos\theta\sin\beta$$
$$= \left(\tfrac{24}{25}\right)\left(-\tfrac{3}{5}\right) - \left(\tfrac{7}{25}\right)\bullet\left(-\tfrac{4}{5}\right)$$
$$= -\tfrac{72}{125} + \tfrac{28}{125}$$
$$= -\tfrac{44}{125}$$

(b)   Using the identity for $\sin(\theta + \beta)$, we have:
$$\sin(\theta + \beta) = \sin\theta\cos\beta + \cos\theta\sin\beta$$
$$= \left(\tfrac{24}{25}\right)\bullet\left(-\tfrac{3}{5}\right) + \left(\tfrac{7}{25}\right)\bullet\left(-\tfrac{4}{5}\right)$$
$$= -\tfrac{72}{125} - \tfrac{28}{125}$$
$$= -\tfrac{100}{125}$$
$$= -\tfrac{4}{5}$$

29.   (a)   Since $0 < \theta < \tfrac{\pi}{2}$, then $\cos\theta > 0$ and thus:
$$\cos\theta = \sqrt{1-\sin^2\theta} = \sqrt{1-\left(\tfrac{1}{5}\right)^2} = \sqrt{1-\tfrac{1}{25}} = \sqrt{\tfrac{24}{25}} = \tfrac{2\sqrt{6}}{5}$$

(b)   Since $\sin 2\theta = \sin(\theta + \theta)$, we use the addition formula for $\sin(s + t)$ to obtain:
$$\sin 2\theta = \sin\theta\cos\theta + \sin\theta\cos\theta = 2\sin\theta\cos\theta = 2\bullet\tfrac{1}{5}\bullet\tfrac{2\sqrt{6}}{5} = \tfrac{4\sqrt{6}}{25}$$

31.   We first find $\sin\theta$, $\cos\theta$, $\sin\beta$ and $\cos\beta$. Since $\tfrac{\pi}{2} < \theta < \pi$, then $\sec\theta < 0$, so using the identity $\sec^2\theta = 1 + \tan^2\theta$ we have:
$$\sec\theta = -\sqrt{1+\tan^2\theta} = -\sqrt{1+\left(-\tfrac{2}{3}\right)^2} = -\sqrt{1+\tfrac{4}{9}} = -\sqrt{\tfrac{13}{9}} = -\tfrac{\sqrt{13}}{3}$$

Thus $\cos\theta = -\tfrac{3}{\sqrt{13}} = -\tfrac{3\sqrt{13}}{13}$. Since $\tfrac{\pi}{2} < \theta < \pi$, then $\sin\theta > 0$ and thus:
$$\sin\theta = \sqrt{1-\cos^2\theta} = \sqrt{1-\left(-\tfrac{3\sqrt{13}}{13}\right)^2} = \sqrt{1-\tfrac{9}{13}} = \sqrt{\tfrac{4}{13}} = \tfrac{2}{\sqrt{13}} = \tfrac{2\sqrt{13}}{13}$$

Since $\csc\beta = 2$, then $\sin\beta = \frac{1}{2}$. Since $0 < \beta < \frac{\pi}{2}$, then $\cos\beta > 0$ and thus:

$$\cos\beta = \sqrt{1 - \sin^2\beta} = \sqrt{1 - \left(\frac{1}{2}\right)^2} = \sqrt{1 - \frac{1}{4}} = \sqrt{\frac{3}{4}} = \frac{\sqrt{3}}{2}$$

Now using the addition formula for $\sin(s + t)$, we have:

$$\sin(\theta + \beta) = \sin\theta\cos\beta + \cos\theta\sin\beta$$
$$= \frac{2\sqrt{13}}{13} \cdot \frac{\sqrt{3}}{2} - \frac{3\sqrt{13}}{13} \cdot \frac{1}{2}$$
$$= \frac{2\sqrt{39}}{26} - \frac{3\sqrt{13}}{26}$$
$$= \frac{2\sqrt{39} - 3\sqrt{13}}{26}$$

Finally using the addition formula for $\cos(s - t)$, we have:

$$\cos(\beta - \theta) = \cos\beta\cos\theta + \sin\beta\sin\theta$$
$$= \frac{\sqrt{3}}{2} \cdot \left(-\frac{3\sqrt{13}}{13}\right) + \frac{1}{2} \cdot \frac{2\sqrt{13}}{13}$$
$$= -\frac{3\sqrt{39}}{26} + \frac{2\sqrt{13}}{26}$$
$$= \frac{2\sqrt{13} - 3\sqrt{39}}{26}$$

33. Using the addition formula for $\sin(s + t)$, we have:

$$\sin\left(t + \frac{\pi}{4}\right) = \sin t\cos\frac{\pi}{4} + \cos t\sin\frac{\pi}{4} = \frac{1}{\sqrt{2}}\sin t + \frac{1}{\sqrt{2}}\cos t = \frac{\sin t + \cos t}{\sqrt{2}}$$

35. Using the results from Exercises 33 and 34, we have:

$$\sin\left(t + \frac{\pi}{4}\right) + \cos\left(t + \frac{\pi}{4}\right) = \frac{\sin t + \cos t}{\sqrt{2}} + \frac{\cos t - \sin t}{\sqrt{2}} = \frac{2\cos t}{\sqrt{2}} = \sqrt{2}\cos t$$

37. (a) Using the addition formula for $\sin(s + t)$, we have:

$$\sin\left(x + \frac{\pi}{2}\right) = \sin x\cos\frac{\pi}{2} + \cos x\sin\frac{\pi}{2} = \sin x \bullet 0 + \cos x \bullet 1 = \cos x$$

(b) Since $\cos x = \sin\left(x + \frac{\pi}{2}\right)$, then the graph of $y = \cos x$ is the graph of $y = \sin x$ shifted $\frac{\pi}{2}$ units to the left.

39. Using the addition formula for $\sin(s + t)$, we have:

$$\frac{\sin(s + t)}{\cos s\cos t} = \frac{\sin s\cos t + \cos s\sin t}{\cos s\cos t} = \frac{\sin s}{\cos s} + \frac{\sin t}{\cos t} = \tan s + \tan t$$

41. Using the addition formulas for $\cos(s-t)$ and $\cos(s+t)$, we have:

$$\cos(A-B) - \cos(A+B) = (\cos A \cos B + \sin A \sin B) - (\cos A \cos B - \sin A \sin B)$$
$$= \cos A \cos B + \sin A \sin B - \cos A \cos B + \sin A \sin B$$
$$= 2\sin A \sin B$$

43. Using the addition formulas for $\cos(s+t)$ and $\cos(s-t)$, as well as the identities $\sin^2 A = 1 - \cos^2 A$ and $\sin^2 B = 1 - \cos^2 B$, we have:

$$\cos(A+B)\cos(A-B) = (\cos A \cos B - \sin A \sin B)(\cos A \cos B + \sin A \sin B)$$
$$= (\cos A \cos B)^2 - (\sin A \sin B)^2$$
$$= \cos^2 A \cos^2 B - \sin^2 A \sin^2 B$$
$$= \cos^2 A(1 - \sin^2 B) - (1 - \cos^2 A)\sin^2 B$$
$$= \cos^2 A - \cos^2 A \sin^2 B - \sin^2 B + \cos^2 A \sin^2 B$$
$$= \cos^2 A - \sin^2 B$$

45. Using the addition formulas for $\cos(s+t)$ and $\sin(s+t)$, as well as the identity $\cos^2 \beta + \sin^2 \beta = 1$, we have:

$$\cos(\alpha+\beta)\cos\beta + \sin(\alpha+\beta)\sin\beta$$
$$= (\cos\alpha\cos\beta - \sin\alpha\sin\beta)\cos\beta + (\sin\alpha\cos\beta + \cos\alpha\sin\beta)\sin\beta$$
$$= \cos\alpha\cos^2\beta - \sin\alpha\sin\beta\cos\beta + \sin\alpha\sin\beta\cos\beta + \cos\alpha\sin^2\beta$$
$$= \cos\alpha\cos^2\beta + \cos\alpha\sin^2\beta$$
$$= \cos\alpha(\cos^2\beta + \sin^2\beta)$$
$$= \cos\alpha$$

47. (a) We complete the table:

t	1	2	3	4
$f(t)$	1.5	1.5	1.5	1.5

(b) Conjecture: $f(t) = 1.5$
To prove this, we first simplify the expressions:

$$\cos\left(t + \tfrac{2\pi}{3}\right) = \cos t \cos\tfrac{2\pi}{3} - \sin t \sin\tfrac{2\pi}{3} = -\tfrac{1}{2}\cos t - \tfrac{\sqrt{3}}{2}\sin t$$
$$\cos\left(t - \tfrac{2\pi}{3}\right) = \cos t \cos\tfrac{2\pi}{3} + \sin t \sin\tfrac{2\pi}{3} = -\tfrac{1}{2}\cos t + \tfrac{\sqrt{3}}{2}\sin t$$

So:

$$\cos^2\left(t + \tfrac{2\pi}{3}\right) = \left(-\tfrac{1}{2}\cos t - \tfrac{\sqrt{3}}{2}\sin t\right)^2 = \tfrac{1}{4}\cos^2 t + \tfrac{\sqrt{3}}{2}\sin t \cos t + \tfrac{3}{4}\sin^2 t$$
$$\cos^2\left(t - \tfrac{2\pi}{3}\right) = \left(-\tfrac{1}{2}\cos t + \tfrac{\sqrt{3}}{2}\sin t\right)^2 = \tfrac{1}{4}\cos^2 t - \tfrac{\sqrt{3}}{2}\sin t \cos t + \tfrac{3}{4}\sin^2 t$$

Thus:

$$f(t) = \cos^2 t + \left(\tfrac{1}{4}\cos^2 t + \tfrac{\sqrt{3}}{2}\sin t \cos t + \tfrac{3}{4}\sin^2 t\right)$$
$$+ \left(\tfrac{1}{4}\cos^2 t - \tfrac{\sqrt{3}}{2}\sin t \cos t + \tfrac{3}{4}\sin^2 t\right)$$
$$= \tfrac{3}{2}\cos^2 t + \tfrac{3}{2}\sin^2 t$$
$$= \tfrac{3}{2}\left(\cos^2 t + \sin^2 t\right)$$
$$= \tfrac{3}{2}$$

49. To prove this identity, we will show an alternate identity:

$$1 - \cos^2 C - \cos^2 A - \cos^2 B - 2\cos A \cos B \cos C$$
$$= \sin^2 C - \cos^2 A - \cos^2 B - 2\cos A \cos B \cos\left[\pi - (A+B)\right]$$
$$= \sin^2\left[\pi - (A+B)\right] - \cos^2 A - \cos^2 B + 2\cos A \cos B \cos(A+B)$$
$$= \sin^2 (A+B) - \cos^2 A - \cos^2 B + 2\cos A \cos B(\cos A \cos B - \sin A \sin B)$$
$$= (\sin A \cos B + \sin B \cos A)^2 - \cos^2 A - \cos^2 B + 2\cos^2 A \cos^2 B$$
$$\qquad\qquad -2\cos A \cos B \sin A \sin B$$
$$= \sin^2 A \cos^2 B + \sin^2 B \cos^2 A - \cos^2 A - \cos^2 B + 2\cos^2 A \cos^2 B$$
$$= \cos^2 B\left(\sin^2 A - 1\right) + \cos^2 A\left(\sin^2 B - 1\right) + 2\cos^2 A \cos^2 B$$
$$= -\cos^2 B \cos^2 A - \cos^2 A \cos^2 B + 2\cos^2 A \cos^2 B$$
$$= 0$$

Hence $\cos^2 A + \cos^2 B + \cos^2 C + 2\cos A \cos B \cos C = 1$.

51. Since $a^2 + b^2 = 1$ and $c^2 + d^2 = 1$, there is some angle θ for which $a = \cos\theta$ and $b = \sin\theta$ and some angle φ for which $c = \cos\varphi$ and $d = \sin\varphi$. Thus:

$$|ac + bd| = |\cos\theta\cos\varphi + \sin\theta\sin\varphi| = |\cos(\theta - \varphi)| \le 1$$

53. Let $A = \tfrac{\pi}{3} - t$ and $B = \tfrac{\pi}{3} + t$, then we have:

$$\sin A \cos B + \cos A \sin B = \sin(A+B)$$

But $A + B = \tfrac{2\pi}{3}$, so $\sin(A+B) = \sin\tfrac{2\pi}{3} = \tfrac{\sqrt{3}}{2}$.

55. (a) Using $\triangle ABH$, $\cos(\alpha + \beta) = \tfrac{AB}{1}$, so $\cos(\alpha + \beta) = AB$.

(b) Using $\triangle ACF$, $\cos\alpha = \tfrac{AC}{AF} = \tfrac{AC}{\cos\beta}$ from Exercise 54(e), so $AC = \cos\alpha\cos\beta$.

(c) Using $\triangle EFH$, $\sin(\angle EHF) = \tfrac{EF}{HF}$. But $\angle EHF = \alpha$ from Exercise 54(c), and

$HF = \sin\beta$ from Exercise 54(b), so $\sin\alpha = \tfrac{EF}{\sin\beta}$, and thus $EF = \sin\alpha\sin\beta$.

(d) From part (a), $\cos(\alpha + \beta) = AB = AC - BC$. But $AC = \cos\alpha\cos\beta$ from part (b), and $BC = EF = \sin\alpha\sin\beta$ from part (c), so $\cos(\alpha + \beta) = \cos\alpha\cos\beta - \sin\alpha\sin\beta$.

4.2 Further Identities

1. (a) Using the identity for $\tan(s + t)$, we have:

$$\tan(t+s) = \frac{\tan t + \tan s}{1 - \tan t \tan s} = \frac{\frac{3}{4} + \frac{7}{24}}{1 - \frac{3}{4} \cdot \frac{7}{24}} = \frac{\frac{25}{24}}{1 - \frac{7}{32}} = \frac{\frac{25}{24}}{\frac{25}{32}} = \frac{4}{3}$$

(b) Using the identity for $\tan(s - t)$, we have:

$$\tan(t-s) = \frac{\tan t - \tan s}{1 + \tan t \tan s} = \frac{\frac{3}{4} - \frac{7}{24}}{1 + \frac{3}{4} \cdot \frac{7}{24}} = \frac{\frac{11}{24}}{1 + \frac{7}{32}} = \frac{\frac{11}{24}}{\frac{39}{32}} = \frac{44}{117}$$

3. (a) Using the identity for $\tan(s + t)$ and rationalizing denominators, we have:

$$\tan\left(x + \frac{3\pi}{4}\right) = \frac{\tan x + \tan\frac{3\pi}{4}}{1 - \tan x \tan\frac{3\pi}{4}}$$
$$= \frac{\sqrt{2} + (-1)}{1 - (\sqrt{2})(-1)}$$
$$= \frac{\sqrt{2} - 1}{1 + \sqrt{2}} \cdot \frac{1 - \sqrt{2}}{1 - \sqrt{2}}$$
$$= \frac{-3 + 2\sqrt{2}}{1 - 2}$$
$$= 3 - 2\sqrt{2}$$

(b) Using the identity for $\tan(s - t)$ and rationalizing denominators, we have:

$$\tan\left(x - \frac{5\pi}{4}\right) = \frac{\tan x - \tan\frac{5\pi}{4}}{1 + \tan x \tan\frac{5\pi}{4}}$$
$$= \frac{\sqrt{2} - 1}{1 + (\sqrt{2})(1)}$$
$$= \frac{\sqrt{2} - 1}{1 + \sqrt{2}} \cdot \frac{1 - \sqrt{2}}{1 - \sqrt{2}}$$
$$= \frac{-3 + 2\sqrt{2}}{1 - 2}$$
$$= 3 - 2\sqrt{2}$$

5. Since $0° < \phi < 90°$, $\sin\phi > 0$ and thus:
$$\sin\phi = \sqrt{1 - \cos^2\phi} = \sqrt{1 - \left(\frac{7}{25}\right)^2} = \sqrt{1 - \frac{49}{625}} = \sqrt{\frac{576}{625}} = \frac{24}{25}$$

(a) Using the double-angle identity for $\sin 2\phi$, we have:
$$\sin 2\phi = 2\sin\phi\cos\phi = 2 \cdot \frac{24}{25} \cdot \frac{7}{25} = \frac{336}{625}$$

(b)   Using the double-angle identity for $\cos 2\phi$, we have:

$$\cos 2\phi = \cos^2 \phi - \sin^2 \phi = \left(\tfrac{7}{25}\right)^2 - \left(\tfrac{24}{25}\right)^2 = \tfrac{49}{625} - \tfrac{576}{625} = -\tfrac{527}{625}$$

(c)   Using parts (a) and (b), we have:

$$\tan 2\phi = \frac{\sin 2\phi}{\cos 2\phi} = \frac{\tfrac{336}{625}}{-\tfrac{527}{625}} = -\frac{336}{527}$$

7.   Since $\tfrac{3\pi}{2} < u < 2\pi$, $\sec u > 0$ and thus:

$$\sec u = \sqrt{1 + \tan^2 u} = \sqrt{1 + (-4)^2} = \sqrt{1 + 16} = \sqrt{17}$$

Thus $\cos u = \tfrac{1}{\sqrt{17}}$. Since $\tfrac{3\pi}{2} < u < 2\pi$, $\sin u < 0$ and thus:

$$\sin u = -\sqrt{1 - \cos^2 u} = -\sqrt{1 - \tfrac{1}{17}} = -\sqrt{\tfrac{16}{17}} = -\frac{4}{\sqrt{17}}$$

(a)   Using the double-angle identity for $\sin 2u$, we have:

$$\sin 2u = 2 \sin u \cos u = 2 \bullet \left(-\tfrac{4}{\sqrt{17}}\right) \bullet \left(\tfrac{1}{\sqrt{17}}\right) = -\tfrac{8}{17}$$

(b)   Using the double-angle identity for $\cos 2u$, we have:

$$\cos 2u = \cos^2 u - \sin^2 u = \left(\tfrac{1}{\sqrt{17}}\right)^2 - \left(-\tfrac{4}{\sqrt{17}}\right)^2 = \tfrac{1}{17} - \tfrac{16}{17} = -\tfrac{15}{17}$$

(c)   Using parts (a) and (b), we have:

$$\tan 2u = \frac{\sin 2u}{\cos 2u} = \frac{-\tfrac{8}{17}}{-\tfrac{15}{17}} = \frac{8}{15}$$

9.   Since $0° < \alpha < 90°$, $\cos \alpha > 0$ and thus:

$$\cos \alpha = \sqrt{1 - \sin^2 \alpha} = \sqrt{1 - \left(\tfrac{\sqrt{3}}{2}\right)^2} = \sqrt{1 - \tfrac{3}{4}} = \sqrt{\tfrac{1}{4}} = \tfrac{1}{2}$$

(a)   Since $0° < \alpha < 90°$, then $0° < \tfrac{\alpha}{2} < 45°$ and thus $\sin \tfrac{\alpha}{2} > 0$. Therefore:

$$\sin \tfrac{\alpha}{2} = \sqrt{\frac{1 - \cos \alpha}{2}} = \sqrt{\frac{1 - \tfrac{1}{2}}{2}} = \sqrt{\tfrac{1}{4}} = \frac{1}{2}$$

(b)   Since $0° < \alpha < 90°$, then $0° < \tfrac{\alpha}{2} < 45°$ and thus $\cos \tfrac{\alpha}{2} > 0$. Therefore:

$$\cos \tfrac{\alpha}{2} = \sqrt{\frac{1 + \cos \alpha}{2}} = \sqrt{\frac{1 + \tfrac{1}{2}}{2}} = \sqrt{\tfrac{3}{4}} = \frac{\sqrt{3}}{2}$$

(c)   Using parts (a) and (b), we have:

$$\tan \tfrac{\alpha}{2} = \frac{\sin \tfrac{\alpha}{2}}{\cos \tfrac{\alpha}{2}} = \frac{\tfrac{1}{2}}{\tfrac{\sqrt{3}}{2}} = \frac{1}{\sqrt{3}} = \frac{\sqrt{3}}{3}$$

Notice that an alternate solution is to spot that since $\sin\alpha = \frac{\sqrt{3}}{2}$ and $0° < \alpha < 90°$, then $\alpha = 60°$ and thus $\frac{\alpha}{2} = 30°$. Thus $\sin\frac{\alpha}{2} = \frac{1}{2}$, $\cos\frac{\alpha}{2} = \frac{\sqrt{3}}{2}$ and $\tan\frac{\alpha}{2} = \frac{\sqrt{3}}{3}$.

11. (a) Since $\frac{\pi}{2} < \theta < \pi$, then $\frac{\pi}{4} < \frac{\theta}{2} < \frac{\pi}{2}$ and thus $\sin\frac{\theta}{2} > 0$. Therefore:

$$\sin\frac{\theta}{2} = \sqrt{\frac{1-\cos\theta}{2}} = \sqrt{\frac{1+\frac{7}{9}}{2}} = \sqrt{\frac{8}{9}} = \frac{2\sqrt{2}}{3}$$

(b) Since $\frac{\pi}{2} < \theta < \pi$, then $\frac{\pi}{4} < \frac{\theta}{2} < \frac{\pi}{2}$ and thus $\cos\frac{\theta}{2} > 0$. Therefore:

$$\cos\frac{\theta}{2} = \sqrt{\frac{1+\cos\theta}{2}} = \sqrt{\frac{1-\frac{7}{9}}{2}} = \sqrt{\frac{1}{9}} = \frac{1}{3}$$

(c) Using parts (a) and (b), we have:

$$\tan\frac{\theta}{2} = \frac{\sin\frac{\theta}{2}}{\cos\frac{\theta}{2}} = \frac{\frac{2\sqrt{2}}{3}}{\frac{1}{3}} = 2\sqrt{2}$$

13. Since $\frac{\pi}{2} < \theta < \pi$, $\cos\theta < 0$ and thus:

$$\cos\theta = -\sqrt{1-\sin^2\theta} = -\sqrt{1-\left(\frac{3}{4}\right)^2} = -\sqrt{1-\frac{9}{16}} = -\sqrt{\frac{7}{16}} = -\frac{\sqrt{7}}{4}$$

(a) Using the double-angle identity for $\sin 2\theta$, we have:

$$\sin 2\theta = 2\sin\theta\cos\theta = 2\bullet\left(\frac{3}{4}\right)\bullet\left(-\frac{\sqrt{7}}{4}\right) = -\frac{3\sqrt{7}}{8}$$

(b) Using the double-angle identity for $\cos 2\theta$, we have:

$$\cos 2\theta = \cos^2\theta - \sin^2\theta = \left(-\frac{\sqrt{7}}{4}\right)^2 - \left(\frac{3}{4}\right)^2 = \frac{7}{16} - \frac{9}{16} = -\frac{1}{8}$$

(c) Since $\frac{\pi}{4} < \frac{\theta}{2} < \frac{\pi}{2}$, $\sin\frac{\theta}{2} > 0$ and thus:

$$\sin\frac{\theta}{2} = \sqrt{\frac{1-\cos\theta}{2}} = \sqrt{\frac{1-\left(-\frac{\sqrt{7}}{4}\right)}{2}} = \sqrt{\frac{4+\sqrt{7}}{8}} = \frac{\sqrt{8+2\sqrt{7}}}{4}$$

(d) Since $\frac{\pi}{4} < \frac{\theta}{2} < \frac{\pi}{2}$, $\cos\frac{\theta}{2} > 0$ and thus:

$$\cos\frac{\theta}{2} = \sqrt{\frac{1+\cos\theta}{2}} = \sqrt{\frac{1-\frac{\sqrt{7}}{4}}{2}} = \sqrt{\frac{4-\sqrt{7}}{8}} = \frac{\sqrt{8-2\sqrt{7}}}{4}$$

15. Since $180° < \theta < 270°$, $\sin\theta < 0$ and thus:

$$\sin\theta = -\sqrt{1-\cos^2\theta} = -\sqrt{1-\left(-\frac{1}{3}\right)^2} = -\sqrt{1-\frac{1}{9}} = -\sqrt{\frac{8}{9}} = -\frac{2\sqrt{2}}{3}$$

(a)   Using the double-angle identity for $\sin 2\theta$, we have:

$$\sin 2\theta = 2 \sin \theta \cos \theta = 2 \bullet \left(-\frac{2\sqrt{2}}{3}\right) \bullet \left(-\frac{1}{3}\right) = \frac{4\sqrt{2}}{9}$$

(b)   Using the double-angle identity for $\cos 2\theta$, we have:

$$\cos 2\theta = \cos^2 \theta - \sin^2 \theta = \left(-\frac{1}{3}\right)^2 - \left(-\frac{2\sqrt{2}}{3}\right)^2 = \frac{1}{9} - \frac{8}{9} = -\frac{7}{9}$$

(c)   Since $90° < \frac{\theta}{2} < 135°$, $\sin \frac{\theta}{2} > 0$ and thus:

$$\sin \frac{\theta}{2} = \sqrt{\frac{1 - \cos \theta}{2}} = \sqrt{\frac{1 + \frac{1}{3}}{2}} = \sqrt{\frac{2}{3}} = \frac{\sqrt{6}}{3}$$

(d)   Since $90° < \frac{\theta}{2} < 135°$, $\cos \frac{\theta}{2} < 0$ and thus:

$$\cos \frac{\theta}{2} = -\sqrt{\frac{1 + \cos \theta}{2}} = -\sqrt{\frac{1 - \frac{1}{3}}{2}} = -\sqrt{\frac{1}{3}} = -\frac{\sqrt{3}}{3}$$

17. (a)   Since $0 < \frac{\pi}{12} < \frac{\pi}{2}$, then $\sin \frac{\pi}{12} > 0$. Using the half-angle formula for $\sin \frac{1}{2} x$, we have:

$$\sin \frac{\pi}{12} = \sqrt{\frac{1 - \cos \frac{\pi}{6}}{2}} = \sqrt{\frac{1 - \frac{\sqrt{3}}{2}}{2}} = \sqrt{\frac{2 - \sqrt{3}}{4}} = \frac{\sqrt{2 - \sqrt{3}}}{2}$$

(b)   Since $0 < \frac{\pi}{12} < \frac{\pi}{2}$, then $\cos \frac{\pi}{12} > 0$. Using the half-angle formula for $\cos \frac{1}{2} x$, we have:

$$\cos \frac{\pi}{12} = \sqrt{\frac{1 + \cos \frac{\pi}{6}}{2}} = \sqrt{\frac{1 + \frac{\sqrt{3}}{2}}{2}} = \frac{\sqrt{2 + \sqrt{3}}}{2}$$

(c)   Using the half-angle formula for $\tan \frac{1}{2} x$, we have:

$$\tan \frac{\pi}{12} = \frac{\sin \frac{\pi}{6}}{1 + \cos \frac{\pi}{6}} = \frac{\frac{1}{2}}{1 + \frac{\sqrt{3}}{2}} = \frac{1}{2 + \sqrt{3}} \bullet \frac{2 - \sqrt{3}}{2 - \sqrt{3}} = \frac{2 - \sqrt{3}}{4 - 3} = 2 - \sqrt{3}$$

Note: We could also use parts (a) and (b), as follows:

$$\tan \frac{\pi}{12} = \frac{\sin \frac{\pi}{12}}{\cos \frac{\pi}{12}}$$

$$= \frac{\frac{\sqrt{2 - \sqrt{3}}}{2}}{\frac{\sqrt{2 + \sqrt{3}}}{2}}$$

$$= \frac{\sqrt{2 - \sqrt{3}}}{\sqrt{2 + \sqrt{3}}} \bullet \frac{\sqrt{2 + \sqrt{3}}}{\sqrt{2 + \sqrt{3}}}$$

$$= \frac{\sqrt{4 - 3}}{2 + \sqrt{3}} \bullet \frac{2 - \sqrt{3}}{2 - \sqrt{3}}$$

$$= \frac{2 - \sqrt{3}}{4 - 3}$$

$$= 2 - \sqrt{3}$$

19. **(a)** Since $90° < 105° < 180°$, then $\sin 105° > 0$. Using the half-angle formula for $\sin \frac{1}{2}x$, we have:

$$\sin 105° = \sqrt{\frac{1-\cos 210°}{2}} = \sqrt{\frac{1+\frac{\sqrt{3}}{2}}{2}} = \sqrt{\frac{2+\sqrt{3}}{4}} = \frac{\sqrt{2+\sqrt{3}}}{2}$$

(b) Since $90° < 105° < 180°$, then $\cos 105° < 0$. Using the half-angle formula for $\cos \frac{1}{2}x$, we have:

$$\cos 105° = -\sqrt{\frac{1+\cos 210°}{2}} = -\sqrt{\frac{1-\frac{\sqrt{3}}{2}}{2}} = -\sqrt{\frac{2-\sqrt{3}}{4}} = -\frac{\sqrt{2-\sqrt{3}}}{2}$$

(c) Using the half-angle formula for $\tan \frac{1}{2}x$, we have:

$$\tan 105° = \frac{\sin 210°}{1+\cos 210°} = \frac{-\frac{1}{2}}{1-\frac{\sqrt{3}}{2}} = -\frac{1}{2-\sqrt{3}} \cdot \frac{2+\sqrt{3}}{2+\sqrt{3}} = -\frac{2+\sqrt{3}}{4-3} = -2-\sqrt{3}$$

21. From the first triangle we have $\sin\theta = \frac{3}{5}$, $\cos\theta = \frac{4}{5}$ and $\tan\theta = \frac{3}{4}$.

(a) By the double-angle identity for $\sin 2\theta$, we have:

$$\sin 2\theta = 2\sin\theta\cos\theta = 2 \cdot \frac{3}{5} \cdot \frac{4}{5} = \frac{24}{25}$$

(b) By the double-angle identity for $\cos 2\theta$, we have:

$$\cos 2\theta = \cos^2\theta - \sin^2\theta = \left(\frac{4}{5}\right)^2 - \left(\frac{3}{5}\right)^2 = \frac{16}{25} - \frac{9}{25} = \frac{7}{25}$$

(c) By the double-angle identity for $\tan 2\theta$, we have:

$$\tan 2\theta = \frac{2\tan\theta}{1-\tan^2\theta} = \frac{2 \cdot \frac{3}{4}}{1-\left(\frac{3}{4}\right)^2} = \frac{\frac{3}{2}}{\frac{7}{16}} = \frac{24}{7}$$

Note: An easier approach, after doing (a) and (b), would be to say:

$$\tan 2\theta = \frac{\sin 2\theta}{\cos 2\theta} = \frac{\frac{24}{25}}{\frac{7}{25}} = \frac{24}{7}$$

23. From the first triangle we have $\sin\beta = \frac{4}{5}$ and $\cos\beta = \frac{3}{5}$.

(a) By the double-angle identity for $\sin 2\beta$, we have:

$$\sin 2\beta = 2\sin\beta\cos\beta = 2 \cdot \frac{4}{5} \cdot \frac{3}{5} = \frac{24}{25}$$

(b) By the double-angle identity for $\cos 2\beta$, we have:

$$\cos 2\beta = \cos^2\beta - \sin^2\beta = \left(\frac{3}{5}\right)^2 - \left(\frac{4}{5}\right)^2 = \frac{9}{25} - \frac{16}{25} = -\frac{7}{25}$$

(c) Using the identity for $\tan x$, we have:

$$\tan 2\beta = \frac{\sin 2\beta}{\cos 2\beta} = \frac{\frac{24}{25}}{-\frac{7}{25}} = -\frac{24}{7}$$

Note: We could also have used the double-angle formula for $\tan 2\beta$.

25. From the first triangle we have $\sin\theta = \frac{3}{5}$ and $\cos\theta = \frac{4}{5}$.

(a) Since $\sin\frac{\theta}{2} > 0$, we use the half-angle formula for $\sin\frac{\theta}{2}$ to obtain:

$$\sin\frac{\theta}{2} = \sqrt{\frac{1-\cos\theta}{2}} = \sqrt{\frac{1-\frac{4}{5}}{2}} = \sqrt{\frac{1}{10}} = \frac{\sqrt{10}}{10}$$

(b) Since $\cos\frac{\theta}{2} > 0$, we use the half-angle formula for $\cos\frac{\theta}{2}$ to obtain:

$$\cos\frac{\theta}{2} = \sqrt{\frac{1+\cos\theta}{2}} = \sqrt{\frac{1+\frac{4}{5}}{2}} = \sqrt{\frac{9}{10}} = \frac{3\sqrt{10}}{10}$$

(c) Using the half-angle formula for $\tan\frac{\theta}{2}$, we obtain:

$$\tan\frac{\theta}{2} = \frac{\sin\theta}{1+\cos\theta} = \frac{\frac{3}{5}}{1+\frac{4}{5}} = \frac{3}{9} = \frac{1}{3}$$

Note: We could also have computed this directly after parts (a) and (b) as:

$$\tan\frac{\theta}{2} = \frac{\sin\frac{\theta}{2}}{\cos\frac{\theta}{2}} = \frac{\frac{\sqrt{10}}{10}}{\frac{3\sqrt{10}}{10}} = \frac{1}{3}$$

27. From the first triangle we have $\cos\beta = \frac{3}{5}$.

(a) Since $\sin\frac{\beta}{2} > 0$, we use the half-angle formula for $\sin\frac{\beta}{2}$ to obtain:

$$\sin\frac{\beta}{2} = \sqrt{\frac{1-\cos\beta}{2}} = \sqrt{\frac{1-\frac{3}{5}}{2}} = \sqrt{\frac{1}{5}} = \frac{\sqrt{5}}{5}$$

(b) Since $\cos\frac{\beta}{2} > 0$, we use the half-angle formula for $\cos\frac{\beta}{2}$ to obtain:

$$\cos\frac{\beta}{2} = \sqrt{\frac{1+\cos\beta}{2}} = \sqrt{\frac{1+\frac{3}{5}}{2}} = \sqrt{\frac{4}{5}} = \frac{2\sqrt{5}}{5}$$

(c) Using the identity for $\tan x$ and parts (a) and (b), we have:

$$\tan\frac{\beta}{2} = \frac{\sin\frac{\beta}{2}}{\cos\frac{\beta}{2}} = \frac{\frac{\sqrt{5}}{5}}{\frac{2\sqrt{5}}{5}} = \frac{1}{2}$$

29. Since $0 < \theta < \frac{\pi}{2}$, then $\cos\theta > 0$. Since $\sin\theta = \frac{x}{5}$, we have:

$$\cos\theta = \sqrt{1-\sin^2\theta} = \sqrt{1-\left(\frac{x}{5}\right)^2} = \sqrt{1-\frac{x^2}{25}} = \frac{\sqrt{25-x^2}}{5}$$

We now apply the double angle formulas:

$$\sin 2\theta = 2\sin\theta\cos\theta = 2\cdot\frac{x}{5}\cdot\frac{\sqrt{25-x^2}}{5} = \frac{2x\sqrt{25-x^2}}{25}$$

$$\cos 2\theta = \cos^2\theta - \sin^2\theta = \left(\frac{\sqrt{25-x^2}}{5}\right)^2 - \left(\frac{x}{5}\right)^2 = \frac{25-x^2}{25} - \frac{x^2}{25} = \frac{25-2x^2}{25}$$

31. Since $0 < \theta < \frac{\pi}{2}$, then $\cos\theta > 0$. Since $\sin\theta = \frac{x-1}{2}$, we have:

$$\cos\theta = \sqrt{1-\sin^2\theta} = \sqrt{1-\left(\frac{x-1}{2}\right)^2} = \sqrt{1-\frac{x^2-2x+1}{4}} = \frac{\sqrt{3+2x-x^2}}{2}$$

We now apply the double-angle formulas:

$$\sin 2\theta = 2\sin\theta\cos\theta = 2\cdot\frac{x-1}{2}\cdot\frac{\sqrt{3+2x-x^2}}{2} = \frac{(x-1)\sqrt{3+2x-x^2}}{2}$$

$$\cos 2\theta = \cos^2\theta - \sin^2\theta$$

$$= \left(\frac{\sqrt{3+2x-x^2}}{2}\right)^2 - \left(\frac{x-1}{2}\right)^2$$

$$= \frac{3+2x-x^2}{4} - \frac{x^2-2x+1}{4}$$

$$= \frac{2+4x-2x^2}{4}$$

$$= \frac{1+2x-x^2}{2}$$

33. Using the identity $\sin^2\theta = \frac{1-\cos 2\theta}{2}$, we have:

$$\sin^4\theta = \left(\sin^2\theta\right)^2$$

$$= \left(\frac{1-\cos 2\theta}{2}\right)^2$$

$$= \frac{1-2\cos 2\theta + (\cos 2\theta)^2}{4}$$

$$= \frac{1-2\cos 2\theta + \frac{1+\cos 4\theta}{2}}{4}$$

$$= \frac{2-4\cos 2\theta + 1 + \cos 4\theta}{8}$$

$$= \frac{3-4\cos 2\theta + \cos 4\theta}{8}$$

35. Using the identities $\sin^2\frac{\theta}{2} = \frac{1-\cos\theta}{2}$ and $\cos^2\theta = \frac{1+\cos 2\theta}{2}$, we have:

$$\sin^4\frac{\theta}{2} = \left(\sin^2\frac{\theta}{2}\right)^2$$

$$= \left(\frac{1-\cos\theta}{2}\right)^2$$

$$= \frac{1-2\cos\theta + \cos^2\theta}{4}$$

$$= \frac{1-2\cos\theta + \frac{1+\cos 2\theta}{2}}{4}$$

$$= \frac{2-4\cos\theta + 1 + \cos 2\theta}{8}$$

$$= \frac{3-4\cos\theta + \cos 2\theta}{8}$$

37. (a) Replacing 2θ with $\theta + \theta$ and using the addition formula for $\cos(s+t)$, we have:

$$\cos 2\theta = \cos(\theta + \theta) = \cos\theta\cos\theta - \sin\theta\sin\theta = \cos^2\theta - \sin^2\theta$$

(b) Replacing 2θ with $\theta + \theta$ and using the addition formula for $\tan(s+t)$, we have:

$$\tan 2\theta = \tan(\theta + \theta) = \frac{\tan\theta + \tan\theta}{1 - \tan\theta\tan\theta} = \frac{2\tan\theta}{1 - \tan^2\theta}$$

39. Working from the right-hand side, we have:

$$\frac{1-\tan^2 s}{1+\tan^2 s} = \frac{1 - \frac{\sin^2 s}{\cos^2 s}}{1 + \frac{\sin^2 s}{\cos^2 s}} = \frac{\cos^2 s - \sin^2 s}{\cos^2 s + \sin^2 s} = \frac{\cos 2s}{1} = \cos 2s$$

41. Writing $\theta = 2 \bullet \frac{\theta}{2}$, we apply the double-angle formula:

$$\cos\theta = \cos\left(2 \bullet \frac{\theta}{2}\right) = \cos^2\frac{\theta}{2} - \sin^2\frac{\theta}{2} = \cos^2\frac{\theta}{2} - \left(1 - \cos^2\frac{\theta}{2}\right) = 2\cos^2\frac{\theta}{2} - 1$$

43. Using the identities $\sin^2\theta = \frac{1-\cos 2\theta}{2}$ and $\cos^2 2\theta = \frac{1+\cos 4\theta}{2}$, we have:

$$\sin^4\theta = \left(\sin^2\theta\right)^2$$

$$= \left(\frac{1-\cos 2\theta}{2}\right)^2$$

$$= \frac{1-2\cos 2\theta + \cos^2 2\theta}{4}$$

$$= \frac{1-2\cos 2\theta + \frac{1+\cos 4\theta}{2}}{4}$$

$$= \frac{2-4\cos 2\theta + 1 + \cos 4\theta}{8}$$

$$= \frac{3-4\cos 2\theta + \cos 4\theta}{8}$$

45. Working from the right-hand side, we have:

$$\frac{2\tan\theta}{1+\tan^2\theta} = \frac{2\bullet\frac{\sin\theta}{\cos\theta}}{1+\frac{\sin^2\theta}{\cos^2\theta}} = \frac{2\sin\theta\cos\theta}{\cos^2\theta+\sin^2\theta} = \frac{\sin 2\theta}{1} = \sin 2\theta$$

47. Working from the right-hand side, we have:

$$2\sin^3\theta\cos\theta + 2\sin\theta\cos^3\theta = (2\sin\theta\cos\theta)(\sin^2\theta+\cos^2\theta) = \sin 2\theta\bullet 1 = \sin 2\theta$$

49. Working from the left-hand side, we have:

$$\frac{1+\tan\frac{\theta}{2}}{1-\tan\frac{\theta}{2}} = \frac{1+\dfrac{\sin\frac{\theta}{2}}{\cos\frac{\theta}{2}}}{1-\dfrac{\sin\frac{\theta}{2}}{\cos\frac{\theta}{2}}}$$

$$= \frac{\cos\frac{\theta}{2}+\sin\frac{\theta}{2}}{\cos\frac{\theta}{2}-\sin\frac{\theta}{2}}$$

$$= \frac{\left(\cos\frac{\theta}{2}+\sin\frac{\theta}{2}\right)^2}{\cos^2\frac{\theta}{2}-\sin^2\frac{\theta}{2}}$$

$$= \frac{1+2\cos\frac{\theta}{2}\sin\frac{\theta}{2}}{\cos\theta}$$

$$= \frac{1+\sin\theta}{\cos\theta}$$

$$= \tan\theta+\sec\theta$$

51. Working from the left-hand side and using the addition formula for $\sin(s-t)$, we have:

$$2\sin^2(45°-\theta) = 2(\sin 45°\cos\theta - \cos 45°\sin\theta)^2$$

$$= 2\left(\frac{\cos\theta-\sin\theta}{\sqrt{2}}\right)^2$$

$$= 2\bullet\frac{\cos^2\theta-2\sin\theta\cos\theta+\sin^2\theta}{2}$$

$$= 1-2\cos\theta\sin\theta$$

$$= 1-\sin 2\theta$$

53. Simplifying the left-hand side, we have:

$$1+\tan\theta\tan 2\theta = 1+\tan\theta\bullet\frac{2\tan\theta}{1-\tan^2\theta}$$

$$= 1+\frac{2\tan^2\theta}{1-\tan^2\theta}$$

$$= \frac{1-\tan^2\theta+2\tan^2\theta}{1-\tan^2\theta}$$

$$= \frac{1+\tan^2\theta}{1-\tan^2\theta}$$

Simplifying the right-hand side, we have:

$$\tan 2\theta \cot \theta - 1 = \frac{2\tan\theta}{1-\tan^2\theta} \bullet \frac{1}{\tan\theta} - 1 = \frac{2}{1-\tan^2\theta} - 1 = \frac{2-1+\tan^2\theta}{1-\tan^2\theta} = \frac{1+\tan^2\theta}{1-\tan^2\theta}$$

Since both sides simplify to the same quantity, the original equation is an identity.

55. Writing $\alpha + 2\beta = (\alpha + \beta) + \beta$, we apply the addition formula for $\sin(s+t)$ to obtain:

$$\sin(\alpha + 2\beta) = \sin\left[(\alpha + \beta) + \beta\right]$$
$$= \sin(\alpha + \beta)\cos\beta + \sin\beta\cos(\alpha + \beta)$$
$$= \left[\sin\alpha\cos\beta + \sin\beta\cos\alpha\right]\cos\beta + \sin\beta \bullet 0$$
$$= \sin\alpha\cos^2\beta + \cos\alpha\sin\beta\cos\beta$$
$$= (\sin\alpha)(1 - \sin^2\beta) + \cos\alpha\sin\beta\cos\beta$$
$$= \sin\alpha + \sin\beta(\cos\alpha\cos\beta - \sin\alpha\sin\beta)$$
$$= \sin\alpha + \sin\beta\cos(\alpha + \beta)$$
$$= \sin\alpha + \sin\beta \bullet 0$$
$$= \sin\alpha$$

57. Since $0 < \theta < \frac{\pi}{2}$, then $\cos\theta > 0$ and thus (since $a > 0$, $b > 0$):

$$\cos\theta = \sqrt{1 - \sin^2\theta}$$
$$= \sqrt{1 - \left(\frac{a^2 - b^2}{a^2 + b^2}\right)^2}$$
$$= \sqrt{\frac{\left(a^2 + b^2\right)^2 - \left(a^2 - b^2\right)^2}{\left(a^2 + b^2\right)^2}}$$
$$= \sqrt{\frac{4a^2b^2}{\left(a^2 + b^2\right)^2}}$$
$$= \frac{2ab}{a^2 + b^2}$$

Now using the half-angle formula for $\tan\frac{\theta}{2}$, we have:

$$\tan\frac{\theta}{2} = \frac{\sin\theta}{1+\cos\theta} = \frac{a^2 - b^2}{a^2 + b^2 + 2ab} = \frac{a^2 - b^2}{(a+b)^2} = \frac{a-b}{a+b}$$

59. (a) Since P is a point on the unit circle located θ radians from the x-axis, then the coordinates of P are $P(\cos\theta, \sin\theta)$. Similarly, since Q is a point on the unit circle located 2θ radians from the x-axis, then the coordinates of Q are $Q(\cos 2\theta, \sin 2\theta)$.

(b) Using the coordinates found in part (a), we compute $(PQ)^2$ by using the distance formula:

$$\begin{aligned}(PQ)^2 &= (\cos 2\theta - \cos \theta)^2 + (\sin 2\theta - \sin \theta)^2\\ &= \cos^2 2\theta - 2\cos 2\theta \cos \theta + \cos^2 \theta + \sin^2 2\theta - 2\sin 2\theta \sin \theta + \sin^2 \theta\\ &= \left(\cos^2 2\theta + \sin^2 2\theta\right) + \left(\cos^2 \theta + \sin^2 \theta\right) - 2\cos 2\theta \cos \theta - 2\sin 2\theta \sin \theta\\ &= 2 - 2\cos \theta \cos 2\theta - 2\sin \theta \sin 2\theta\end{aligned}$$

Since the coordinates of R are $R\,(1,0)$, we compute $(PR)^2$ by using the distance formula:

$$\begin{aligned}(PR)^2 &= (\cos \theta - 1)^2 + (\sin \theta - 0)^2\\ &= \cos^2 \theta - 2\cos \theta + 1 + \sin^2 \theta\\ &= \left(\cos^2 \theta + \sin^2 \theta\right) + 1 - 2\cos \theta\\ &= 2 - 2\cos \theta\end{aligned}$$

(c) Since $OR = OQ$ (they are both radii of the unit circle) and $\angle QOP \approx \angle POR$, then the two triangles are congruent (side-angle-side postulate).

(d) Equating the two expressions, we have:

$$\begin{aligned}2 - 2\cos \theta \cos 2\theta - 2\sin \theta \sin 2\theta &= 2 - 2\cos \theta\\ \cos \theta \cos 2\theta + \sin \theta \sin 2\theta &= \cos \theta\\ \cos \theta \cos 2\theta &= \cos \theta - \sin \theta \sin 2\theta\end{aligned}$$

(e) Squaring and replacing $\cos^2 2\theta$ with $1 - \sin^2 2\theta$, we have:

$$\begin{aligned}\cos^2 \theta \cos^2 2\theta &= \cos^2 \theta - 2\cos \theta \sin \theta \sin 2\theta + \sin^2 \theta \sin^2 2\theta\\ \cos^2 \theta\left(1 - \sin^2 2\theta\right) &= \cos^2 \theta - 2\cos \theta \sin \theta \sin 2\theta + \sin^2 \theta \sin^2 2\theta\\ -\cos^2 \theta \sin^2 2\theta &= -2\cos \theta \sin \theta \sin 2\theta + \sin^2 \theta \sin^2 2\theta\\ 0 &= \left(\cos^2 \theta + \sin^2 \theta\right)\sin^2 2\theta - 2\cos \theta \sin \theta \sin 2\theta\\ 0 &= \sin^2 2\theta - 2\sin \theta \cos \theta \sin 2\theta\end{aligned}$$

(f) Factoring, we have:

$$(\sin 2\theta)(\sin 2\theta - 2\sin \theta \cos \theta) = 0$$

Since $0 < \theta < \frac{\pi}{2}$, then $\sin 2\theta \neq 0$. Thus $\sin 2\theta - 2\sin \theta \cos \theta = 0$, or $\sin 2\theta = 2\sin \theta \cos \theta$.

61. (a) Since $\triangle AOC$ is isosceles, then its base angles are congruent. Thus $\angle ACO = \theta$.

(b) We have $\angle DAO = \angle AOC + \angle ACO = \theta + \theta = 2\theta$.

(c) Since $OD = OA$ (they are both radial lines), then $\triangle DOA$ is isosceles, and thus its base angles are congruent. Thus $\angle ODA = 2\theta$.

(d)   Since $\angle DOE$ is an exterior angle to $\triangle DOC$, then:
$$\angle DOE = \angle ODA + \angle DCO = 2\theta + \theta = 3\theta$$

(e)   For G, we have the following figure:

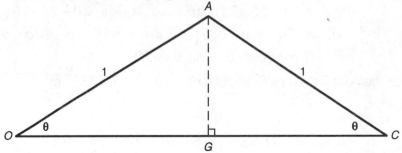

Thus $\cos\theta = \frac{OG}{1} = OG$ and $\cos\theta = \frac{GC}{1} = GC$, and thus:
$$GC = OG = \cos\theta$$
For F, we have the following figure:

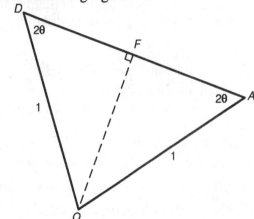

Thus $\cos 2\theta = \frac{DF}{1} = DF$ and $\cos 2\theta = \frac{FA}{1} = FA$, and thus:
$$FA = DF = \cos 2\theta$$

(f)   Using the right triangle CFO, we have:
$$\cos\theta = \frac{CF}{CO} = \frac{1 + \cos 2\theta}{2\cos\theta}$$

(g)   Solving for $\cos 2\theta$:
$$\cos\theta = \frac{1 + \cos 2\theta}{2\cos\theta}$$
$$2\cos^2\theta = 1 + \cos 2\theta$$
$$\cos 2\theta = 2\cos^2\theta - 1$$

(h)   Using the right triangle DFO, we have:
$$\sin 2\theta = \frac{OF}{1} = OF$$

(i) Using the right triangle CFO, we have:
$$\sin\theta = \frac{OF}{OC} = \frac{\sin 2\theta}{2\cos\theta}$$
Solving for $\sin 2\theta$:
$$\sin 2\theta = 2\sin\theta\cos\theta$$

(j) We have $DC = DF + FA + AC = \cos 2\theta + \cos 2\theta + 1 = 1 + 2\cos 2\theta$. Using right triangle DEO, we have:
$$\cos 3\theta = \frac{EO}{1} = EO$$

(k) Using $\triangle CDE$:
$$\cos\theta = \frac{EC}{DC} = \frac{\cos 3\theta + 2\cos\theta}{1 + 2\cos 2\theta}$$

(l) Using the results of parts (k) and (g):
$$\cos\theta = \frac{\cos 3\theta + 2\cos\theta}{1 + 2\cos 2\theta} = \frac{\cos 3\theta + 2\cos\theta}{1 + 4\cos^2\theta - 2} = \frac{\cos 3\theta + 2\cos\theta}{4\cos^2\theta - 1}$$
Cross-multiplying, we have:
$$4\cos^3\theta - \cos\theta = \cos 3\theta + 2\cos\theta$$
$$\cos 3\theta = 4\cos^3\theta - 3\cos\theta$$

(m) Using right triangle DEO:
$$\sin 3\theta = \frac{DE}{DO} = \frac{DE}{1} = DE$$

(n) Using $\triangle CDE$:
$$\sin\theta = \frac{DE}{DC} = \frac{\sin 3\theta}{1 + 2\cos 2\theta}$$
Cross-multiplying, we have:
$$\sin 3\theta = (\sin\theta)(1 + 2\cos 2\theta)$$
$$= (\sin\theta)(1 + 4\cos^2\theta - 2) \quad \text{using the identity from (g)}$$
$$= 4\sin\theta\cos^2\theta - \sin\theta$$
$$= (4\sin\theta)(1 - \sin^2\theta) - \sin\theta$$
$$= 4\sin\theta - 4\sin^3\theta - \sin\theta$$
$$= 3\sin\theta - 4\sin^3\theta$$

63. (a) Following the suggestion and applying the double-angle formulas for $\sin 4\theta$ and $\sin 2\theta$, we have:
$$\frac{\sin 4\theta}{4\sin\theta} = \frac{2\sin 2\theta\cos 2\theta}{4\sin\theta} = \frac{4\sin\theta\cos\theta\cos 2\theta}{4\sin\theta} = \cos\theta\cos 2\theta$$

(b) Following the suggestion and applying the double-angle formulas for $\sin 8\theta$, $\sin 4\theta$ and $\sin 2\theta$, we have:

$$\frac{\sin 8\theta}{8\sin\theta} = \frac{2\sin 4\theta\cos 4\theta}{8\sin\theta}$$
$$= \frac{4\sin 2\theta\cos 2\theta\cos 4\theta}{8\sin\theta}$$
$$= \frac{8\sin\theta\cos\theta\cos 2\theta\cos 4\theta}{8\sin\theta}$$
$$= \cos\theta\cos 2\theta\cos 4\theta$$

(c) Following the suggestion and applying the double-angle formulas for $\sin 16\theta$, $\sin 8\theta$, $\sin 4\theta$ and $\sin 2\theta$, we have:

$$\frac{\sin 16\theta}{16\sin\theta} = \frac{2\sin 8\theta\cos 8\theta}{16\sin\theta}$$
$$= \frac{4\sin 4\theta\cos 4\theta\cos 8\theta}{16\sin\theta}$$
$$= \frac{8\sin 2\theta\cos 2\theta\cos 4\theta\cos 8\theta}{16\sin\theta}$$
$$= \frac{16\sin\theta\cos\theta\cos 2\theta\cos 4\theta\cos 8\theta}{16\sin\theta}$$
$$= \cos\theta\cos 2\theta\cos 4\theta\cos 8\theta$$

65. (a) Using the addition and double-angle formulas for $\cos 3\theta$ and $\cos 2\theta$, we have:

$$\cos 3\theta = \cos(\theta + 2\theta)$$
$$= \cos\theta\cos 2\theta - \sin\theta\sin 2\theta$$
$$= \cos\theta\left(2\cos^2\theta - 1\right) - \sin\theta(2\sin\theta\cos\theta)$$
$$= 2\cos^3\theta - \cos\theta - 2\cos\theta\sin^2\theta$$
$$= 2\cos^3\theta - \cos\theta - 2\cos\theta\left(1 - \cos^2\theta\right)$$
$$= 2\cos^3\theta - \cos\theta - 2\cos\theta + 2\cos^3\theta$$
$$= 4\cos^3\theta - 3\cos\theta$$

(b) (i) Since $36° + 54° = 90°$, we have:
$$\sin 36° = \cos(90° - 36°) = \cos 54°$$

(ii) Since $2\sin\theta\cos\theta = \sin 2\theta$, we have:
$$2\sin 18°\cos 18° = \sin 36° = \cos 54° = \cos 3(18°) = 4\cos^3 18° - 3\cos 18°$$

(iii) Dividing each side of the last equation in (ii) by $\cos 18°$ yields:
$$2\sin 18° = 4\cos^2 18° - 3$$

(c)   Making the substitution $\cos^2 18° = 1 - \sin^2 18°$, we have:
$$2\sin 18° = 4\cos^2 18° - 3 = 4(1 - \sin^2 18°) - 3 = 4 - 4\sin^2 18° - 3$$
$$0 = 4\sin^2 18° + 2\sin 18° - 1$$

This is a quadratic equation in $\sin 18°$. Applying the quadratic formula yields:
$$\sin 18° = \frac{-2 \pm \sqrt{4 - 4(4)(-1)}}{2(4)} = \frac{-2 \pm \sqrt{20}}{8} = \frac{-2 \pm 2\sqrt{5}}{8} = \frac{-1 \pm \sqrt{5}}{4}$$

We choose the positive root here because $\sin 18° > 0$. Thus:
$$\sin 18° = \frac{-1 + \sqrt{5}}{4}$$

(d)   Since $\cos 18° > 0$, we have:
$$\cos 18° = \sqrt{1 - \sin^2 18°}$$
$$= \sqrt{1 - \left(\frac{\sqrt{5} - 1}{4}\right)^2}$$
$$= \sqrt{1 - \frac{6 - 2\sqrt{5}}{16}}$$
$$= \sqrt{\frac{10 + 2\sqrt{5}}{16}}$$
$$= \frac{\sqrt{10 + 2\sqrt{5}}}{4}$$

(e)   We first find $\sin 15°$ and $\cos 15°$:
$$\sin 15° = \sin(45° - 30°)$$
$$= \sin 45° \cos 30° - \sin 30° \cos 45°$$
$$= \frac{\sqrt{2}}{2} \cdot \frac{\sqrt{3}}{2} - \frac{1}{2} \cdot \frac{\sqrt{2}}{2}$$
$$= \frac{\sqrt{6} - \sqrt{2}}{4}$$
$$\cos 15° = \cos(45° - 30°)$$
$$= \cos 45° \cos 30° + \sin 45° \sin 30°$$
$$= \frac{\sqrt{2}}{2} \cdot \frac{\sqrt{3}}{2} + \frac{\sqrt{2}}{2} \cdot \frac{1}{2}$$
$$= \frac{\sqrt{6} + \sqrt{2}}{4}$$

Now using the hint, we have:
$$\sin 3° = \sin(18° - 15°)$$
$$= \sin 18° \cos 15° - \sin 15° \cos 18°$$
$$= \frac{\sqrt{5} - 1}{4} \cdot \frac{\sqrt{6} + \sqrt{2}}{4} - \frac{\sqrt{10 + 2\sqrt{5}}}{4} \cdot \frac{\sqrt{6} - \sqrt{2}}{4}$$
$$= \frac{1}{16}(\sqrt{5} - 1)(\sqrt{6} + \sqrt{2}) - \frac{\sqrt{2}}{16} \cdot \sqrt{5 + \sqrt{5}} \cdot (\sqrt{6} - \sqrt{2})$$
$$= \frac{1}{16}\left[(\sqrt{5} - 1)(\sqrt{6} + \sqrt{2}) - 2\sqrt{5 + \sqrt{5}}(\sqrt{3} - 1)\right]$$

(f)   Using a calculator, we have sin 18° ≈ 0.3090, cos 18° ≈ 0.9511 and sin 3° ≈ 0.0523. These results check the radical values.

4.3 Product-to-Sum and Sum-to-Product Formulas

1.   Using the product-to-sum formula for $\cos A \cos B$, we have:
$$\cos 70° \cos 20° = \tfrac{1}{2}\left[\cos(70°-20°) + \cos(70°+20°)\right]$$
$$= \tfrac{1}{2}\left[\cos 50° + \cos 90°\right]$$
$$= \tfrac{1}{2}\left[\cos 50° + 0\right]$$
$$= \tfrac{1}{2}\cos 50°$$

3.   Using the product-to-sum formula for $\sin A \sin B$, we have:
$$\sin 5° \sin 85° = \tfrac{1}{2}\left[\cos(5°-85°) - \cos(5°+85°)\right]$$
$$= \tfrac{1}{2}\left[\cos(-80°) - \cos 90°\right]$$
$$= \tfrac{1}{2}\left[\cos 80° - 0\right]$$
$$= \tfrac{1}{2}\cos 80°$$

5.   Using the product-to-sum formula for $\sin A \cos B$, we have:
$$\sin 20° \cos 10° = \tfrac{1}{2}\left[\sin(20°-10°) + \sin(20°+10°)\right]$$
$$= \tfrac{1}{2}\left[\sin 10° + \sin 30°\right]$$
$$= \tfrac{1}{2}\left[\sin 10° + \tfrac{1}{2}\right]$$
$$= \tfrac{1}{2}\sin 10° + \tfrac{1}{4}$$

7.   Using the product-to-sum formula for $\cos A \cos B$, we have:
$$\cos\tfrac{\pi}{5}\cos\tfrac{4\pi}{5} = \tfrac{1}{2}\left[\cos\left(\tfrac{\pi}{5}-\tfrac{4\pi}{5}\right) + \cos\left(\tfrac{\pi}{5}+\tfrac{4\pi}{5}\right)\right]$$
$$= \tfrac{1}{2}\left[\cos\left(-\tfrac{3\pi}{5}\right) + \cos\pi\right]$$
$$= \tfrac{1}{2}\left[\cos\tfrac{3\pi}{5} - 1\right]$$
$$= \tfrac{1}{2}\cos\tfrac{3\pi}{5} - \tfrac{1}{2}$$

9.   Using the product-to-sum formula for $\sin A \sin B$, we have:
$$\sin\tfrac{2\pi}{7}\sin\tfrac{5\pi}{7} = \tfrac{1}{2}\left[\cos\left(\tfrac{2\pi}{7}-\tfrac{5\pi}{7}\right) - \cos\left(\tfrac{2\pi}{7}+\tfrac{5\pi}{7}\right)\right]$$
$$= \tfrac{1}{2}\left[\cos\left(-\tfrac{3\pi}{7}\right) - \cos\pi\right]$$
$$= \tfrac{1}{2}\left[\cos\tfrac{3\pi}{7} - (-1)\right]$$
$$= \tfrac{1}{2}\cos\tfrac{3\pi}{7} + \tfrac{1}{2}$$

11. Using the product-to-sum formula for $\sin A \cos B$, we have:

$$\sin \frac{7\pi}{12} \cos \frac{\pi}{12} = \frac{1}{2}\left[\sin\left(\frac{7\pi}{12} - \frac{\pi}{12}\right) + \sin\left(\frac{7\pi}{12} + \frac{\pi}{12}\right)\right]$$
$$= \frac{1}{2}\left[\sin \frac{\pi}{2} + \sin \frac{2\pi}{3}\right]$$
$$= \frac{1}{2}\left[1 + \frac{\sqrt{3}}{2}\right]$$
$$= \frac{1}{2} + \frac{\sqrt{3}}{4}$$

13. Using the product-to-sum formula for $\sin A \sin B$, we have:

$$\sin 3x \sin 4x = \frac{1}{2}\left[\cos(3x - 4x) - \cos(3x + 4x)\right]$$
$$= \frac{1}{2}\left[\cos(-x) - \cos(7x)\right]$$
$$= \frac{1}{2}\left[\cos x - \cos 7x\right]$$
$$= \frac{1}{2}\cos x - \frac{1}{2}\cos 7x$$

15. Using the product-to-sum formula for $\sin A \cos B$, we have:

$$\sin 6\theta \cos 5\theta = \frac{1}{2}\left[\sin(6\theta - 5\theta) + \sin(6\theta + 5\theta)\right]$$
$$= \frac{1}{2}\left[\sin \theta + \sin 11\theta\right]$$
$$= \frac{1}{2}\sin \theta + \frac{1}{2}\sin 11\theta$$

17. Using the product-to-sum formula for $\sin A \cos B$, we have:

$$\sin \frac{3\theta}{2} \cos \frac{\theta}{2} = \frac{1}{2}\left[\sin\left(\frac{3\theta}{2} - \frac{\theta}{2}\right) + \sin\left(\frac{3\theta}{2} + \frac{\theta}{2}\right)\right]$$
$$= \frac{1}{2}\left[\sin \theta + \sin 2\theta\right]$$
$$= \frac{1}{2}\sin \theta + \frac{1}{2}\sin 2\theta$$

19. Using the product-to-sum formula for $\sin A \sin B$, we have:

$$\sin(2x + y)\sin(2x - y) = \frac{1}{2}\left[\cos(2x + y - 2x + y) - \cos(2x + y + 2x - y)\right]$$
$$= \frac{1}{2}\left[\cos 2y - \cos 4x\right]$$
$$= \frac{1}{2}\cos 2y - \frac{1}{2}\cos 4x$$

21. Using the product-to-sum formula for $\sin A \cos B$, we have:

$$\sin 2t \cos(s - t) = \frac{1}{2}\left[\sin(2t - s + t) + \sin(2t + s - t)\right]$$
$$= \frac{1}{2}\left[\sin(3t - s) + \sin(t + s)\right]$$
$$= \frac{1}{2}\sin(3t - s) + \frac{1}{2}\sin(t + s)$$

23. Using the sum-to-product formula for $\cos\alpha + \cos\beta$, we have:

$$\cos 35° + \cos 55° = 2\cos\frac{35°+55°}{2}\cos\frac{35°-55°}{2}$$
$$= 2\cos 45°\cos(-10°)$$
$$= 2\cdot\frac{\sqrt{2}}{2}\cdot\cos 10°$$
$$= \sqrt{2}\cos 10°$$

25. Using the sum-to-product formula for $\sin\alpha - \sin\beta$, we have:

$$\sin\frac{\pi}{5} - \sin\frac{3\pi}{10} = 2\cos\frac{\frac{\pi}{5}+\frac{3\pi}{10}}{2}\sin\frac{\frac{\pi}{5}-\frac{3\pi}{10}}{2}$$
$$= 2\cos\frac{\pi}{4}\sin\left(-\frac{\pi}{20}\right)$$
$$= -2\cdot\frac{\sqrt{2}}{2}\sin\frac{\pi}{20}$$
$$= -\sqrt{2}\sin\frac{\pi}{20}$$

27. Using the sum-to-product formula for $\cos\alpha - \cos\beta$, we have:

$$\cos 5\theta - \cos 3\theta = -2\sin\frac{5\theta+3\theta}{2}\sin\frac{5\theta-3\theta}{2} = -2\sin 4\theta\sin\theta$$

29. Using the hint, we have $\cos 65° = \sin(90° - 65°) = \sin 25°$. Now using the sum-to-product formula for $\sin\alpha + \sin\beta$, we have:

$$\sin 35° + \sin 25° = 2\sin\frac{35°+25°}{2}\cos\frac{35°-25°}{2} = 2\sin 30°\cos 5° = 2\cdot\frac{1}{2}\cos 5° = \cos 5°$$

31. Using the sum-to-product formula for $\sin\alpha - \sin\beta$, we have:

$$\sin\left(\frac{\pi}{3}+2\theta\right) - \sin\left(\frac{\pi}{3}-2\theta\right) = 2\cos\frac{2\pi/3}{2}\sin\frac{4\theta}{2} = 2\cos\frac{\pi}{3}\sin 2\theta = 2\cdot\frac{1}{2}\sin 2\theta = \sin 2\theta$$

33. We will simplify the numerator and denominator separately. For the numerator, we first use the identity $\cos\theta = \sin\left(\frac{\pi}{2}-\theta\right)$ so that $\cos\frac{5\pi}{12} = \sin\frac{\pi}{12}$. Now using the sum-to-product formula for $\sin\alpha + \sin\beta$, we have:

$$\sin\frac{\pi}{12} + \sin\frac{5\pi}{12} = 2\sin\frac{\frac{\pi}{12}+\frac{5\pi}{12}}{2}\cos\frac{\frac{\pi}{12}-\frac{5\pi}{12}}{2} = 2\sin\frac{\pi}{4}\cos\left(-\frac{\pi}{6}\right) = 2\cdot\frac{\sqrt{2}}{2}\cdot\frac{\sqrt{3}}{2} = \frac{\sqrt{6}}{2}$$

For the denominator, we first use the identity $\cos\theta = \sin\left(\frac{\pi}{2}-\theta\right)$ so that $\cos\frac{\pi}{12} = \sin\frac{5\pi}{12}$. Now using the sum-to-product formula for $\sin\alpha - \sin\beta$, we have:

$$\sin\frac{5\pi}{12} - \sin\frac{\pi}{12} = 2\cos\frac{\frac{5\pi}{12}+\frac{\pi}{12}}{2}\sin\frac{\frac{5\pi}{12}-\frac{\pi}{12}}{2} = 2\cos\frac{\pi}{4}\sin\frac{\pi}{6} = 2\cdot\frac{\sqrt{2}}{2}\cdot\frac{1}{2} = \frac{\sqrt{2}}{2}$$

Thus, the original problem becomes:

$$\frac{\cos\frac{5\pi}{12} + \sin\frac{5\pi}{12}}{\cos\frac{\pi}{12} - \sin\frac{\pi}{12}} = \frac{\frac{\sqrt{6}}{2}}{\frac{\sqrt{2}}{2}} = \frac{\sqrt{6}}{\sqrt{2}} = \sqrt{3}$$

35. Using the sum-to-product formulas for $\sin s + \sin t$ and $\cos s + \cos t$, we have:

$$\frac{\sin s + \sin t}{\cos s + \cos t} = \frac{2\sin\frac{s+t}{2}\cos\frac{s-t}{2}}{2\cos\frac{s+t}{2}\cos\frac{s-t}{2}} = \frac{\sin\frac{s+t}{2}}{\cos\frac{s+t}{2}} = \tan\frac{s+t}{2}$$

37. Using the sum-to-product formulas for $\sin 2x + \sin 2y$ and $\cos 2x + \cos 2y$, we have:

$$\frac{\sin 2x + \sin 2y}{\cos 2x + \cos 2y} = \frac{2\sin\frac{2x+2y}{2}\cos\frac{2x-2y}{2}}{2\cos\frac{2x+2y}{2}\cos\frac{2x-2y}{2}} = \frac{\sin(x+y)}{\cos(x+y)} = \tan(x+y)$$

Note that we can also use the result from Exercise 35 where $s = 2x$ and $t = 2y$.

39. Using the hint, we use the sum-to-product formulas to obtain:

$$\cos 7\theta + \cos 5\theta = 2\cos\frac{7\theta+5\theta}{2}\cos\frac{7\theta-5\theta}{2} = 2\cos 6\theta\cos\theta$$

$$\cos 3\theta + \cos\theta = 2\cos\frac{3\theta+\theta}{2}\cos\frac{3\theta-\theta}{2} = 2\cos 2\theta\cos\theta$$

Thus, we have:

$$\begin{aligned}
\cos 7\theta + \cos 5\theta + \cos 3\theta + \cos\theta &= 2\cos 6\theta\cos\theta + 2\cos 2\theta\cos\theta \\
&= (2\cos\theta)(\cos 6\theta + \cos 2\theta) \\
&= (2\cos\theta)\left(2\cos\frac{6\theta+2\theta}{2}\cos\frac{6\theta-2\theta}{2}\right) \\
&= 4\cos\theta\cos 4\theta\cos 2\theta
\end{aligned}$$

41. (a) Since $\cos\frac{x}{2} = \sin\left(\frac{\pi}{2} - \frac{x}{2}\right)$, we can use the sum-to-product identity for $\sin\alpha + \sin\beta$ to obtain:

$$\begin{aligned}
\sqrt{2}\left[\sin\frac{x}{2} + \cos\frac{x}{2}\right] &= \sqrt{2}\left[\sin\frac{x}{2} + \sin\left(\frac{\pi}{2} - \frac{x}{2}\right)\right] \\
&= \sqrt{2}\left[2\sin\frac{\frac{\pi}{2}}{2}\cos\frac{x - \frac{\pi}{2}}{2}\right] \\
&= \sqrt{2}\left[2\sin\frac{\pi}{4}\cos\left(\frac{x}{2} - \frac{\pi}{4}\right)\right] \\
&= \sqrt{2}\cdot 2\cdot\frac{\sqrt{2}}{2}\cos\left(\frac{x}{2} - \frac{\pi}{4}\right) \\
&= 2\cos\left(\frac{x}{2} - \frac{\pi}{4}\right)
\end{aligned}$$

(b)   Since $f(x) = 2\cos\left(\frac{x}{2} - \frac{\pi}{4}\right)$, we have an amplitude of 2, a period of $\frac{2\pi}{1/2} = 4\pi$, and a

phase shift of $\frac{\pi/4}{1/2} = \frac{\pi}{2}$. We sketch the graph:

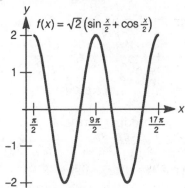

$f(x) = \sqrt{2}\left(\sin\frac{x}{2} + \cos\frac{x}{2}\right)$

43.   We begin with the product-to-sum formula:
$$\cos A \cos B = \tfrac{1}{2}\left[\cos(A - B) + \cos(A + B)\right]$$
If we let $A + B = \alpha$ and $A - B = \beta$, then:
$$A = \tfrac{\alpha+\beta}{2} \text{ and } B = \tfrac{\alpha-\beta}{2}$$
Substituting, we have:
$$\cos\tfrac{\alpha+\beta}{2}\cos\tfrac{\alpha-\beta}{2} = \tfrac{1}{2}\left[\cos\beta + \cos\alpha\right]$$
Multiplying by 2, we have the desired identity:
$$\cos\alpha + \cos\beta = 2\cos\tfrac{\alpha+\beta}{2}\cos\tfrac{\alpha-\beta}{2}$$

45.   Using the identity for $\sin\alpha + \sin\beta$ (derived in Exercise 44), we replace β with $-\beta$ and
note that $\sin(-\beta) = -\sin\beta$:
$$\sin\alpha + \sin(-\beta) = 2\sin\tfrac{\alpha-\beta}{2}\cos\tfrac{\alpha+\beta}{2}$$
$$\sin\alpha - \sin\beta = 2\cos\tfrac{\alpha+\beta}{2}\sin\tfrac{\alpha-\beta}{2}$$

47.   Obtaining a common denominator in the expression, and noting that
$\sin 10° = \cos(90° - 10°) = \cos 80°$, we have:
$$\frac{1}{2\sin 10°} - 2\sin 70° = \frac{1 - 4\sin 10°\sin 70°}{2\sin 10°} = \frac{1 - 4\sin 10°\sin 70°}{2\cos 80°}$$

Now using the product-to-sum formula for $\sin A \sin B$, we have:
$$\sin 10°\sin 70° = \tfrac{1}{2}\left[\cos(10° - 70°) - \cos(10° + 70°)\right]$$
$$= \tfrac{1}{2}\left[\cos(-60°) - \cos 80°\right]$$
$$= \tfrac{1}{2}\left[\tfrac{1}{2} - \cos 80°\right]$$
$$= \tfrac{1}{4} - \tfrac{1}{2}\cos 80°$$

Thus, we have:

$$\frac{1}{2\sin 10°} - 2\sin 70° = \frac{1-4\sin 10°\sin 70°}{2\cos 80°}$$

$$= \frac{1-4\left(\frac{1}{4}-\frac{1}{2}\cos 80°\right)}{2\cos 80°}$$

$$= \frac{1-1+2\cos 80°}{2\cos 80°}$$

$$= \frac{2\cos 80°}{2\cos 80°}$$

$$= 1$$

49. First, note that $A + B + C = 180°$ implies that $A + B = 180° - C$, $B + C = 180° - A$ and $A + C = 180° - B$. Also, note the identities $\sin(180° - \alpha) = \sin\alpha$ and $\cos(180° - \alpha) = -\cos\alpha$. Working from the right-hand side, note that:

$$\sin\tfrac{A}{2}\sin\tfrac{B}{2} = \tfrac{1}{2}\cos\left(\tfrac{A}{2}-\tfrac{B}{2}\right)-\tfrac{1}{2}\cos\left(\tfrac{A}{2}+\tfrac{B}{2}\right)$$

Thus:

$$\sin\tfrac{A}{2}\sin\tfrac{B}{2}\sin\tfrac{C}{2}$$

$$= \tfrac{1}{2}\cos\left(\tfrac{A}{2}-\tfrac{B}{2}\right)\sin\tfrac{C}{2}-\tfrac{1}{2}\cos\left(\tfrac{A}{2}+\tfrac{B}{2}\right)\sin\tfrac{C}{2}$$

$$= \tfrac{1}{2}\sin\left[90°-\tfrac{A}{2}+\tfrac{B}{2}\right]\sin\tfrac{C}{2}-\tfrac{1}{2}\sin\left[90°-\tfrac{A}{2}-\tfrac{B}{2}\right]\sin\tfrac{C}{2}$$

$$= \tfrac{1}{4}\left[\cos\left(90°-\tfrac{A}{2}+\tfrac{B}{2}-\tfrac{C}{2}\right)-\cos\left(90°-\tfrac{A}{2}+\tfrac{B}{2}+\tfrac{C}{2}\right)\right]$$

$$\quad -\tfrac{1}{4}\left[\cos\left(90°-\tfrac{A}{2}-\tfrac{B}{2}-\tfrac{C}{2}\right)-\cos\left(90°-\tfrac{A}{2}-\tfrac{B}{2}+\tfrac{C}{2}\right)\right]$$

$$= \tfrac{1}{4}\left[\cos\tfrac{180°+B-(A+C)}{2}-\cos\tfrac{180°-A+(B+C)}{2}-\cos\tfrac{180°-(A+B+C)}{2}+\cos\tfrac{180°+C-(A+B)}{2}\right]$$

$$= \tfrac{1}{4}\left[\cos\tfrac{180°+B-(180°-B)}{2}-\cos\tfrac{180°-A+(180°-A)}{2}-\cos\tfrac{180°-180°}{2}+\cos\tfrac{180°+C-(180°-C)}{2}\right]$$

$$= \tfrac{1}{4}\left[\cos B - \cos(180°-A)-\cos 0+\cos C\right]$$

$$= \tfrac{1}{4}\left[\cos B + \cos A - 1 + \cos C\right]$$

Now we prove the identity:

$$1+4\sin\tfrac{A}{2}\sin\tfrac{B}{2}\sin\tfrac{C}{2} = 1+\cos B+\cos A-1+\cos C = \cos A+\cos B+\cos C$$

4.4 Trigonometric Equations

1. For $\theta = \frac{\pi}{2}$, $2\cos^2\theta - 3\cos\theta = 2(0)^2 - 3(0) = 0$, so $\theta = \frac{\pi}{2}$ is a solution.

3. For $x = \frac{3\pi}{4}$, $\tan^2 x - 3\tan x + 2 = (-1)^2 + 3 + 2 = 6$, so $x = \frac{3\pi}{4}$ is not a solution.

5.   Since $\sin\theta=\frac{\sqrt{3}}{2}$, then $\theta=\frac{\pi}{3}$ and $\theta=\frac{2\pi}{3}$ are the primary solutions. All solutions will be of the form $\theta=\frac{\pi}{3}+2\pi k$ or $\theta=\frac{2\pi}{3}+2\pi k$, where k is any integer.

7.   Since $\sin\theta=-\frac{1}{2}$, then $\theta=\frac{7\pi}{6}$ and $\theta=\frac{11\pi}{6}$ are the primary solutions. All solutions will be of the form $\theta=\frac{7\pi}{6}+2\pi k$ or $\theta=\frac{11\pi}{6}+2\pi k$, where k is any integer.

9.   Since $\cos\theta=-1$, then $\theta=\pi$ is the primary solution. All solutions will be of the form $\theta=\pi+2\pi k$, where k is any integer.

11. Since $\tan\theta=\sqrt{3}$, then $\theta=\frac{\pi}{3}$ is the primary solution. All solutions will be of the form $\theta=\frac{\pi}{3}+\pi k$, where k is any integer.

13. Since $\tan x=0$, then $x=0$ is the primary solution. All solutions will be of the form $x=0+\pi k$, or $x=\pi k$, where k is any integer.

15. Since $2\cos^2\theta+\cos\theta=\cos\theta(2\cos\theta+1)=0$, then the primary solutions are the solutions of $\cos\theta=0$ or $\cos\theta=-\frac{1}{2}$, which are $\theta=\frac{\pi}{2}$, $\theta=\frac{3\pi}{2}$, $\theta=\frac{2\pi}{3}$ or $\theta=\frac{4\pi}{3}$. All solutions will be of the form $\theta=\frac{\pi}{2}+\pi k$, $\theta=\frac{2\pi}{3}+2\pi k$ or $\theta=\frac{4\pi}{3}+2\pi k$, where k is any integer.

17. Since $\cos^2 t\sin t-\sin t=\sin t(\cos^2 t-1)=0$, then $\sin t=0$ or $\cos t=\pm1$. Therefore primary solutions are $t=0$ or $t=\pi$. All solutions are of the form $t=\pi k$, where k is any integer.

19. Using the identity $\cos^2 x=1-\sin^2 x$, we have:
$$2\cos^2 x-\sin x-1=2\left(1-\sin^2 x\right)-\sin x-1$$
$$=-2\sin^2 x-\sin x+1$$
$$=(-2\sin x+1)(\sin x+1)$$

So $\sin x=\frac{1}{2}$ or $\sin x=-1$. Thus the primary solutions are $x=\frac{\pi}{6}$, $x=\frac{5\pi}{6}$ or $x=\frac{3\pi}{2}$. All solutions are of the form $x=\frac{\pi}{6}+2\pi k$, $x=\frac{5\pi}{6}+2\pi k$ or $x=\frac{3\pi}{2}+2\pi k$, where k is any integer.

21. Since $\sqrt{3}\sin t - \sqrt{1+\sin^2 t} = 0$ is equivalent to $3\sin^2 t = 1 + \sin^2 t$ by squaring each side, then $2\sin^2 t = 1$ and thus $\sin^2 t = \frac{1}{2}$ and $\sin t = \pm\frac{\sqrt{2}}{2}$. This would have primary solutions of $\frac{\pi}{4}$, $\frac{3\pi}{4}$, $\frac{5\pi}{4}$ or $\frac{7\pi}{4}$, but $\frac{5\pi}{4}$ and $\frac{7\pi}{4}$ do not work in the original equation. So the primary solutions are $t = \frac{\pi}{4}$ or $t = \frac{3\pi}{4}$. All solutions are of the form $t = \frac{\pi}{4} + 2\pi k$ or $t = \frac{3\pi}{4} + 2\pi k$, where k is any integer.

23. Since $\cos 3\theta = 1$, then $3\theta = 2\pi k$ for any integer k. So $\theta = \frac{2\pi k}{3}$. Thus the values of θ in the interval $[0°, 360°)$ are $0°$, $120°$ and $240°$.

25. Since $\sin 3\theta = -\frac{\sqrt{2}}{2}$, then $3\theta = \frac{5\pi}{4} + 2\pi k$ or $\frac{7\pi}{4} + 2\pi k$. Thus $\theta = \frac{5\pi}{12} + \frac{2\pi k}{3}$ or $\frac{7\pi}{12} + \frac{2\pi k}{3}$. So the primary solutions are $75°$, $105°$, $195°$, $225°$, $315°$ or $345°$.

27. Using the hint, we have:
$$\sin\theta = \cos\frac{\theta}{2}$$
$$2\sin\frac{\theta}{2}\cos\frac{\theta}{2} = \cos\frac{\theta}{2}$$
$$\left(\cos\frac{\theta}{2}\right)\left(2\sin\frac{\theta}{2} - 1\right) = 0$$

Thus $\cos\frac{\theta}{2} = 0$ or $\sin\frac{\theta}{2} = \frac{1}{2}$. Now $\sin\frac{\theta}{2} = \frac{1}{2}$ when $\frac{\theta}{2} = 30°$ or $\frac{\theta}{2} = 150°$, and therefore when $\theta = 60°$ or $300°$. When $\cos\frac{\theta}{2} = 0$ we have $\frac{\theta}{2} = 90° + 360°k$ or $270° + 360°k$, so $\theta = 180°$. Combining we have $\theta = 60°$, $180°$ or $300°$.

29. Dividing each side by $\cos 2\theta$ results in $\tan 2\theta = \sqrt{3}$, thus $2\theta = 60° + 180°k$ and so $\theta = 30° + 90°k$. So the solutions in the interval $[0°, 360°)$ are $\theta = 30°$, $120°$, $210°$ or $300°$.

31. Since $\sin\theta = \frac{1}{4}$, then $\theta \approx 14.5°$ or $165.5°$.

33. Since $2\tan\theta = -4$, then $\tan\theta = -2$ and thus $\theta \approx 116.6° + 180°k$. So the solutions in the interval $[0°, 360°)$ are $\theta \approx 116.6°$ or $296.6°$.

35. Using the quadratic formula, we have:
$$\cos x = \frac{1 \pm \sqrt{(-1)^2 - 4(-1)}}{2(1)} = \frac{1 \pm \sqrt{1+4}}{2} = \frac{1 \pm \sqrt{5}}{2}$$

Since $\frac{1+\sqrt{5}}{2} > 1$, the only solutions are those for which $\cos x = \frac{1-\sqrt{5}}{2} \approx -0.62$, and thus $x \approx 128.2°$ or $231.8°$.

37. Since $\cos x = 0.184$, then $x \approx 1.39$ or $x \approx 4.90$.

39. Since $\sin x = \frac{1}{\sqrt{5}}$, then $x \approx 0.46$ or $x \approx 2.68$.

41. Since $\tan x = 6$, then $x \approx 1.41$ or $x \approx 4.55$.

43. Dividing through by $\cos t$ results in $\tan t = 5$, thus $t \approx 1.37$ or $t \approx 4.51$.

45. Since $\sec t = 2.24$, then $\cos t = \frac{1}{2.24} \approx 0.45$. Thus $t \approx 1.11$ or $t \approx 5.18$.

47. We use the sum-to-product formula for $\sin \alpha - \sin \beta$:
$$\sin 5x = \sin 3x$$
$$\sin 5x - \sin 3x = 0$$
$$2 \cos \tfrac{5x+3x}{2} \sin \tfrac{5x-3x}{2} = 0$$
$$2 \cos 4x \sin x = 0$$
$$\cos 4x \sin x = 0$$
$$\cos 4x = 0 \text{ or } \sin x = 0$$

So $4x = \frac{\pi}{2}, \frac{3\pi}{2}, \frac{5\pi}{2}, \frac{7\pi}{2}, \frac{9\pi}{2}, \frac{11\pi}{2}, \frac{13\pi}{2}, \frac{15\pi}{2}$, or $x = 0, \pi, 2\pi$. So the solutions are

$x = \frac{\pi}{8}, \frac{3\pi}{8}, \frac{5\pi}{8}, \frac{7\pi}{8}, \frac{9\pi}{8}, \frac{11\pi}{8}, \frac{13\pi}{8}, \frac{15\pi}{8}, 0, \pi, 2\pi$.

49. We have $\sin 3x = \cos 2x = \sin\left(\frac{\pi}{2} - 2x\right)$, so:
$$\sin 3x - \sin\left(\tfrac{\pi}{2} - 2x\right) = 2 \cos\left(\tfrac{x}{2} + \tfrac{\pi}{4}\right) \sin\left(\tfrac{5x}{2} - \tfrac{\pi}{4}\right) = 0$$

Hence $\frac{x}{2} + \frac{\pi}{4} = \frac{\pi}{2}, \frac{3\pi}{2}$, or $\frac{5x}{2} - \frac{\pi}{4} = 0, \pi, 2\pi, 3\pi, 4\pi, 5\pi$. So $\frac{x}{2} = \frac{\pi}{4}, \frac{5\pi}{4}$ and thus

$x = \frac{\pi}{2}, \frac{5\pi}{2}$, or $\frac{5x}{2} = \frac{\pi}{4}, \frac{5\pi}{4}, \frac{9\pi}{4}, \frac{13\pi}{4}, \frac{17\pi}{4}, \frac{21\pi}{4}$ and thus $x = \frac{\pi}{10}, \frac{\pi}{2}, \frac{9\pi}{10}, \frac{13\pi}{10}, \frac{17\pi}{10}, \frac{21\pi}{10}$. Now

$\frac{5\pi}{2}$ and $\frac{21\pi}{10}$ are not in the interval, so the solutions are $x = \frac{\pi}{10}, \frac{\pi}{2}, \frac{9\pi}{10}, \frac{13\pi}{10}$ or $\frac{17\pi}{10}$.

51. Using the sum-to-product formula for $\cos \alpha + \cos \beta$, we have:
$$\cos x + \cos 7x = 2 \cos \tfrac{x+7x}{2} \cos \tfrac{x-7x}{2} = 2 \cos 4x \cos(-3x) = 2 \cos 4x \cos 3x$$
Therefore the equation becomes:
$$2 \cos 4x \cos 3x = \cos 4x$$
$$2 \cos 4x \cos 3x - \cos 4x = 0$$
$$\cos 4x (2 \cos 3x - 1) = 0$$
$$\cos 4x = 0 \text{ or } \cos 3x = \tfrac{1}{2}$$

For the first factor we must have $4x = \frac{\pi}{2}$ or $\frac{3\pi}{2}$, so $x = \frac{\pi}{8}$ or $\frac{3\pi}{8}$. For the second factor we

must have $3x = \frac{\pi}{3}$, so $x = \frac{\pi}{9}$.

53. Using the product-to-sum formula for $\sin A \cos B$, we have:

$$\sin 5x \cos 3x = \tfrac{1}{2}\big[\sin(5x-3x)+\sin(5x+3x)\big] = \tfrac{1}{2}\sin 2x + \tfrac{1}{2}\sin 8x$$

$$\sin 9x \cos 7x = \tfrac{1}{2}\big[\sin(9x-7x)+\sin(9x+7x)\big] = \tfrac{1}{2}\sin 2x + \tfrac{1}{2}\sin 16x$$

Now since $\sin 16x = 2\sin 8x \cos 8x$, we have:

$$\sin 5x \cos 3x = \sin 9x \cos 7x$$
$$\tfrac{1}{2}\sin 2x + \tfrac{1}{2}\sin 8x = \tfrac{1}{2}\sin 2x + \tfrac{1}{2}\sin 16x$$
$$\sin 8x = \sin 16x$$
$$\sin 8x = 2\sin 8x \cos 8x$$
$$0 = \sin 8x(2\cos 8x - 1)$$
$$\sin 8x = 0 \text{ or } \cos 8x = \tfrac{1}{2}$$

For the first factor we must have $8x = 0,\ \pi,\ 2\pi,\ 3\pi$ or 4π, so $x = 0,\ \frac{\pi}{8},\ \frac{\pi}{4},\ \frac{3\pi}{8}$ or $\frac{\pi}{2}$. For

the second factor we must have $8x = \frac{\pi}{3},\ \frac{5\pi}{3},\ \frac{7\pi}{3}$ or $\frac{11\pi}{3}$, so $x = \frac{\pi}{24},\ \frac{5\pi}{24},\ \frac{7\pi}{24}$ or $\frac{11\pi}{24}$.

55. Using the identity $\sin\theta = \cos\!\left(\frac{\pi}{2}-\theta\right)$, we have:

$$\sin\!\left(x+\tfrac{\pi}{18}\right) = \cos\!\left(\tfrac{\pi}{2}-x-\tfrac{\pi}{18}\right) = \cos\!\left(\tfrac{4\pi}{9}-x\right)$$

Therefore the original equation becomes:

$$\cos\!\left(\tfrac{4\pi}{9}-x\right) = \cos\!\left(x-\tfrac{2\pi}{9}\right),\ \text{or}\ \cos\!\left(\tfrac{4\pi}{9}-x\right)-\cos\!\left(x-\tfrac{2\pi}{9}\right) = 0$$

Using the sum-to-product formula for $\cos\alpha - \cos\beta$, we have:

$$\cos\!\left(\tfrac{4\pi}{9}-x\right)-\cos\!\left(x-\tfrac{2\pi}{9}\right) = -2\sin\!\left(\tfrac{2\pi/9}{2}\right)\sin\!\left(\tfrac{2\pi/3-2x}{2}\right)$$
$$= -2\sin\tfrac{\pi}{9}\sin\!\left(\tfrac{\pi}{3}-x\right)$$
$$= 2\sin\tfrac{\pi}{9}\sin\!\left(x-\tfrac{\pi}{3}\right)$$

Since $2\sin\frac{\pi}{9} \neq 0$, we must have $\sin\!\left(x-\frac{\pi}{3}\right) = 0$ and thus $x-\frac{\pi}{3} = 0,\ \pi$, therefore $x = \frac{\pi}{3}$ or $\frac{4\pi}{3}$.

57. Following the hint, we use the addition formula and double-angle formula for tangent to obtain:

$$\tan 3x = \tan(2x+x) = \frac{\tan 2x + \tan x}{1-\tan x \tan 2x} = \frac{\frac{2\tan x}{1-\tan^2 x}+\tan x}{1-\tan x\frac{2\tan x}{1-\tan^2 x}} = \frac{3\tan x - \tan^3 x}{1-3\tan^2 x}$$

So the left-hand side of the equation becomes:

$$\tan 3x - \tan x = \frac{3\tan x - \tan^3 x}{1 - 3\tan^2 x} - \tan x$$

$$= \frac{3\tan x - \tan^3 x - \tan x + 3\tan^3 x}{1 - 3\tan^2 x}$$

$$= \frac{2\tan x + 2\tan^3 x}{1 - 3\tan^2 x}$$

$$= \frac{2\tan x\left(1 + \tan^2 x\right)}{1 - 3\tan^2 x}$$

Since $1 + \tan^2 x \neq 0$, we must have $\tan x = 0$ and thus $x = 0$ or $x = \pi$.

59. By the half-angle formula for cosine, we have:

$$\cos\tfrac{x}{2} = \pm\sqrt{\frac{1 + \cos x}{2}}$$

Writing the original equation, then squaring, we have:

$$\pm\sqrt{\frac{1 + \cos x}{2}} = 1 + \cos x$$

$$\frac{1 + \cos x}{2} = 1 + 2\cos x + \cos^2 x$$

$$1 + \cos x = 2 + 4\cos x + 2\cos^2 x$$

$$0 = 2\cos^2 x + 3\cos x + 1$$

$$0 = (2\cos x + 1)(\cos x + 1)$$

$$\cos x = -\tfrac{1}{2} \text{ or } \cos x = -1$$

These equations have solutions of $x = \frac{2\pi}{3}, \frac{4\pi}{3}$ or π. Upon checking we find that $x = \frac{4\pi}{3}$ is

not a solution, and thus the solutions are $x = \frac{2\pi}{3}$ or $x = \pi$.

61. Using the identity for $\sec 4\theta$ and the double-angle identity for sine, we have:

$$\sec 4\theta + 2\sin 4\theta = 0$$

$$\frac{1}{\cos 4\theta} + 2\sin 4\theta = 0$$

$$1 + 2\sin 4\theta \cos 4\theta = 0$$

$$\sin 8\theta = -1$$

Thus $8\theta = \frac{3\pi}{2} + 2\pi k$, and so $\theta = \frac{3\pi}{16} + \frac{k\pi}{4}$, where k is any integer.

63. Following the hint, we have:

$$4\sin\theta - 3\cos\theta = 2$$
$$4\sin\theta = 3\cos\theta + 2$$
$$16\sin^2\theta = 9\cos^2\theta + 12\cos\theta + 4$$
$$16\left(1 - \cos^2\theta\right) = 9\cos^2\theta + 12\cos\theta + 4$$
$$16 - 16\cos^2\theta = 9\cos^2\theta + 12\cos\theta + 4$$
$$25\cos^2\theta + 12\cos\theta - 12 = 0$$

This will not factor, so we use the quadratic formula:

$$\cos\theta = \frac{-12 \pm \sqrt{(12)^2 - 4(25)(-12)}}{2(25)} = \frac{-12 \pm \sqrt{1344}}{50} = \frac{-12 \pm 8\sqrt{21}}{50} = \frac{-6 \pm 4\sqrt{21}}{25}$$

So $\cos\theta = 0.4932$ or $\cos\theta = -0.9732$, and thus $\theta = 60.45°$ (the other solution is not in the required interval).

65. (a) Squaring each side, we have:

$$\sin^2 x\cos^2 x = 1$$
$$\sin^2 x\left(1 - \sin^2 x\right) = 1$$
$$\sin^2 x - \sin^4 x = 1$$
$$\sin^4 x - \sin^2 x + 1 = 0$$

(b) Using the quadratic formula:

$$\sin^2 x = \frac{1 \pm \sqrt{(-1)^2 - 4(1)(1)}}{2(1)} = \frac{1 \pm \sqrt{1 - 4}}{2} = \frac{1 \pm \sqrt{-3}}{2}$$

Thus the original equation has no real-number solutions.

67. (a) Since $\cos x = 0.412$, then $x \approx 1.146$ or $x \approx -1.146$. We must now add multiples of 2π on to these values until we reach an x-value greater than 1000. Note that $\frac{1000}{2\pi} \approx 159$ so we check $x = 1.146 + 159(2\pi)$ as a starting point. The first such value is at $x \approx 1000.173$.

(b) Since $\cos x = -0.412$, then $x \approx 1.995$ or $x \approx -1.995$. Again, we add multiples of 2π on to these values until we reach an x-value greater than 1000. See the note from part (a). The first such value is at $x \approx 1001.022$.

Section 4.4 TI-81 Graphing Calculator Exercises

1. (a) The graphs of $Y_1 = \cos x$ and $Y_2 = 0.351$ with the indicated settings appear as:

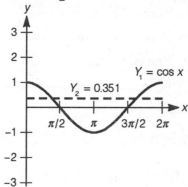

   (b) The graphs of $Y_1 = \cos x$ and $Y_2 = 0.351$ with the indicated settings appear as:

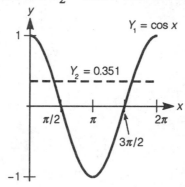

   (c) The approximation $x \approx 1.19$ is confirmed.

   (d) The approximation $x \approx 1.215$ is confirmed.

   (e) The approximation $x \approx 1.212$ is confirmed.

Note: For Exercises 3-9 be sure to set your calculator to the radian mode and use the settings from Exercise 1.

3. Using the trace and zoom technique, the two solutions are $x \approx 0.90$ and $x \approx 5.38$.

5. Using the trace and zoom technique, the two solutions are $x \approx 0.43$ and $x \approx 2.71$.

7. (a) Using the trace and zoom technique, there are no solutions to this equation.

   (b) Using the trace and zoom technique, the four solutions are $x \approx 1.93$, $x \approx 2.28$, $x \approx 5.07$ and $x \approx 5.42$.

9. Using the trace and zoom technique, the three solutions are $x = 0$, $x \approx 0.69$ and $x \approx 4.26$.

Note: For Exercises 11-15 be sure to set your calculator to the degree mode and use the settings from Exercise 2.

11. Using the trace and zoom technique, the two solutions are $x \approx 67.5°$ and $x \approx 247.5°$.

13. Using the trace and zoom technique, the two solutions are $x \approx 120.8°$ and $x \approx 329.2°$.

15. Using the trace and zoom technique, the three solutions are $x = 0°$, $x \approx 1.0°$ and $x \approx 193.9°$.

4.5 The Inverse Sine and the Inverse Cosine Functions

1. We are asked to find the number x in the interval $\left[-\frac{\pi}{2}, \frac{\pi}{2}\right]$ such that $\sin x = \frac{\sqrt{3}}{2}$. Since

$x = \frac{\pi}{3}$ is that number, then $\sin^{-1}\left(\frac{\sqrt{3}}{2}\right) = \frac{\pi}{3}$.

3. We are asked to find the number x in the interval $[0, \pi]$ such that $\cos x = \frac{\sqrt{3}}{2}$. Since

$x = \frac{\pi}{6}$ is that number, then $\cos^{-1}\left(\frac{\sqrt{3}}{2}\right) = \frac{\pi}{6}$.

5. We are asked to find the number x in the interval $\left[-\frac{\pi}{2}, \frac{\pi}{2}\right]$ such that $\sin x = -\frac{\sqrt{2}}{2}$. Since

$x = -\frac{\pi}{4}$ is that number, then $\arcsin\left(-\frac{\sqrt{2}}{2}\right) = -\frac{\pi}{4}$.

7. We are asked to find the number x in the interval $[0, \pi]$ such that $\cos x = 1$. Since $x = 0$ is that number, then $\cos^{-1}1 = 0$.

9. We are asked to find the number x in the interval $[0, \pi]$ such that $\cos x = 2\pi$. Since $\cos x \le 1$ for all x, this value for x does not exist, thus $\cos^{-1}2\pi$ is undefined.

11. If $x = \sin^{-1}\left(\frac{1}{4}\right)$, then $\sin x = \frac{1}{4}$. Thus $\sin\left[\sin^{-1}\frac{1}{4}\right] = \frac{1}{4}$.

13. If $x = \cos^{-1}\left(\frac{3}{4}\right)$, then $\cos x = \frac{3}{4}$. Thus $\cos\left[\cos^{-1}\frac{3}{4}\right] = \frac{3}{4}$.

15. Since $\sin\frac{\pi}{2} = 1$, we must find $\arcsin 1$. We are asked to find the number x in the interval $\left[-\frac{\pi}{2}, \frac{\pi}{2}\right]$ such that $\sin x = 1$. Since $x = \frac{\pi}{2}$ is that number, then $\arcsin\left[\sin\frac{\pi}{2}\right] = \frac{\pi}{2}$.

17. Since $\cos 2\pi = 1$, we must find arccos 1. We are asked to find the number x in the interval $[0, \pi]$ such that $\cos x = 1$. Since $x = 0$ is that number, then $\arccos(\cos 2\pi) = 0$.

19. If $x = \sin^{-1}\left(\frac{4}{5}\right)$, then $\sin x = \frac{4}{5}$ and thus $\cos x = \frac{3}{5}$. Therefore:

$$\tan\left[\sin^{-1}\left(\tfrac{4}{5}\right)\right] = \tan x = \frac{\sin x}{\cos x} = \frac{\frac{4}{5}}{\frac{3}{5}} = \frac{4}{3}$$

21. If $x = \sin^{-1} 1$, then $\sin x = 1$ and x is in the interval $\left[-\frac{\pi}{2}, \frac{\pi}{2}\right]$, and thus $x = \frac{\pi}{2}$. Therefore:

$$\sin\left(\sin^{-1} 1\right) = \sin\tfrac{\pi}{2} = 1$$

23. If $x = \arccos\frac{5}{13}$, then $\cos x = \frac{5}{13}$ and thus $\sin x = \frac{12}{13}$. Therefore:

$$\tan\left(\arccos\tfrac{5}{13}\right) = \tan x = \frac{\sin x}{\cos x} = \frac{\frac{12}{13}}{\frac{5}{13}} = \frac{12}{5}$$

25. Since $\sin^{-1} x + \cos^{-1} x = \frac{\pi}{2}$ for every x in the interval $[-1, 1]$, then:

$$\sin^{-1} 0.1 + \cos^{-1} 0.1 = \tfrac{\pi}{2}$$

27. If $x = \arccos\left(-\frac{1}{3}\right)$, then $\cos x = -\frac{1}{3}$ and since x is in the interval $[0, \pi]$, then x must lie in the second quadrant, so:

$$\sin x = \sqrt{1 - \cos^2 x} = \sqrt{1 - \left(-\tfrac{1}{3}\right)^2} = \sqrt{1 - \tfrac{1}{9}} = \sqrt{\tfrac{8}{9}} = \tfrac{2\sqrt{2}}{3}$$

Therefore:

$$\sin\left[\arccos\left(-\tfrac{1}{3}\right)\right] = \sin x = \tfrac{2\sqrt{2}}{3}$$

29. (a) Using a calculator, we have $\sin^{-1}\left(\frac{4}{5}\right) \approx 0.927$.

    (b) Using a calculator, we have $\cos^{-1}\left(\frac{5}{8}\right) \approx 0.896$.

31. We first compute $\sin^{-1}\left(\frac{1}{2}\right)$ and $\cos^{-1}\left(\frac{1}{2}\right)$:

$$\sin^{-1}\left(\tfrac{1}{2}\right) = \tfrac{\pi}{6}, \text{ since } \sin\tfrac{\pi}{6} = \tfrac{1}{2} \text{ and } \tfrac{\pi}{6} \text{ is in the interval } \left[-\tfrac{\pi}{2}, \tfrac{\pi}{2}\right]$$

$$\cos^{-1}\left(\tfrac{1}{2}\right) = \tfrac{\pi}{3}, \text{ since } \cos\tfrac{\pi}{3} = \tfrac{1}{2} \text{ and } \tfrac{\pi}{3} \text{ is in the interval } [0, \pi]$$

Therefore, we have:

$$\csc\left[\sin^{-1}\left(\tfrac{1}{2}\right) - \cos^{-1}\left(\tfrac{1}{2}\right)\right] = \csc\left(\tfrac{\pi}{6} - \tfrac{\pi}{3}\right) = \csc\left(-\tfrac{\pi}{6}\right) = \frac{1}{\sin\left(-\tfrac{\pi}{6}\right)} = \frac{1}{-\frac{1}{2}} = -2$$

33. We first compute $\cos^{-1}\left(-\frac{1}{2}\right)$ and $\cos^{-1}(0)$:

$\cos^{-1}\left(-\frac{1}{2}\right) = \frac{2\pi}{3}$, since $\cos\frac{2\pi}{3} = -\frac{1}{2}$ and $\frac{2\pi}{3}$ is in the interval $[0, \pi]$

$\cos^{-1} 0 = \frac{\pi}{2}$, since $\cos\frac{\pi}{2} = 0$ and $\frac{\pi}{2}$ is in the interval $[0, \pi]$

Therefore, we have:

$$\cot\left[\cos^{-1}\left(-\frac{1}{2}\right) + \cos^{-1}(0)\right] = \cot\left(\frac{2\pi}{3} + \frac{\pi}{2}\right) = \cot\frac{7\pi}{6} = \frac{\cos\frac{7\pi}{6}}{\sin\frac{7\pi}{6}} = \frac{-\frac{\sqrt{3}}{2}}{-\frac{1}{2}} = \sqrt{3}$$

35. Since $\cos 2\theta = 1 - 2\sin^2\theta$ by a double-angle identity for cosine, and since $\sin\theta = 2x$

then $\theta = \sin^{-1}(2x)$, thus:

$$\theta + \cos 2\theta = \theta + 1 - 2\sin^2\theta = \sin^{-1}(2x) + 1 - 2(2x)^2 = 1 + \sin^{-1}(2x) - 8x^2$$

37. Since $\cos 2\theta = 2\cos^2\theta - 1$ by a double-angle identity for cosine, and since

$\cos\theta = x - 1$ then $\theta = \cos^{-1}(x - 1)$, thus:

$$\begin{aligned}
2\theta - \cos 2\theta &= 2\theta - 2\cos^2\theta + 1 \\
&= 2\cos^{-1}(x-1) - 2(x-1)^2 + 1 \\
&= 2\cos^{-1}(x-1) - 2x^2 + 4x - 2 + 1 \\
&= 2\cos^{-1}(x-1) - 2x^2 + 4x - 1
\end{aligned}$$

39. Let $\theta = \sin^{-1}\left(\frac{5}{13}\right)$, so $\sin\theta = \frac{5}{13}$ and θ is in the interval $\left[-\frac{\pi}{2}, \frac{\pi}{2}\right]$. Then $\cos\theta > 0$, so:

$$\cos\theta = \sqrt{1 - \sin^2\theta} = \sqrt{1 - \left(\frac{5}{13}\right)^2} = \sqrt{1 - \frac{25}{169}} = \sqrt{\frac{144}{169}} = \frac{12}{13}$$

Using a double-angle identity for $\cos 2\theta$, we have:

$$\cos\left[2\sin^{-1}\left(\frac{5}{13}\right)\right] = \cos 2\theta = \cos^2\theta - \sin^2\theta = \left(\frac{12}{13}\right)^2 - \left(\frac{5}{13}\right)^2 = \frac{144}{169} - \frac{25}{169} = \frac{119}{169}$$

41. (a) Since $\alpha = \sin^{-1} x$, then $-\frac{\pi}{2} \le \alpha \le \frac{\pi}{2}$ and since $\beta = \cos^{-1} x$, then $0 \le \beta \le \pi$. Thus:

$$-\frac{\pi}{2} + 0 \le \alpha + \beta \le \frac{\pi}{2} + \pi$$
$$-\frac{\pi}{2} \le \alpha + \beta \le \frac{3\pi}{2}$$

(b) Since $\alpha = \sin^{-1} x$, then $\sin\alpha = x$ and thus:

$$\cos\alpha = \sqrt{1 - \sin^2\alpha} = \sqrt{1 - x^2}$$

Similarly, since $\beta = \cos^{-1} x$, then $\cos\beta = x$ and thus:

$$\sin\beta = \sqrt{1 - \cos^2\beta} = \sqrt{1 - x^2}$$

Using the addition formula for $\sin(\alpha + \beta)$, we have:
$$\sin(\alpha + \beta) = \sin\alpha\cos\beta + \cos\alpha\sin\beta$$
$$= x \bullet x + \sqrt{1-x^2} \bullet \sqrt{1-x^2}$$
$$= x^2 + 1 - x^2$$
$$= 1$$

But if $-\frac{\pi}{2} \le \alpha + \beta \le \frac{3\pi}{2}$ (from part (a)) and $\sin(\alpha + \beta) = 1$, then $\alpha + \beta = \frac{\pi}{2}$.

43. We take the sine of each side to get:
$$\sin\left(\sin^{-1}(3t-2)\right) = \sin\left(\sin^{-1}t - \cos^{-1}t\right)$$
$$3t - 2 = \sin\left(\sin^{-1}t - \cos^{-1}t\right)$$

Let $u = \sin^{-1}t$, so $\sin u = t$ and $\cos u = \sqrt{1-t^2}$. Also, let $v = \cos^{-1}t$, so $\cos v = t$ and $\sin v = \sqrt{1-t^2}$. Thus:
$$\sin\left(\sin^{-1}t - \cos^{-1}t\right) = \sin(u-v)$$
$$= \sin u\cos v - \cos u\sin v$$
$$= t \bullet t - \sqrt{1-t^2}\sqrt{1-t^2}$$
$$= t^2 - 1 + t^2$$
$$= 2t^2 - 1$$

Thus, we have the equation:
$$3t - 2 = 2t^2 - 1$$
$$0 = 2t^2 - 3t + 1$$
$$0 = (2t-1)(t-1)$$
$$t = \tfrac{1}{2}, 1$$

45. Let $x = \cos^{-1}A$, $y = \cos^{-1}B$ and $z = \cos^{-1}C$. Then $\cos x = A$, $\cos y = B$, and $\cos z = C$, and $x + y + z = \pi$. Thus we want to prove that:
$$\cos^2 x + \cos^2 y + \cos^2 z + 2\cos x\cos y\cos z = 1$$
First note that $z = \pi - (x + y)$, thus:
$$\cos z = \cos\left[\pi - (x+y)\right] = -\cos(x+y) = -\cos x\cos y + \sin x\sin y$$
Squaring, we have:
$$\cos^2 z = \left(-\cos x\cos y + \sin x\sin y\right)^2$$
$$= \cos^2 x\cos^2 y - 2\sin x\cos x\sin y\cos y + \sin^2 x\sin^2 y$$

Working from the left-hand side, we have:

$$\cos^2 x + \cos^2 y + \cos^2 z + 2\cos x \cos y \cos z$$
$$= \cos^2 x + \cos^2 y + \left(\cos^2 x \cos^2 y - 2\sin x \cos x \sin y \cos y + \sin^2 x \sin^2 y\right)$$
$$\quad + 2\cos x \cos y(-\cos x \cos y + \sin x \sin y)$$
$$= \cos^2 x + \cos^2 y + \cos^2 x \cos^2 y - 2\sin x \cos x \sin y \cos y + \sin^2 x \sin^2 y$$
$$\quad - 2\cos^2 x \cos^2 y + 2\sin x \cos x \sin y \cos y$$
$$= \cos^2 x + \cos^2 y - \cos^2 x \cos^2 y + \sin^2 x \sin^2 y$$
$$= \cos^2 x(1 - \cos^2 y) + \cos^2 y + \sin^2 x \sin^2 y$$
$$= \cos^2 x \sin^2 y + \cos^2 y + \sin^2 x \sin^2 y$$
$$= \sin^2 y(\cos^2 x + \sin^2 x) + \cos^2 y$$
$$= \sin^2 y + \cos^2 y$$
$$= 1$$

4.6 The Inverse Tangent and Inverse Secant Functions

1.   (a) C
      (b) D
      (c) A
      (d) B

3.   We are asked to find the number x in the interval $\left(-\frac{\pi}{2}, \frac{\pi}{2}\right)$ such that $\tan x = \sqrt{3}$. Since $x = \frac{\pi}{3}$ is that number, then $\tan^{-1}\sqrt{3} = \frac{\pi}{3}$.

5.   We are asked to find the number x in the interval $\left(-\frac{\pi}{2}, \frac{\pi}{2}\right)$ such that $\tan x = -\frac{\sqrt{3}}{3}$. Since $x = -\frac{\pi}{6}$ is that number, then $\tan^{-1}\left(-\frac{\sqrt{3}}{3}\right) = -\frac{\pi}{6}$.

7.   We are asked to find the number x in the interval $\left(-\frac{\pi}{2}, \frac{\pi}{2}\right)$ such that $\tan x = 1$. Since $x = \frac{\pi}{4}$ is that number, then $\arctan 1 = \frac{\pi}{4}$.

9.   We are asked to find the number x in the interval $\left[0, \frac{\pi}{2}\right) \cup \left(\frac{\pi}{2}, \pi\right]$ such that $\sec x = 2$. Since $x = \frac{\pi}{3}$ is that number, then $\sec^{-1} 2 = \frac{\pi}{3}$.

11. We are asked to find the number x in the interval $\left[0,\frac{\pi}{2}\right)\cup\left(\frac{\pi}{2},\pi\right]$ such that $\sec x = \frac{1}{2}$. But then $\cos x = 2$, which is impossible. Thus $\sec^{-1}\left(\frac{1}{2}\right)$ is undefined.

13. We are asked to find the number x in the interval $\left[0,\frac{\pi}{2}\right)\cup\left(\frac{\pi}{2},\pi\right]$ such that $\sec x = -\frac{2}{\sqrt{3}}$. Since $x = \frac{5\pi}{6}$ is that number, then $\operatorname{arcsec}\left(-\frac{2}{\sqrt{3}}\right) = \frac{5\pi}{6}$.

15. We are asked to find the number x in the interval $\left[0,\frac{\pi}{2}\right)\cup\left(\frac{\pi}{2},\pi\right]$ such that $\sec x = \sqrt{2}$. Since $x = \frac{\pi}{4}$ is that number, then $\operatorname{arcsec}\sqrt{2} = \frac{\pi}{4}$.

17. (a) Since $\tan\left(\tan^{-1}x\right) = x$ for every real number x, then $\tan\left(\tan^{-1}0.3\right) = 0.3$.

    (b) Since $\tan\left(\tan^{-1}x\right) = x$ for every real number x, then $\tan\left(\tan^{-1}(-10)\right) = -10$.

19. (a) Since $\tan\frac{\pi}{6} = \frac{\sqrt{3}}{3}$, we are asked to find $x = \tan^{-1}\left(\frac{\sqrt{3}}{3}\right)$. Then $\tan x = \frac{\sqrt{3}}{3}$ and x is in the interval $\left(-\frac{\pi}{2},\frac{\pi}{2}\right)$, and so $x = \frac{\pi}{6}$. Thus $\tan^{-1}\left(\tan\frac{\pi}{6}\right) = \frac{\pi}{6}$. Note that we could also use the identity $\tan^{-1}(\tan x) = x$, since $\frac{\pi}{6}$ is in the interval $\left(-\frac{\pi}{2},\frac{\pi}{2}\right)$.

    (b) Since $\tan\frac{5\pi}{6} = -\frac{\sqrt{3}}{3}$, we are asked to find $x = \tan^{-1}\left(-\frac{\sqrt{3}}{3}\right)$. Then $\tan x = -\frac{\sqrt{3}}{3}$ and x is in the interval $\left(-\frac{\pi}{2},\frac{\pi}{2}\right)$, and so $x = -\frac{\pi}{6}$. Thus $\tan^{-1}\left(\tan\frac{5\pi}{6}\right) = -\frac{\pi}{6}$. Note that we cannot use the identity $\tan^{-1}(\tan x) = x$, since $\frac{5\pi}{6}$ is not in the interval $\left(-\frac{\pi}{2},\frac{\pi}{2}\right)$.

21. (a) Since $\sec\left(\sec^{-1}x\right) = x$ for every x in the set $(-\infty,-1]\cup[1,\infty)$, then $\sec\left(\sec^{-1}2\right) = 2$.

    (b) Since $\sec\left(\sec^{-1}x\right) = x$ for every x in the set $(-\infty,-1]\cup[1,\infty)$, then $\sec\left(\sec^{-1}(-10\pi)\right) = -10\pi$.

23. (a) Since $\sec\frac{\pi}{3} = 2$, we are asked to find $x = \sec^{-1}2$. Then $\sec x = 2$ and x is in the

interval $\left[0,\frac{\pi}{2}\right)\cup\left(\frac{\pi}{2},\pi\right]$, and so $x = \frac{\pi}{3}$. Thus $\sec^{-1}\left(\sec\frac{\pi}{3}\right) = \frac{\pi}{3}$. Note that we could

also use the identity $\sec^{-1}(\sec x) = x$, since $\frac{\pi}{3}$ is in the interval $\left[0,\frac{\pi}{2}\right)\cup\left(\frac{\pi}{2},\pi\right]$.

(b) Since $\sec\frac{4\pi}{3} = -2$, we are asked to find $x = \sec^{-1}(-2)$. Then $\sec x = -2$ and x is in

the interval $\left[0,\frac{\pi}{2}\right)\cup\left(\frac{\pi}{2},\pi\right]$, and so $x = \frac{2\pi}{3}$. Thus $\sec^{-1}\left(\sec\frac{4\pi}{3}\right) = \frac{2\pi}{3}$. Note that we

cannot use the identity $\sec^{-1}(\sec x) = x$, since $\frac{4\pi}{3}$ is not in the interval

$\left[0,\frac{\pi}{2}\right)\cup\left(\frac{\pi}{2},\pi\right]$.

25. (a) Since $\tan\theta = \frac{x}{2}$ and $0 < \theta < \frac{\pi}{2}$, then $\theta = \tan^{-1}\left(\frac{x}{2}\right)$. Thus:
$$2\theta - \tan\theta = 2\tan^{-1}\left(\frac{x}{2}\right) - \frac{x}{2}$$

(b) By the double-angle formula for $\tan 2\theta$, we have:
$$\tan 2\theta = \frac{2\tan\theta}{1-\tan^2\theta} = \frac{2\bullet\frac{x}{2}}{1-\left(\frac{x}{2}\right)^2} = \frac{x}{1-\frac{x^2}{4}} = \frac{4x}{4-x^2}$$

Therefore:
$$2\theta - \tan 2\theta = 2\tan^{-1}\left(\frac{x}{2}\right) - \frac{4x}{4-x^2}$$

27. Let $\theta = \tan^{-1}\left(\frac{1}{3}\right)$, so $\tan\theta = \frac{1}{3}$. Since θ lies in the first quadrant, we construct the
triangle:

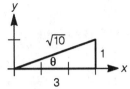

Therefore:
$$\sin\left(\tan^{-1}\left(\tfrac{1}{3}\right)\right) = \sin\theta = \frac{1}{\sqrt{10}} = \frac{\sqrt{10}}{10}$$

29. Let $\theta = \sec^{-1}\left(\frac{5}{3}\right)$, so $\sec\theta = \frac{5}{3}$. Since θ lies in the first quadrant, we construct the triangle:

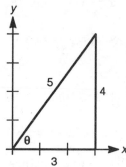

Therefore:

$$\tan\left(\sec^{-1}\left(\frac{5}{3}\right)\right) = \tan\theta = \frac{4}{3}$$

31. Let $\theta = \tan^{-1}x$, so $\tan\theta = x$. Since $x > 0$, θ must lie in the first quadrant, so we construct the triangle:

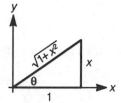

Therefore:

$$\sin\left(\tan^{-1}x\right) = \sin\theta = \frac{x}{\sqrt{1+x^2}} = \frac{x\sqrt{1+x^2}}{1+x^2}$$

33. (a) Using a calculator, we have $\sec^{-1}4.2 \approx 1.330$.

(b) Using a calculator, we have $\sec^{-1}(-4.2) \approx 1.811$.

35. Let $s = \arccos\frac{3}{5}$, so $\cos s = \frac{3}{5}$. Drawing a triangle:

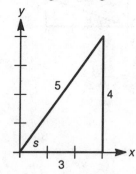

Let $t = \arctan\frac{7}{13}$, so $\tan t = \frac{7}{13}$. Drawing a triangle:

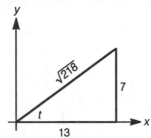

Using the addition formula for $\sin(s - t)$, we have:

$$\sin\left(\arccos\frac{3}{5} - \arctan\frac{7}{13}\right) = \sin(s - t)$$
$$= \sin s \cos t - \cos s \sin t$$
$$= \frac{4}{5} \cdot \frac{13}{\sqrt{218}} - \frac{3}{5} \cdot \frac{7}{\sqrt{218}}$$
$$= \frac{31}{5\sqrt{218}}$$
$$= \frac{31\sqrt{218}}{1090}$$

37. (a) Let $s = \tan^{-1}\left(\frac{4}{3}\right)$, so $\tan s = \frac{4}{3}$. Let $t = \tan^{-1}\left(\frac{1}{7}\right)$, so $\tan t = \frac{1}{7}$. Using the addition formula for $\tan(s - t)$, we have:

$$\tan\left(\tan^{-1}\left(\frac{4}{3}\right) - \tan^{-1}\left(\frac{1}{7}\right)\right) = \tan(s - t) = \frac{\tan s - \tan t}{1 + \tan s \tan t} = \frac{\frac{4}{3} - \frac{1}{7}}{1 + \frac{4}{3} \cdot \frac{1}{7}} = \frac{\frac{25}{21}}{\frac{25}{21}} = 1$$

(b) Since $\tan\left(\tan^{-1}\left(\frac{4}{3}\right) - \tan^{-1}\left(\frac{1}{7}\right)\right) = 1$ and $\tan^{-1}\left(\frac{4}{3}\right) - \tan^{-1}\left(\frac{1}{7}\right)$ is in the first quadrant, then $\tan^{-1}\left(\frac{4}{3}\right) - \tan^{-1}\left(\frac{1}{7}\right) = \frac{\pi}{4}$.

(c) Using a calculator, we find $\tan^{-1}\left(\frac{4}{3}\right) - \tan^{-1}\left(\frac{1}{7}\right) \approx 0.785398$, which is the same value as $\frac{\pi}{4}$.

39. If we write $0.8 = \frac{4}{5}$ and $0.75 = \frac{3}{4}$, we are asked to find $\sin^{-1}\left(\frac{4}{5}\right) + \tan^{-1}\left(\frac{3}{4}\right)$. Let

$\alpha = \sin^{-1}\left(\frac{4}{5}\right)$ and $\beta = \tan^{-1}\left(\frac{3}{4}\right)$, so $\sin\alpha = \frac{4}{5}$ and $\tan\beta = \frac{3}{4}$. Note the following triangle:

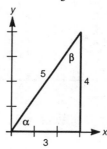

Thus $\alpha + \beta = \frac{\pi}{2}$.

41. Let $\alpha = \arctan 1$, so $\tan\alpha = 1$ and thus $\alpha = \frac{\pi}{4}$. Let $\beta = \arctan 2$, so $\tan\beta = 2$. We draw the triangle:

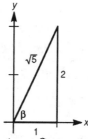

Using the addition formula for $\cos(\alpha + \beta)$, we have:
$$\cos(\arctan 1 + \arctan 2) = \cos(\alpha + \beta)$$
$$= \cos\alpha\cos\beta - \sin\alpha\sin\beta$$
$$= \frac{\sqrt{2}}{2} \cdot \frac{1}{\sqrt{5}} - \frac{\sqrt{2}}{2} \cdot \frac{2}{\sqrt{5}}$$
$$= -\frac{\sqrt{2}}{2\sqrt{5}}$$
$$= -\frac{\sqrt{10}}{10}$$

43. Let $\theta = \sec^{-1} 4$, so $\sec\theta = 4$. We draw the triangle:

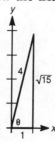

Therefore:
$$\sin\left(\sec^{-1} 4\right) = \sin\theta = \frac{\sqrt{15}}{4}$$

45. Let $s = \arctan x$ and $t = \arctan y$, so $\tan s = x$ and $\tan t = y$. Using the addition formula for $\tan(s+t)$, we have:

$$\tan(s+t) = \frac{\tan s + \tan t}{1 - \tan s \tan t} = \frac{x+y}{1-xy}$$

Then:

$$\arctan(\tan(s+t)) = \arctan\left(\tfrac{x+y}{1-xy}\right)$$

So $s + t = \arctan x + \arctan y = \arctan\left(\tfrac{x+y}{1-xy}\right)$.

47. Following the hint, we take the tangent of each side to obtain:

$$\tan\left(2\tan^{-1}x\right) = \tan\left[\tan^{-1}\left(\tfrac{1}{4x}\right)\right]$$
$$\tan\left(2\tan^{-1}x\right) = \tfrac{1}{4x}$$

Let $\alpha = \tan^{-1}x$, so using the double-angle formula for $\tan 2\alpha$:

$$\frac{2\tan\alpha}{1-\tan^2\alpha} = \frac{1}{4x}$$
$$\frac{2x}{1-x^2} = \frac{1}{4x}$$
$$8x^2 = 1-x^2$$
$$9x^2 = 1$$
$$x^2 = \tfrac{1}{9}$$
$$x = \pm\tfrac{1}{3}$$

49. We take the tangent of each side of the equation to obtain:

$$\tan\left(2\tan^{-1}\sqrt{t-t^2}\right) = \tan\left(\tan^{-1}t + \tan^{-1}(1-t)\right)$$

Let $\alpha = \tan^{-1}\sqrt{t-t^2}$, so $\tan\alpha = \sqrt{t-t^2}$. Let $\beta = \tan^{-1}t$, so $\tan\beta = t$. Let $\gamma = \tan^{-1}(1-t)$, so $\tan\gamma = 1-t$. Now simplify each side of the equation, using the double-angle and sum-angle identities for tangent:

$$\tan\left(2\tan^{-1}\sqrt{t-t^2}\right) = \tan 2\alpha = \frac{2\tan\alpha}{1-\tan^2\alpha} = \frac{2\sqrt{t-t^2}}{1-\left(t-t^2\right)} = \frac{2\sqrt{t-t^2}}{1-t+t^2}$$

$$\tan\left(\tan^{-1}+\tan^{-1}(1-t)\right) = \tan(\beta+\gamma) = \frac{\tan\beta+\tan\gamma}{1-\tan\beta\tan\gamma} = \frac{t+1-t}{1-t(1-t)} = \frac{1}{1-t+t^2}$$

So, our original equation becomes:

$$\frac{2\sqrt{t-t^2}}{1-t+t^2} = \frac{1}{1-t+t^2}$$
$$2\sqrt{t-t^2} = 1$$

Squaring each side, we have:

$$4\left(t - t^2\right) = 1$$
$$4t - 4t^2 = 1$$
$$0 = 4t^2 - 4t + 1$$
$$0 = (2t - 1)^2$$
$$t = \tfrac{1}{2}$$

51. We must solve the equation $\sec^{-1} x = \cos^{-1}(x - 1)$. Taking the cosine of each side of the equation, we have:

$$\cos\left(\sec^{-1} x\right) = \cos\left(\cos^{-1}(x - 1)\right)$$
$$\cos\left(\sec^{-1} x\right) = x - 1$$

Let $\theta = \sec^{-1} x$, so $\sec \theta = x$ and thus $\cos\left(\sec^{-1} x\right) = \cos\theta = \tfrac{1}{x}$. Substituting, we solve the equation:

$$\tfrac{1}{x} = x - 1$$
$$1 = x^2 - x$$
$$0 = x^2 - x - 1$$

Using the quadratic formula:

$$x = \frac{-(-1) \pm \sqrt{(-1)^2 - 4(-1)}}{2} = \frac{1 \pm \sqrt{1 + 4}}{2} = \frac{1 \pm \sqrt{5}}{2}$$

Since $x \geq 0$, the negative root is discarded and thus $x = \frac{1 + \sqrt{5}}{2} \approx 1.618$ and $y \approx \sec^{-1} 1.618 \approx 0.905$. The intersection point is approximately $(1.618, 0.905)$, which is consistent with the figure.

53. (a) We complete the table:

x	$f(x)$
0.1	3.141592654
0.2	3.141592654
0.3	3.141592654
0.4	3.141592654
0.5	3.141592654

(b) It would appear that these values always equal π. To prove this, let $A = \tan^{-1}\left(\frac{1+x}{1-x}\right)$, so $\tan A = \frac{1+x}{1-x}$. We draw the triangle:

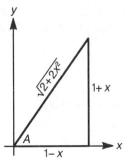

Also let $B = \sin^{-1}\left(\frac{1-x^2}{1+x^2}\right)$, so $\sin B = \frac{1-x^2}{1+x^2}$. We draw the triangle:

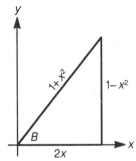

Now using the hint and the addition and double-angle formulas, we have:

$$\sin[f(x)] = \sin(2A+B)$$
$$= \sin 2A \cos B + \cos 2A \sin B$$
$$= 2\sin A \cos A \cos B + \left(\cos^2 A - \sin^2 A\right)\sin B$$
$$= 2 \cdot \frac{1+x}{\sqrt{2+2x^2}} \cdot \frac{1-x}{\sqrt{2+2x^2}} \cdot \frac{2x}{1+x^2} + \left(\frac{(1-x)^2}{2+2x^2} - \frac{(1+x)^2}{2+2x^2}\right) \cdot \frac{1-x^2}{1+x^2}$$
$$= \frac{4x(1+x)(1-x)}{2\left(1+x^2\right)^2} + \frac{-4x\left(1-x^2\right)}{2\left(1+x^2\right)^2}$$
$$= \frac{4x\left(1-x^2\right) - 4x\left(1-x^2\right)}{2\left(1+x^2\right)^2}$$
$$= 0$$

So $f(x) = 0,\ \pi,\ 2\pi,...$ Since $0 < x < 1$, only $f(x) = \pi$ is a legitimate value.

55. We are asked to find the number x in the interval $(0, \pi)$ such that $\cot x = \sqrt{3}$. Since $x = \frac{\pi}{6}$ is that number, then $\cot^{-1}\sqrt{3} = \frac{\pi}{6}$.

57. We are asked to find the number x in the interval $(0, \pi)$ such that $\cot x = 0$. Since $x = \frac{\pi}{2}$ is that number, then $\cot^{-1} 0 = \frac{\pi}{2}$.

59. Since $\cot(\cot^{-1} x) = x$ for every real number x, then $\cot(\cot^{-1} 2) = 2$.

61. Since $\cot \frac{\pi}{6} = \sqrt{3}$, we are asked to find $\cot^{-1} \sqrt{3}$. If $x = \cot^{-1} \sqrt{3}$, then $\cot x = \sqrt{3}$ and x is in the interval $(0, \pi)$, so $x = \frac{\pi}{6}$. Thus $\cot^{-1}\left[\cot \frac{\pi}{6}\right] = \frac{\pi}{6}$. Note that we could also use the identity $\cot^{-1}(\cot x) = x$, since $\frac{\pi}{6}$ is in the interval $(0, \pi)$.

63. Let $\theta = \cot^{-1}\left(\frac{2}{3}\right)$, so $\cot \theta = \frac{2}{3}$. We draw the triangle:

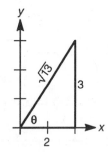

Therefore:
$$\cos\left[\cot^{-1}\left(\tfrac{2}{3}\right)\right] = \cos\theta = \frac{2}{\sqrt{13}} = \frac{2\sqrt{13}}{13}$$

65. Let $\theta = \cot^{-1}\left(-\frac{4}{3}\right)$, so $\cot \theta = -\frac{4}{3}$ and θ lies in the second quadrant. We draw the triangle:

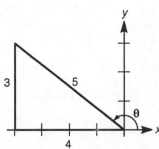

Therefore:
$$\sin\left[\cot^{-1}\left(-\tfrac{4}{3}\right)\right] = \sin\theta = \tfrac{3}{5}$$

67. **(a)** Using a calculator and the identity in Exercise 66(b), we have:
$$\cot^{-1} 6 = \tan^{-1}\left(\tfrac{1}{6}\right) \approx 0.165$$

(b)   Using a calculator and the identity in Exercise 66(a), we have:
$$\cot^{-1}(-6) = \tfrac{\pi}{2} - \tan^{-1}(-6) \approx 2.976$$

(c)   Using a calculator and the identity in Exercise 66(b), we have:
$$\cot^{-1}\!\left(\tfrac{\pi}{5}\right) = \tan^{-1}\!\left(\tfrac{5}{\pi}\right) \approx 1.010$$

(d)   Using a calculator and the identity in Exercise 66(a), we have:
$$\cot^{-1}\!\left(-\tfrac{\pi}{5}\right) = \tfrac{\pi}{2} - \tan^{-1}\!\left(-\tfrac{\pi}{5}\right) \approx 2.132$$

69.   We are asked to find the number x in the interval $\left[-\tfrac{\pi}{2},0\right)\cup\left(0,\tfrac{\pi}{2}\right]$ such that $\csc x = \sqrt{2}$, thus $\sin x = \tfrac{1}{\sqrt{2}}$. Since $x = \tfrac{\pi}{4}$ is that number, then $\csc^{-1}\sqrt{2} = \tfrac{\pi}{4}$.

71.   We are asked to find the number x in the interval $\left[-\tfrac{\pi}{2},0\right)\cup\left(0,\tfrac{\pi}{2}\right]$ such that $\csc x = -1$, thus $\sin x = -1$. Since $x = -\tfrac{\pi}{2}$ is that number, then $\csc^{-1}(-1) = -\tfrac{\pi}{2}$.

73.   We are asked to find the number x in the interval $\left[-\tfrac{\pi}{2},0\right)\cup\left(0,\tfrac{\pi}{2}\right]$ such that $\csc x = \tfrac{2\sqrt{3}}{3}$, thus $\sin x = \tfrac{\sqrt{3}}{2}$. Since $x = \tfrac{\pi}{3}$ is that number, then $\csc^{-1}\!\left(\tfrac{2\sqrt{3}}{3}\right) = \tfrac{\pi}{3}$.

75.   Since $\csc\tfrac{5\pi}{4} = -\sqrt{2}$, we are asked to find $\csc^{-1}(-\sqrt{2})$. If $x = \csc^{-1}(-\sqrt{2})$, then $\csc x = -\sqrt{2}$ and x is in the interval $\left[-\tfrac{\pi}{2},0\right)\cup\left(0,\tfrac{\pi}{2}\right]$, so $x = -\tfrac{\pi}{4}$. Thus $\csc^{-1}\!\left(\csc\tfrac{5\pi}{4}\right) = -\tfrac{\pi}{4}$. Note that we cannot use the identity $\csc^{-1}(\csc x) = x$, since $\tfrac{5\pi}{4}$ is not in the interval $\left[-\tfrac{\pi}{2},0\right)\cup\left(0,\tfrac{\pi}{2}\right]$.

77.   Let $\theta = \csc^{-1}\!\left(\tfrac{3}{2}\right)$, so $\csc\theta = \tfrac{3}{2}$. We draw the triangle:

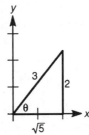

Therefore:
$$\tan\!\left[\csc^{-1}\!\left(\tfrac{3}{2}\right)\right] = \tan\theta = \tfrac{2}{\sqrt{5}} = \tfrac{2\sqrt{5}}{5}$$

79. Let $\theta = \csc^{-1}\left(-\frac{13}{12}\right)$, so $\csc\theta = -\frac{13}{12}$. We draw the triangle:

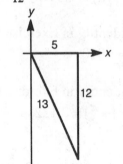

Therefore:
$$\sec\left[\csc^{-1}\left(-\tfrac{13}{12}\right)\right] = \sec\theta = \tfrac{13}{5}$$

81. **(a)** If A corresponds to $A(0,0)$, then the coordinates of the remaining points are $B(9,3)$, $C(12,0)$, $D(10,0)$ and $E(11,1)$. Following the hint, we compute the slopes:
$$m_{\overline{DE}} = \tfrac{1-0}{11-10} = \tfrac{1}{1} = 1$$
$$m_{\overline{BC}} = \tfrac{0-3}{12-9} = \tfrac{-3}{3} = -1$$
$$m_{\overline{AB}} = \tfrac{3-0}{9-0} = \tfrac{3}{9} = \tfrac{1}{3}$$
$$m_{\overline{BD}} = \tfrac{0-3}{10-9} = \tfrac{-3}{1} = -3$$

Since $m_{\overline{DE}} \bullet m_{\overline{BC}} = -1$, then $\overline{DE}$ is perpendicular to $\overline{BC}$. Since $m_{\overline{AB}} \bullet m_{\overline{BD}} = -1$, then $\overline{AB}$ is perpendicular to $\overline{BD}$.

(b) We compute each distance using the coordinates specified in part (a) and the distance formula:
$$DE = \sqrt{(11-10)^2 + (1-0)^2} = \sqrt{1+1} = \sqrt{2}$$
$$CE = \sqrt{(11-12)^2 + (1-0)^2} = \sqrt{1+1} = \sqrt{2}$$
$$BE = \sqrt{(11-9)^2 + (1-3)^2} = \sqrt{4+4} = 2\sqrt{2}$$
$$AB = \sqrt{(9-0)^2 + (3-0)^2} = \sqrt{81+9} = 3\sqrt{10}$$
$$BD = \sqrt{(10-9)^2 + (0-3)^2} = \sqrt{1+9} = \sqrt{10}$$

(c) We compute each tangent specified:
$$\tan\alpha = \tfrac{CE}{DE} = \tfrac{\sqrt{2}}{\sqrt{2}} = 1$$
$$\tan\beta = \tfrac{BE}{DE} = \tfrac{2\sqrt{2}}{\sqrt{2}} = 2$$
$$\tan\gamma = \tfrac{AB}{BD} = \tfrac{3\sqrt{10}}{\sqrt{10}} = 3$$
Thus $\alpha = \tan^{-1}1$, $\beta = \tan^{-1}2$ and $\gamma = \tan^{-1}3$. Since $\alpha + \beta + \gamma = \pi$, then:
$$\tan^{-1}1 + \tan^{-1}2 + \tan^{-1}3 = \pi$$

Section 4.6 TI-81 Graphing Calculator Exercises

1. (a) The graphs of $Y_1 = x$ and $Y_2 = \sin^{-1}(\sin x)$ are the same using the indicated settings.

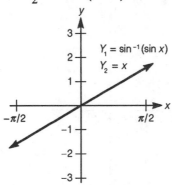

(b) The results demonstrate that $\sin^{-1}(\sin x) = x$, and thus $\sin^{-1} x$ is the inverse function of $\sin x$ on the interval $-\frac{\pi}{2} \leq x \leq \frac{\pi}{2}$.

3. (a) The graph of $Y_1 = \sin^{-1} x + \cos^{-1} x$ appears to be constant.

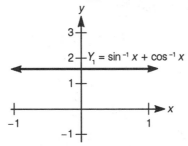

(b) The y-intercept is $\frac{\pi}{2}$. This demonstrates the identity $\sin^{-1} x + \cos^{-1} x = \frac{\pi}{2}$.

5. (a) Using the Zoom-7 settings, the graphs of $Y_1 = \tan\left(\tan^{-1} x\right)$ and $Y_2 = x$ are identical. This demonstrates that $\tan\left(\tan^{-1} x\right) = x$, at least for the values of x that are graphed.

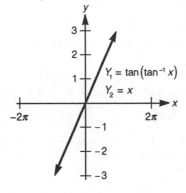

(b)   The graphs of $Y_1 = \tan^{-1}(\tan x)$ and $Y_2 = x$ are identical. This demonstrates that $\tan^{-1}(\tan x) = x$ for $-\frac{\pi}{2} < x < \frac{\pi}{2}$.

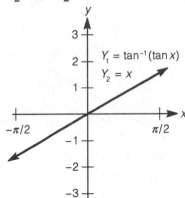

(c)   It was necessary to change the range settings since $\tan x$ is one-to-one on the interval $-\frac{\pi}{2} < x < \frac{\pi}{2}$. Using the interval $-2\pi < x < 2\pi$ would not produce the same result (try it!).

7.   Using the settings $X_{\min} = -8$, $X_{\max} = 8$, $Y_{\min} = -1$, $Y_{\max} = 4$, we obtain the graph of $Y_1 = \sec^{-1} x$ using the identity $\sec^{-1} x = \cos^{-1}\left(\frac{1}{x}\right)$.

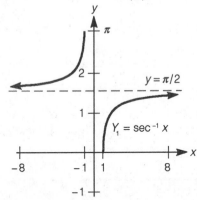

9.   Using the trace and zoom technique, the single solution is $x \approx 0.74$.

11.   (a)   Using the trace and zoom technique, the single solution is $x \approx 0.96$.
      (b)   Using the trace and zoom technique, the single solution is $x \approx 0.96$.

13.   (a)   Using the trace and zoom technique, the single solution is $x \approx 0.24$.
      (b)   Using the trace and zoom technique, there are two solutions of $x \approx 0.19$ and $x \approx 0.68$.

15.   (a)   Using the trace and zoom technique, the single solution is $x \approx 0.56$.
      (b)   Using the trace and zoom technique, there are two solutions of $x \approx 0.51$ and $x \approx 0.84$.

Chapter Four Review Exercises

1. Using the addition formula for sin $(s + t)$, we have:
$$\sin\left(x + \tfrac{3\pi}{2}\right) = \sin x \cos\tfrac{3\pi}{2} + \cos x \sin\tfrac{3\pi}{2} = \sin x \bullet 0 + \cos x \bullet (-1) = -\cos x$$

3. Using the addition formula for cos $(s - t)$, we have:
$$\cos(\pi - x) = \cos\pi\cos x + \sin\pi\sin x = -1 \bullet \cos x + 0 \bullet \sin x = -\cos x$$

5. Using the addition formula for cos $(s - t)$, we have:
$$\cos 175° \cos 25° + \sin 175° \sin 25° = \cos\left(175° - 25°\right) = \cos 150° = -\tfrac{\sqrt{3}}{2}$$

7. Using the addition formulas for cos $(s - t)$ and cos $(s + t)$, we have:
$$\cos\left(x - \tfrac{2\pi}{3}\right) = \cos x \cos\tfrac{2\pi}{3} + \sin x \sin\tfrac{2\pi}{3}$$
$$= \cos x \bullet \left(-\tfrac{1}{2}\right) + \sin x \bullet \tfrac{\sqrt{3}}{2}$$
$$= -\tfrac{1}{2}\cos x + \tfrac{\sqrt{3}}{2}\sin x$$
$$\cos\left(x + \tfrac{2\pi}{3}\right) = \cos x \cos\tfrac{2\pi}{3} - \sin x \sin\tfrac{2\pi}{3}$$
$$= \cos x \bullet \left(-\tfrac{1}{2}\right) - \sin x \bullet \tfrac{\sqrt{3}}{2}$$
$$= -\tfrac{1}{2}\cos x - \tfrac{\sqrt{3}}{2}\sin x$$

Therefore:
$$\cos\left(x - \tfrac{2\pi}{3}\right) - \cos\left(x + \tfrac{2\pi}{3}\right) = \left(-\tfrac{1}{2}\cos x + \tfrac{\sqrt{3}}{2}\sin x\right) - \left(-\tfrac{1}{2}\cos x - \tfrac{\sqrt{3}}{2}\sin x\right)$$
$$= -\tfrac{1}{2}\cos x + \tfrac{\sqrt{3}}{2}\sin x + \tfrac{1}{2}\cos x + \tfrac{\sqrt{3}}{2}\sin x$$
$$= \sqrt{3}\sin x$$

9. Using the addition formulas for tan $(s + t)$ and tan $(s - t)$, we have:
$$\tan\left(x + 45°\right) = \frac{\tan x + \tan 45°}{1 - \tan x \tan 45°} = \frac{\tan x + 1}{1 - \tan x}$$
$$\tan\left(x - 45°\right) = \frac{\tan x - \tan 45°}{1 + \tan x \tan 45°} = \frac{\tan x - 1}{1 + \tan x}$$

Multiplying, we have:
$$\tan\left(x + 45°\right)\tan\left(x - 45°\right) = \frac{\tan x + 1}{1 - \tan x} \bullet \frac{\tan x - 1}{1 + \tan x} = \frac{\tan^2 x - 1}{1 - \tan^2 x} = -1$$

11. (a)   Using the addition formula for $\tan(s-t)$, we have:

$$\tan\frac{\pi}{12} = \tan\left(\frac{\pi}{4} - \frac{\pi}{6}\right)$$

$$= \frac{\tan\frac{\pi}{4} - \tan\frac{\pi}{6}}{1 + \tan\frac{\pi}{4}\tan\frac{\pi}{6}}$$

$$= \frac{1 - \frac{\sqrt{3}}{3}}{1 + 1 \bullet \frac{\sqrt{3}}{3}}$$

$$= \frac{3 - \sqrt{3}}{3 + \sqrt{3}} \bullet \frac{3 - \sqrt{3}}{3 - \sqrt{3}}$$

$$= \frac{9 - 6\sqrt{3} + 3}{9 - 3}$$

$$= \frac{12 - 6\sqrt{3}}{6}$$

$$= 2 - \sqrt{3}$$

(b)   Using the identity for $\cot x$, we have:

$$\cot\frac{\pi}{12} = \frac{1}{\tan\frac{\pi}{12}} = \frac{1}{2 - \sqrt{3}} \bullet \frac{2 + \sqrt{3}}{2 + \sqrt{3}} = \frac{2 + \sqrt{3}}{4 - 3} = 2 + \sqrt{3}$$

13.   Using the identity for $\cot x$ and the addition formula for $\tan(s+t)$, we have:

$$\cot(x+y) = \frac{1}{\tan(x+y)} = \frac{1 - \tan x \tan y}{\tan x + \tan y} \bullet \frac{\cot x \cot y}{\cot x \cot y} = \frac{\cot x \cot y - 1}{\cot y + \cot x}$$

15.   Working from the right-hand side, we have:

$$\frac{2\tan x}{1 + \tan^2 x} = \frac{\frac{2\sin x}{\cos x}}{1 + \frac{\sin^2 x}{\cos^2 x}} = \frac{2\sin x \cos x}{\cos^2 x + \sin^2 x} = \frac{\sin 2x}{1} = \sin 2x$$

17.   Using the addition formulas for $\sin(x+y)$ and $\sin(x-y)$, we have:

$$\frac{\sin(x+y)\sin(x-y)}{\cos^2 x \cos^2 y} = \frac{(\sin x \cos y + \cos x \sin y)(\sin x \cos y - \cos x \sin y)}{\cos^2 x \cos^2 y}$$

$$= \frac{\sin^2 x \cos^2 y - \cos^2 x \sin^2 y}{\cos^2 x \cos^2 y}$$

$$= \frac{\sin^2 x}{\cos^2 x} - \frac{\sin^2 y}{\cos^2 y}$$

$$= \tan^2 x - \tan^2 y$$

19. Using the half-angle formula for $\tan\frac{x}{2}$, we have:

$$\sin x\left(\tan\frac{x}{2}+\cot\frac{x}{2}\right)=\sin x\left[\frac{\sin x}{1+\cos x}+\frac{1+\cos x}{\sin x}\right]$$

$$=\sin x\left[\frac{\sin^2 x+1+2\cos x+\cos^2 x}{(\sin x)(1+\cos x)}\right]$$

$$=\sin x\left[\frac{2+2\cos x}{\sin x(1+\cos x)}\right]$$

$$=\frac{2(1+\cos x)}{1+\cos x}$$

$$=2$$

21. Using the addition formula for $\tan(s-t)$, we have:

$$\tan\left(\frac{\pi}{4}-x\right)=\frac{\tan\frac{\pi}{4}-\tan x}{1+\tan x\tan\frac{\pi}{4}}=\frac{1-\tan x}{1+\tan x}$$

Now using the result of Exercise 20, we have:

$$\tan\left(\frac{\pi}{4}+x\right)-\tan\left(\frac{\pi}{4}-x\right)=\frac{1+\tan x}{1-\tan x}-\frac{1-\tan x}{1+\tan x}=\frac{4\tan x}{1-\tan^2 x}=2\tan 2x$$

23. Using the double-angle formula for $\sin 2\theta$, we have:

$$2\sin\left(\frac{\pi}{4}-\frac{x}{2}\right)\cos\left(\frac{\pi}{4}-\frac{x}{2}\right)=\sin\left[2\bullet\left(\frac{\pi}{4}-\frac{x}{2}\right)\right]=\sin\left(\frac{\pi}{2}-x\right)=\cos x$$

25. We first simplify $\tan\left(\frac{\pi}{4}-t\right)$ using the addition formula for $\tan(s-t)$:

$$\tan\left(\frac{\pi}{4}-t\right)=\frac{\tan\frac{\pi}{4}-\tan t}{1+\tan\frac{\pi}{4}\tan t}=\frac{1-\tan t}{1+\tan t}$$

Therefore:

$$\frac{1-\tan\left(\frac{\pi}{4}-t\right)}{1+\tan\left(\frac{\pi}{4}-t\right)}=\frac{1-\frac{1-\tan t}{1+\tan t}}{1+\frac{1-\tan t}{1+\tan t}}=\frac{1+\tan t-1+\tan t}{1+\tan t+1-\tan t}=\frac{2\tan t}{2}=\tan t$$

27. Using the addition formula for $\tan(s+t)$, we have:

$$\frac{\tan(\alpha-\beta)+\tan\beta}{1-\tan(\alpha-\beta)\tan\beta}=\tan\left[(\alpha-\beta)+\beta\right]=\tan\alpha$$

29. Using the addition formula for $\tan(s+t)$ and the double-angle formula for $\tan 2\theta$, we
     have:

$$
\begin{aligned}
\tan 3\theta &= \tan(\theta + 2\theta) \\
&= \frac{\tan\theta + \tan 2\theta}{1 - \tan\theta\tan 2\theta} \\
&= \frac{\tan\theta + \frac{2\tan\theta}{1-\tan^2\theta}}{1 - \tan\theta \cdot \frac{2\tan\theta}{1-\tan^2\theta}} \\
&= \frac{\tan\theta(1-\tan^2\theta)+2\tan\theta}{1-\tan^2\theta - 2\tan^2\theta} \\
&= \frac{3\tan\theta - \tan^3\theta}{1 - 3\tan^2\theta} \\
&= \frac{3t - t^3}{1 - 3t^2}, \text{ where } t = \tan\theta
\end{aligned}
$$

31. Working from the right-hand side, we have:

$$
\begin{aligned}
\frac{\cos x + \sin x}{\cos x - \sin x} &= \frac{\cos x + \sin x}{\cos x - \sin x} \cdot \frac{\cos x + \sin x}{\cos x + \sin x} \\
&= \frac{\cos^2 x + 2\sin x\cos x + \sin^2 x}{\cos^2 x - \sin^2 x} \\
&= \frac{1 + \sin 2x}{\cos 2x} \\
&= \frac{1}{\cos 2x} + \frac{\sin 2x}{\cos 2x} \\
&= \tan 2x + \sec 2x
\end{aligned}
$$

33. Using the double-angle formula for $\sin 2x$, we have:

$$
\begin{aligned}
2\sin x + \sin 2x &= 2\sin x + 2\sin x\cos x \\
&= 2\sin x(1 + \cos x) \\
&= 2\sin x(1 + \cos x) \cdot \frac{1 - \cos x}{1 - \cos x} \\
&= \frac{2\sin x(1 - \cos^2 x)}{1 - \cos x} \\
&= \frac{2\sin^3 x}{1 - \cos x}
\end{aligned}
$$

35. Working from the right-hand side, we have:

$$\frac{1-\cos x + \sin x}{1+\cos x + \sin x} = \frac{1+\sin x - \cos x}{1+\sin x + \cos x} \cdot \frac{1+\sin x + \cos x}{1+\sin x + \cos x}$$

$$= \frac{(1+\sin x)^2 - \cos^2 x}{\left[(1+\sin x) + \cos x\right]^2}$$

$$= \frac{1+2\sin x + \sin^2 x - \cos^2 x}{\left(1+2\sin x + \sin^2 x\right) + 2\cos x(1+\sin x) + \cos^2 x}$$

$$= \frac{2\sin x + 2\sin^2 x}{2+2\sin x + 2\cos x(1+\sin x)}$$

$$= \frac{2\sin x(1+\sin x)}{2(1+\sin x) + 2\cos x(1+\sin x)}$$

$$= \frac{2\sin x(1+\sin x)}{2(1+\sin x)(1+\cos x)}$$

$$= \frac{\sin x}{1+\cos x}$$

$$= \tan \tfrac{x}{2}$$

37. Using the addition formula for $\sin(s-t)$, we have:

$$\sin(x+y)\cos y - \cos(x+y)\sin y = \sin\left[(x+y)-y\right] = \sin x$$

39. Working from the left-hand side, we have:

$$\frac{1-\tan^2 \tfrac{x}{2}}{1+\tan^2 \tfrac{x}{2}} = \frac{1-\dfrac{\sin^2 \tfrac{x}{2}}{\cos^2 \tfrac{x}{2}}}{1+\dfrac{\sin^2 \tfrac{x}{2}}{\cos^2 \tfrac{x}{2}}} = \frac{\cos^2 \tfrac{x}{2} - \sin^2 \tfrac{x}{2}}{\cos^2 \tfrac{x}{2} + \sin^2 \tfrac{x}{2}} = \cos^2 \tfrac{x}{2} - \sin^2 \tfrac{x}{2} = \cos\left(2\bullet\tfrac{x}{2}\right) = \cos x$$

41. Using the double-angle formulas for $\sin 2\theta$ and $\cos 2\theta$, we have:

$$\sin 4x = 2\sin 2x\cos 2x$$

$$= 2(2\sin x\cos x)\left(\cos^2 x - \sin^2 x\right)$$

$$= 4\sin x\cos x\left(1-2\sin^2 x\right)$$

$$= 4\sin x\cos x - 8\sin^3 x\cos x$$

43. Using the addition formula for $\sin(s+t)$, we have:

$$\sin 5x = \sin(4x+x) = \sin 4x\cos x + \cos 4x\sin x$$

Simplifying each of these products using the double-angle formulas for $\sin 2\theta$ and $\cos 2\theta$, we have:

$$\sin 4x\cos x = 2\sin 2x\cos 2x\cos x$$
$$= 2(2\sin x\cos x)(\cos^2 x - \sin^2 x)\cos x$$
$$= 4\sin x\cos^2 x(1 - 2\sin^2 x)$$
$$= 4\sin x(1 - \sin^2 x)(1 - 2\sin^2 x)$$
$$= 4\sin x - 12\sin^3 x + 8\sin^5 x$$

$$\cos 4x\sin x = (\cos^2 2x - \sin^2 2x)(\sin x)$$
$$= \left[(\cos^2 x - \sin^2 x)^2 - (2\sin x\cos x)^2\right]\sin x$$
$$= \left[(1 - 2\sin^2 x)^2 - 4\sin^2 x\cos^2 x\right]\sin x$$
$$= \left[1 - 4\sin^2 x + 4\sin^4 x - 4\sin^2 x(1 - \sin^2 x)\right]\sin x$$
$$= \left[1 - 8\sin^2 x + 8\sin^4 x\right]\sin x$$
$$= \sin x - 8\sin^3 x + 8\sin^5 x$$

Therefore, we have:

$$\sin 5x = \sin 4x\cos x + \cos 4x\sin x$$
$$= 4\sin x - 12\sin^3 x + 8\sin^5 x + \sin x - 8\sin^3 x + 8\sin^5 x$$
$$= 16\sin^5 x - 20\sin^3 x + 5\sin x$$

45. Using the sum-to-product formula for $\sin\alpha - \sin\beta$, we have:

$$\sin 80° - \sin 20° = 2\cos\frac{80°+20°}{2}\sin\frac{80°-20°}{2} = 2\cos 50°\sin 30° = 2\cos 50°\cdot\frac{1}{2} = \cos 50°$$

47. Using the sum-to-product formulas for $\cos\alpha - \cos\beta$ and $\sin\alpha + \sin\beta$, we have:

$$\cos x - \cos 3x = -2\sin\frac{x+3x}{2}\sin\frac{x-3x}{2} = -2\sin 2x\sin(-x) = 2\sin 2x\sin x$$
$$\sin x + \sin 3x = 2\sin\frac{x+3x}{2}\cos\frac{x-3x}{2} = 2\sin 2x\cos(-x) = 2\sin 2x\cos x$$

Therefore:

$$\frac{\cos x - \cos 3x}{\sin x + \sin 3x} = \frac{2\sin 2x\sin x}{2\sin 2x\cos x} = \frac{\sin x}{\cos x} = \tan x$$

49. Using the sum-to-product formula for $\sin\alpha + \sin\beta$, we have:

$$\sin\tfrac{5\pi}{12} + \sin\tfrac{\pi}{12} = 2\sin\frac{\frac{5\pi}{12}+\frac{\pi}{12}}{2}\cos\frac{\frac{5\pi}{12}-\frac{\pi}{12}}{2} = 2\sin\tfrac{\pi}{4}\cos\tfrac{\pi}{6} = 2\cdot\frac{\sqrt{2}}{2}\cdot\frac{\sqrt{3}}{2} = \frac{\sqrt{6}}{2}$$

51. Using the sum-to-product formulas for $\cos\alpha + \cos\beta$ and $\sin\alpha + \sin\beta$, we have:

$$\cos 3y + \cos(2x - 3y) = 2\cos\tfrac{2x}{2}\cos\frac{6y-2x}{2} = 2\cos x\cos(3y - x)$$
$$\sin 3y + \sin(2x - 3y) = 2\sin\tfrac{2x}{2}\cos\frac{6y-2x}{2} = 2\sin x\cos(3y - x)$$

Therefore:

$$\frac{\cos 3y + \cos(2x - 3y)}{\sin 3y + \sin(2x - 3y)} = \frac{2\cos x\cos(3y - x)}{2\sin x\cos(3y - x)} = \frac{\cos x}{\sin x} = \cot x$$

53. Using the sum-to-product formulas for $\sin\alpha - \sin\beta$ and $\cos\alpha - \cos\beta$, we have:

$$\sin 40° - \sin 20° = 2\cos\frac{40°+20°}{2}\sin\frac{40°-20°}{2}$$
$$= 2\cos 30°\sin 10°$$
$$= 2\cdot\frac{\sqrt{3}}{2}\sin 10°$$
$$= \sqrt{3}\sin 10°$$

$$\cos 20° - \cos 40° = -2\sin\frac{40°+20°}{2}\sin\frac{40°-20°}{2}$$
$$= -2\sin 30°\sin(-10°)$$
$$= 2\cdot\tfrac{1}{2}\sin 10°$$
$$= \sin 10°$$

Therefore:

$$\frac{\sin 40° - \sin 20°}{\cos 20° - \cos 40°} = \frac{\sqrt{3}\sin 10°}{\sin 10°} = \sqrt{3}$$

Using the result of Exercise 52, we have shown the required identity.

55. (a) Since $a = 1$, the area is $\tan^{-1}1 = \frac{\pi}{4}$.

    (b) (i) Since $\tan^{-1}a = 1.5$, then $a = \tan 1.5 \approx 14$.

        (ii) Since $\tan^{-1}a = 1.56$, then $a = \tan 1.56 \approx 93$. ·

        (iii) Since $\tan^{-1}a = 1.57$, then $a = \tan 1.57 \approx 1256$.

57. The principal solution is $x = \tan^{-1}4.26 \approx 1.34$. Since $\tan x$ is also positive in the third quadrant, the other solution in the interval $[0, 2\pi]$ is $1.34 + \pi \approx 4.48$.

59. Since $\csc x = 2.24$, then $\sin x = \frac{1}{2.24} \approx 0.45$. The principal solution is $x = \sin^{-1}0.45 \approx 0.46$. Since $\sin x$ is also positive in the second quadrant, the other solution in the interval $[0, 2\pi)$ is $\pi - 0.46 \approx 2.68$.

61. We have $\tan^2 x - 3 = 0$, so $\tan^2 x = 3$ and $\tan x = \pm\sqrt{3}$. If $\tan x = \sqrt{3}$ then $x = \frac{\pi}{3}$ or $\frac{4\pi}{3}$ while if $\tan x = -\sqrt{3}$ then $x = \frac{2\pi}{3}$ or $\frac{5\pi}{3}$. So the solutions in the interval $[0, 2\pi)$ are $x = \frac{\pi}{3}, \frac{2\pi}{3}, \frac{4\pi}{3}$ or $\frac{5\pi}{3}$.

63. Squaring each side of the equation, we have:

$$(1 + \sin x)^2 = \cos^2 x$$
$$1 + 2\sin x + \sin^2 x = 1 - \sin^2 x$$
$$2\sin^2 x + 2\sin x = 0$$
$$2\sin x(\sin x + 1) = 0$$

So $\sin x = 0$ or $\sin x = -1$, thus $x = 0$, π or $\frac{3\pi}{2}$. Upon checking we find that $x = \pi$ is not a solution (recall that squaring an equation can produce extraneous roots), so the solutions are $x = 0$ or $\frac{3\pi}{2}$.

65. Using the double-angle formula for $\cos 2x$, we have:
$$\sin x - \left(\cos^2 x - \sin^2 x\right) + 1 = 0$$
$$\sin x - \left(1 - 2\sin^2 x\right) + 1 = 0$$
$$2\sin^2 x + \sin x = 0$$
$$\sin x(2\sin x + 1) = 0$$

So $\sin x = 0$ or $\sin x = -\frac{1}{2}$. Thus the solutions are $x = 0$, π, $\frac{7\pi}{6}$ or $\frac{11\pi}{6}$.

67. Solving the equation, we have:
$$3\csc x - 4\sin x = 0$$
$$\frac{3}{\sin x} = 4\sin x$$
$$\sin^2 x = \frac{3}{4}$$
$$\sin x = \pm\frac{\sqrt{3}}{2}$$

So the solutions are $x = \frac{\pi}{3}$, $\frac{2\pi}{3}$, $\frac{4\pi}{3}$ or $\frac{5\pi}{3}$.

69. Factoring, we have:
$$2\sin^4 x - 3\sin^2 x + 1 = 0$$
$$\left(2\sin^2 x - 1\right)\left(\sin^2 x - 1\right) = 0$$
$$\sin^2 x = \tfrac{1}{2} \qquad \text{or} \qquad \sin^2 x = 1$$
$$\sin x = \pm\tfrac{\sqrt{2}}{2} \qquad \text{or} \qquad \sin x = \pm 1$$

So the solutions are $x = \frac{\pi}{4}$, $\frac{\pi}{2}$, $\frac{3\pi}{4}$, $\frac{5\pi}{4}$, $\frac{3\pi}{2}$ or $\frac{7\pi}{4}$.

71. Using the identity $\sin^2 x = 1 - \cos^2 x$, we have:
$$\left(1 - \cos^2 x\right)^2 + \cos^4 x = \tfrac{5}{8}$$
$$1 - 2\cos^2 x + 2\cos^4 x = \tfrac{5}{8}$$
$$16\cos^4 x - 16\cos^2 x + 3 = 0$$
$$\left(4\cos^2 x - 3\right)\left(4\cos^2 x - 1\right) = 0$$
$$\cos^2 x = \tfrac{3}{4} \qquad \text{or} \qquad \cos^2 x = \tfrac{1}{4}$$
$$\cos x = \pm\tfrac{\sqrt{3}}{2} \qquad \text{or} \qquad \cos x = \pm\tfrac{1}{2}$$

So the solutions are $x = \frac{\pi}{6}$, $\frac{\pi}{3}$, $\frac{2\pi}{3}$, $\frac{5\pi}{6}$, $\frac{7\pi}{6}$, $\frac{4\pi}{3}$, $\frac{5\pi}{3}$ or $\frac{11\pi}{6}$.

73. Using the suggestion, we re-write the equation in terms of sines and cosines:

$$\cot x + \csc x + \sec x = \tan x$$

$$\frac{\cos x}{\sin x} + \frac{1}{\sin x} + \frac{1}{\cos x} = \frac{\sin x}{\cos x}$$

Multiplying each side of the equation by $\sin x \cos x$, we get:

$$\cos^2 x + \cos x + \sin x = \sin^2 x$$

$$\cos^2 x - \sin^2 x + \cos x + \sin x = 0$$

$$(\cos x + \sin x)(\cos x - \sin x + 1) = 0$$

$$\cos x + \sin x = 0 \quad \text{or} \quad \cos x - \sin x + 1 = 0$$

From the first equation we have $\sin x = -\cos x$, so $\tan x = -1$ and thus $x = \frac{3\pi}{4}$ or $\frac{7\pi}{4}$.
From the second equation, we isolate $\cos x$ and square each side:

$$\cos x = \sin x - 1$$

$$\cos^2 x = \sin^2 x - 2\sin x + 1$$

$$1 - \sin^2 x = \sin^2 x - 2\sin x + 1$$

$$0 = 2\sin^2 x - 2\sin x$$

$$0 = 2\sin x(\sin x - 1)$$

Now $\sin x = 0$ when $x = 0$ or π, but then $\csc x$ is undefined. Also $\sin x = 1$ when $x = \frac{\pi}{2}$,

but then $\tan x$ is undefined. So the only solutions are $x = \frac{3\pi}{4}$ or $\frac{7\pi}{4}$.

75. Let $\theta = \tan^{-1}\left(\frac{\sqrt{2}}{2}\right)$, so $\tan \theta = \frac{\sqrt{2}}{2}$. We draw the triangle:

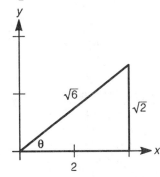

Thus $\sin \theta = \frac{\sqrt{2}}{\sqrt{6}} = \frac{\sqrt{3}}{3}$, so:

$$\cos\left\{\tan^{-1}\left[\sin\left(\tan^{-1}\left(\frac{\sqrt{2}}{2}\right)\right)\right]\right\} = \cos\left\{\tan^{-1}[\sin \theta]\right\} = \cos\left\{\tan^{-1}\left(\frac{\sqrt{3}}{3}\right)\right\} = \cos\frac{\pi}{6} = \frac{\sqrt{3}}{2}$$

77. We are asked to find the number x in $\left(-\frac{\pi}{2}, \frac{\pi}{2}\right)$ such that $\tan x = \frac{\sqrt{3}}{3}$. Since $x = \frac{\pi}{6}$ is that

number, then $\arctan\frac{\sqrt{3}}{3} = \frac{\pi}{6}$.

79. We are asked to find the number x in $\left[-\frac{\pi}{2}, \frac{\pi}{2}\right]$ such that $\sin x = \frac{1}{2}$. Since $x = \frac{\pi}{6}$ is that number, then $\arcsin \frac{1}{2} = \frac{\pi}{6}$.

81. We are asked to find the number x in $[0, \pi]$ such that $\cos x = \frac{1}{2}$. Since $x = \frac{\pi}{3}$ is that number, then $\cos^{-1}\left(\frac{1}{2}\right) = \frac{\pi}{3}$.

83. We are asked to find the number x in $[0, \pi]$ such that $\cos x = -\frac{1}{2}$. Since $x = \frac{2\pi}{3}$ is that number, then $\cos^{-1}\left(-\frac{1}{2}\right) = \frac{2\pi}{3}$.

85. Since $\cos\left(\cos^{-1} x\right) = x$ for every x in the interval $[-1, 1]$, then $\cos\left[\cos^{-1}\left(\frac{2}{7}\right)\right] = \frac{2}{7}$.

87. Let $\theta = \tan^{-1}(-1)$, so $\tan \theta = -1$ and θ is in the interval $\left(-\frac{\pi}{2}, \frac{\pi}{2}\right)$, thus $\theta = -\frac{\pi}{4}$.
     Therefore:
     $$\sin\left[\tan^{-1}(-1)\right] = \sin\left(-\frac{\pi}{4}\right) = -\frac{\sqrt{2}}{2}$$

89. Let $\theta = \cos^{-1}\left(\frac{\sqrt{2}}{3}\right)$, so $\cos \theta = \frac{\sqrt{2}}{3}$ and θ is in the interval $[0, \pi]$. Therefore:
     $$\sec\left[\cos^{-1}\left(\frac{\sqrt{2}}{3}\right)\right] = \sec \theta = \frac{1}{\cos \theta} = \frac{1}{\frac{\sqrt{2}}{3}} = \frac{3}{\sqrt{2}} = \frac{3\sqrt{2}}{2}$$

91. Let $\theta = \sin^{-1}\left(\frac{5}{13}\right)$, so $\sin \theta = \frac{5}{13}$ and θ is in the interval $\left[-\frac{\pi}{2}, \frac{\pi}{2}\right]$. We draw the triangle:

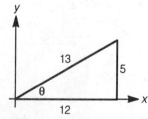

Using the addition formula for $\tan(s + t)$, we have:
$$\tan\left(\frac{\pi}{4} + \theta\right) = \frac{\tan\frac{\pi}{4} + \tan\theta}{1 - \tan\frac{\pi}{4}\tan\theta} = \frac{1 + \frac{5}{12}}{1 - 1 \bullet \frac{5}{12}} = \frac{\frac{17}{12}}{\frac{7}{12}} = \frac{17}{7}$$

93. Let $\theta = \tan^{-1} 2$, so $\tan \theta = 2$ and θ is in the interval $\left(-\frac{\pi}{2}, \frac{\pi}{2}\right)$. Using the double-angle formula for $\tan 2\theta$, we have:

$$\tan(2\theta) = \frac{2\tan\theta}{1-\tan^2\theta} = \frac{2 \bullet 2}{1-(2)^2} = \frac{4}{-3} = -\frac{4}{3}$$

95. Let $\theta = \cos^{-1}\left(\frac{4}{5}\right)$, so $\cos\theta = \frac{4}{5}$ and θ lies in the first quadrant. Thus $\frac{\theta}{2}$ lies in the first quadrant, so using the half-angle formula for $\cos\frac{\theta}{2}$, we have:

$$\cos\frac{\theta}{2} = \sqrt{\frac{1+\cos\theta}{2}} = \sqrt{\frac{1+\frac{4}{5}}{2}} = \sqrt{\frac{9}{10}} = \frac{3}{\sqrt{10}} = \frac{3\sqrt{10}}{10}$$

97. Let $\theta = \sec^{-1}\left(-\sqrt{2}\right)$, so $\sec\theta = -\sqrt{2}$ and θ is in the interval $\left[0,\frac{\pi}{2}\right) \cup \left(\frac{\pi}{2}, \pi\right]$. Since $\cos\theta = -\frac{\sqrt{2}}{2}$, then $\theta = \frac{3\pi}{4}$. So $\sec^{-1}\left(-\sqrt{2}\right) = \frac{3\pi}{4}$.

99. Since $\sec\left(\sec^{-1}x\right) = x$ for every x in $(-\infty,-1] \cup [1,\infty)$, then $\sec\left(\sec^{-1}\sqrt{6}\right) = \sqrt{6}$.

101. Let $\alpha = \tan^{-1}x$ and $\beta = \tan^{-1}y$, so $\tan\alpha = x$ and $\tan\beta = y$. Using the addition formula for $\tan(s+t)$, we have:

$$\tan\left(\tan^{-1}x + \tan^{-1}y\right) = \tan(\alpha+\beta) = \frac{\tan\alpha + \tan\beta}{1-\tan\alpha\tan\beta} = \frac{x+y}{1-xy}$$

103. Let $\theta = \arctan x$, so $\tan\theta = x$. We draw the triangle:

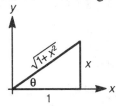

Using the double-angle formula for $\sin 2\theta$, we have:

$$\sin(2\arctan x) = \sin 2\theta = 2\sin\theta\cos\theta = 2 \bullet \frac{x}{\sqrt{x^2+1}} \bullet \frac{1}{\sqrt{x^2+1}} = \frac{2x}{x^2+1}$$

105. Let $\theta = \sin^{-1}(x^2)$, so $\sin\theta = x^2$. Since $\sin\theta \ge 0$, θ must lie in the first quadrant. We draw the triangle:

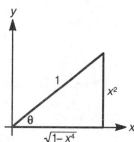

Noting that $\frac{\theta}{2}$ must also lie in the first quadrant, we use the half-angle formula for $\sin\frac{\theta}{2}$ to obtain:

$$\sin\left[\tfrac{1}{2}\sin^{-1}(x^2)\right] = \sin\tfrac{\theta}{2} = \sqrt{\frac{1-\cos\theta}{2}} = \sqrt{\frac{1-\sqrt{1-x^4}}{2}} = \sqrt{\tfrac{1}{2}-\tfrac{1}{2}\sqrt{1-x^4}}$$

107. Let $\alpha = \arcsin\frac{4\sqrt{41}}{41}$ and $\beta = \arcsin\frac{\sqrt{82}}{82}$, so $\sin\alpha = \frac{4\sqrt{41}}{41} = \frac{4}{\sqrt{41}}$ and $\sin\beta = \frac{\sqrt{82}}{82} = \frac{1}{\sqrt{82}}$. We draw the triangles:

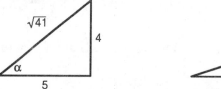

We now find $\tan(\alpha+\beta)$ by the addition formula for $\tan(s+t)$:

$$\tan(\alpha+\beta) = \frac{\tan\alpha + \tan\beta}{1 - \tan\alpha\tan\beta} = \frac{\frac{4}{5}+\frac{1}{9}}{1-\frac{4}{5}\cdot\frac{1}{9}} = \frac{\frac{41}{45}}{\frac{41}{45}} = 1$$

Since $\tan(\alpha+\beta) = 1$ and $0 < \alpha+\beta < \pi$, then $\alpha+\beta = \frac{\pi}{4}$. Therefore:

$$\arcsin\frac{4\sqrt{41}}{41} + \arcsin\frac{\sqrt{82}}{82} = \frac{\pi}{4}$$

109. (a) Using a calculator we find $\cos 20°\cos 40°\cos 60°\cos 80° = 0.0625$.

(b) Since $\cos 60° = \frac{1}{2}$, we have:

$$\cos 20°\cos 40°\cos 60°\cos 80° = \tfrac{1}{2}\cos 20°\cos 40°\cos 80°$$

Using the product-to-sum formula for $\cos A\cos B$, we have:

$$\cos 20°\cos 40° = \tfrac{1}{2}\left[\cos 60°+\cos(-20°)\right] = \tfrac{1}{2}\left[\tfrac{1}{2}+\cos 20°\right] = \tfrac{1}{4}+\tfrac{1}{2}\cos 20°$$

Therefore:

$$\cos 20°\cos 40°\cos 60°\cos 80° = \tfrac{1}{2}\cos 20°\cos 40°\cos 80°$$
$$= \tfrac{1}{2}\left(\tfrac{1}{4}+\tfrac{1}{2}\cos 20°\right)\cos 80°$$
$$= \tfrac{1}{8}\cos 80°+\tfrac{1}{4}\cos 20°\cos 80°$$

Using the product-to-sum formula for $\cos A \cos B$, and the identity $\cos\theta = -\cos(180° - \theta)$, we have:

$$\cos 20°\cos 80° = \tfrac{1}{2}\left[\cos 100° + \cos(-60°)\right]$$
$$= \tfrac{1}{2}\left[-\cos 80° + \tfrac{1}{2}\right]$$
$$= -\tfrac{1}{2}\cos 80° + \tfrac{1}{4}$$

Therefore:

$$\cos 20°\cos 40°\cos 60°\cos 80° = \tfrac{1}{8}\cos 80° + \tfrac{1}{4}\cos 20°\cos 80°$$
$$= \tfrac{1}{8}\cos 80° + \tfrac{1}{4}\left(-\tfrac{1}{2}\cos 80° + \tfrac{1}{4}\right)$$
$$= \tfrac{1}{8}\cos 80° - \tfrac{1}{8}\cos 80° + \tfrac{1}{16}$$
$$= \tfrac{1}{16}$$
$$= 0.0625$$

111. Working from the right-hand side, we have:

$$\frac{2}{\cot\theta + \tan\theta} = \frac{2}{\frac{\cos\theta}{\sin\theta} + \frac{\sin\theta}{\cos\theta}} = \frac{2}{\frac{\cos^2\theta + \sin^2\theta}{\sin\theta\cos\theta}} = \frac{2}{\frac{1}{\sin\theta\cos\theta}} = 2\sin\theta\cos\theta = \sin 2\theta$$

113. Using the double-angle formula for $\tan 2\theta$, we have:

$$\frac{1}{1 + \tan 2\theta\tan\theta} = \frac{1}{1 + \frac{2\tan\theta}{1-\tan^2\theta}\bullet\tan\theta}$$
$$= \frac{1}{1 + \frac{2\tan^2\theta}{1-\tan^2\theta}}$$
$$= \frac{1-\tan^2\theta}{\left(1-\tan^2\theta\right) + 2\tan^2\theta}$$
$$= \frac{1-\tan^2\theta}{1+\tan^2\theta}$$
$$= \frac{1 - \frac{\sin^2\theta}{\cos^2\theta}}{1 + \frac{\sin^2\theta}{\cos^2\theta}}$$
$$= \frac{\cos^2\theta - \sin^2\theta}{\cos^2\theta + \sin^2\theta}$$
$$= \frac{\cos 2\theta}{1}$$
$$= \cos 2\theta$$

115. (a) Working from the right-hand side, we have:

$$\frac{2\tan\theta}{1+\tan^2\theta} = \frac{2\bullet\frac{\sin\theta}{\cos\theta}}{1 + \frac{\sin^2\theta}{\cos^2\theta}} = \frac{2\sin\theta\cos\theta}{\cos^2\theta + \sin^2\theta} = \frac{\sin 2\theta}{1} = \sin 2\theta$$

(b) Working from the right-hand side, we have:

$$\frac{1-\tan^2\theta}{1+\tan^2\theta}=\frac{1-\frac{\sin^2\theta}{\cos^2\theta}}{1+\frac{\sin^2\theta}{\cos^2\theta}}=\frac{\cos^2\theta-\sin^2\theta}{\cos^2\theta+\sin^2\theta}=\frac{\cos2\theta}{1}=\cos2\theta$$

117. Using the product-to-sum formula for $\sin A\sin B$, we have:

$$\sin\theta\sin\left[(n-1)\theta\right]=\tfrac{1}{2}\left[\cos(2\theta-n\theta)-\cos(n\theta)\right]$$
$$=\tfrac{1}{2}\cos\left[(2-n)\theta\right]-\tfrac{1}{2}\cos n\theta$$
$$=\tfrac{1}{2}\cos\left[(n-2)\theta\right]-\tfrac{1}{2}\cos n\theta$$

Therefore, the right-hand side of the original identity becomes:

$$\cos\left[(n-2)\theta\right]-2\sin\theta\sin\left[(n-1)\theta\right]=\cos\left[(n-2)\theta\right]-\cos\left[(n-2)\theta\right]+\cos n\theta$$
$$=\cos n\theta$$

Chapter Four Test

1.   Using the addition formula for $\sin(s+t)$, we have:

$$\sin\left(\theta+\tfrac{3\pi}{2}\right)=\sin\theta\cos\tfrac{3\pi}{2}+\cos\theta\sin\tfrac{3\pi}{2}=\sin\theta\bullet0+\cos\theta\bullet(-1)=-\cos\theta$$

2.   Since $\tfrac{3\pi}{2}<t<2\pi$, we draw the triangle:

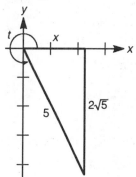

By the Pythagorean theorem:

$$x^2+\left(2\sqrt{5}\right)^2=(5)^2$$
$$x^2+20=25$$
$$x^2=5$$
$$x=\sqrt{5}$$

So $\cos t=\tfrac{\sqrt{5}}{5}$, now use the double-angle formula for $\cos 2t$:

$$\cos2t=\cos^2t-\sin^2t=\left(\tfrac{\sqrt{5}}{5}\right)^2-\left(-\tfrac{2\sqrt{5}}{5}\right)^2=\tfrac{1}{5}-\tfrac{4}{5}=-\tfrac{3}{5}$$

3. Since $\pi < \theta < \frac{3\pi}{2}$, we draw the triangle:

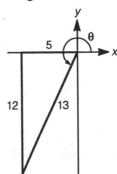

So $\sin \theta = -\frac{12}{13}$, now use the half-angle formula for $\tan \frac{\theta}{2}$:

$$\tan \frac{\theta}{2} = \frac{\sin \theta}{1 + \cos \theta} = \frac{-\frac{12}{13}}{1 - \frac{5}{13}} = \frac{-\frac{12}{13}}{\frac{8}{13}} = -\frac{3}{2}$$

4. Dividing each side of the equation by $\cos x$ yields $\tan x = 3$, which has a principal solution of $x = \tan^{-1} 3 \approx 1.25$. Since $\tan x$ is also positive in the third quadrant, the other solution in the interval $(0, 2\pi)$ is $1.25 + \pi \approx 4.39$.

5. Factoring, we have:
$$2 \sin^2 x + 7 \sin x + 3 = 0$$
$$(2 \sin x + 1)(\sin x + 3) = 0$$
$$\sin x = -\frac{1}{2} \quad \text{or} \quad \sin x = -3 \ \text{(impossible)}$$

So $x = \frac{7\pi}{6}, \frac{11\pi}{6}$.

6. Since $\cos \alpha = \frac{2}{\sqrt{5}}$ and $\frac{3\pi}{2} < \alpha < 2\pi$ (fourth quadrant), then:
$$\sin \alpha = -\sqrt{1 - \cos^2 \alpha} = -\sqrt{1 - \frac{4}{5}} = -\frac{1}{\sqrt{5}}$$

Similarly, since $\sin \beta = \frac{4}{5}$ and $\frac{\pi}{2} < \beta < \pi$ (second quadrant), then:
$$\cos \beta = -\sqrt{1 - \sin^2 \beta} = -\sqrt{1 - \frac{16}{25}} = -\sqrt{\frac{9}{25}} = -\frac{3}{5}$$
Using the addition formula for $\sin (\beta - \alpha)$, we have:
$$\sin (\beta - \alpha) = \sin \beta \cos \alpha - \cos \beta \sin \alpha = \frac{4}{5} \cdot \frac{2}{\sqrt{5}} - \left(-\frac{3}{5}\right)\left(-\frac{1}{\sqrt{5}}\right) = \frac{8}{5\sqrt{5}} - \frac{3}{5\sqrt{5}} = \frac{5}{5\sqrt{5}} = \frac{\sqrt{5}}{5}$$

7. First, we simplify the left-hand side using the addition formula for $\sin (\alpha + \beta)$:
$$\sin (x + 30°) = \sin x \cos 30° + \cos x \sin 30° = \frac{\sqrt{3}}{2} \sin x + \frac{1}{2} \cos x$$

Thus, we have the equation:

$$\frac{\sqrt{3}}{2}\sin x + \frac{1}{2}\cos x = \sqrt{3}\sin x$$
$$\sqrt{3}\sin x + \cos x = 2\sqrt{3}\sin x$$
$$\cos x = \sqrt{3}\sin x$$
$$\tan x = \frac{1}{\sqrt{3}}$$
$$x = 30°$$

8.   If $\csc\theta = -3$, then $\sin\theta = -\frac{1}{3}$. Since $\pi < \theta < \frac{3\pi}{2}$ (third quadrant), we have:

$$\cos\theta = -\sqrt{1-\sin^2\theta} = -\sqrt{1-\frac{1}{9}} = -\sqrt{\frac{8}{9}} = -\frac{2\sqrt{2}}{3}$$

Now $\frac{\pi}{2} < \frac{\theta}{2} < \frac{3\pi}{4}$ (second quadrant), so using the half-angle formula for $\sin\frac{\theta}{2}$ we have:

$$\sin\frac{\theta}{2} = \sqrt{\frac{1-\cos\theta}{2}} = \sqrt{\frac{1+\frac{2\sqrt{2}}{3}}{2}} = \sqrt{\frac{3+2\sqrt{2}}{6}} = \frac{\sqrt{18+12\sqrt{2}}}{6}$$

9.   For the restricted sine function, the domain is $\left[-\frac{\pi}{2},\frac{\pi}{2}\right]$ and the range is $[-1,1]$. For the

function $y = \sin^{-1}x$, the domain is $[-1,1]$ and the range is $\left[-\frac{\pi}{2},\frac{\pi}{2}\right]$.

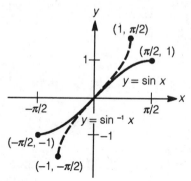

10. (a)   Since $\sin^{-1}(\sin x) = x$ for every x in the interval $\left[-\frac{\pi}{2},\frac{\pi}{2}\right]$, then $\sin^{-1}\left(\sin\frac{\pi}{10}\right) = \frac{\pi}{10}$.

     (b)   Since $\sin 2\pi = 0$, we are asked to find $x = \sin^{-1}0$. Then $\sin x = 0$ and x is in the

           interval $\left[-\frac{\pi}{2},\frac{\pi}{2}\right]$, thus $x = 0$. So $\sin^{-1}(\sin 2\pi) = 0$. Notice that we cannot use the

           identity $\sin^{-1}(\sin x) = x$, since 2π is not in the interval $\left[-\frac{\pi}{2},\frac{\pi}{2}\right]$.

11. Let $\theta = \arcsin \frac{3}{4}$, so $\sin \theta = \frac{3}{4}$ and θ is in the interval $\left[-\frac{\pi}{2}, \frac{\pi}{2}\right]$. We draw the triangle:

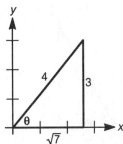

Therefore:
$$\cos\left(\arcsin \frac{3}{4}\right) = \cos \theta = \frac{\sqrt{7}}{4}$$

12. Using the addition formula for $\tan(s + t)$, we have:
$$\tan\left(\frac{\pi}{4} + \frac{\theta}{2}\right) = \frac{\tan \frac{\pi}{4} + \tan \frac{\theta}{2}}{1 - \tan \frac{\pi}{4} \tan \frac{\theta}{2}} = \frac{1 + \tan \frac{\theta}{2}}{1 - \tan \frac{\theta}{2}}$$

Now using the half-angle formula for $\tan \frac{\theta}{2}$, we have:
$$\tan\left(\frac{\pi}{4} + \frac{\theta}{2}\right) = \frac{1 + \tan \frac{\theta}{2}}{1 - \tan \frac{\theta}{2}} = \frac{1 + \frac{\sin \theta}{1 + \cos \theta}}{1 - \frac{\sin \theta}{1 + \cos \theta}} = \frac{1 + \cos \theta + \sin \theta}{1 + \cos \theta - \sin \theta}$$

13. Using the product-to-sum formula for $\sin A \cos B$, we have:
$$\begin{aligned}
\sin \frac{7\pi}{24} \cos \frac{\pi}{24} &= \frac{1}{2}\left[\sin\left(\frac{7\pi}{24} + \frac{\pi}{24}\right) + \sin\left(\frac{7\pi}{24} - \frac{\pi}{24}\right)\right] \\
&= \frac{1}{2}\left[\sin \frac{\pi}{3} + \sin \frac{\pi}{4}\right] \\
&= \frac{1}{2}\left(\frac{\sqrt{3}}{2} + \frac{\sqrt{2}}{2}\right) \\
&= \frac{\sqrt{3} + \sqrt{2}}{4}
\end{aligned}$$

14. Using the sum-to-product formulas for $\sin \alpha + \sin \beta$ and $\cos \alpha + \cos \beta$, we have:
$$\sin 3\theta + \sin 5\theta = 2 \sin \frac{3\theta + 5\theta}{2} \cos \frac{3\theta - 5\theta}{2} = 2 \sin 4\theta \cos(-\theta) = 2 \sin 4\theta \cos \theta$$
$$\cos 3\theta + \cos 5\theta = 2 \cos \frac{3\theta + 5\theta}{2} \cos \frac{3\theta - 5\theta}{2} = 2 \cos 4\theta \cos(-\theta) = 2 \cos 4\theta \cos \theta$$

Therefore:
$$\frac{\sin 3\theta + \sin 5\theta}{\cos 3\theta + \cos 5\theta} = \frac{2 \sin 4\theta \cos \theta}{2 \cos 4\theta \cos \theta} = \frac{\sin 4\theta}{\cos 4\theta} = \tan 4\theta$$

15. **(a)** Let $\theta = \arctan\sqrt{x^2 - 1}$, so $\tan\theta = \sqrt{x^2 - 1}$. We draw the triangle:

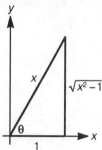

Therefore:

$$\sec\left[\arctan\sqrt{x^2 - 1}\right] = \sec\theta = \tfrac{x}{1} = x$$

(b) Let $\alpha = \sec^{-1}\left(\tfrac{5}{3}\right)$ and $\beta = \tan^{-1}\left(\tfrac{3}{4}\right)$, so $\sec\alpha = \tfrac{5}{3}$ and $\tan\beta = \tfrac{3}{4}$. We draw the triangles:

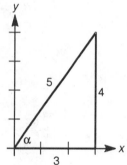 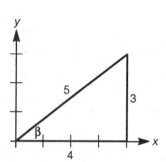

Since α and β are both in the first quadrant, then $\sin\alpha > 0$, $\sin\beta > 0$ and $\cos\beta > 0$. Now use the addition formula for $\sin(s + t)$:

$$
\begin{aligned}
\sin\left[\sec^{-1}\left(\tfrac{5}{3}\right) + \tan^{-1}\left(\tfrac{3}{4}\right)\right] &= \sin(\alpha + \beta) \\
&= \sin\alpha\cos\beta + \cos\alpha\sin\beta \\
&= \tfrac{4}{5}\bullet\tfrac{4}{5} + \tfrac{3}{5}\bullet\tfrac{3}{5} \\
&= \tfrac{16}{25} + \tfrac{9}{25} \\
&= \tfrac{25}{25} \\
&= 1
\end{aligned}
$$

Notice that since $\sin(\alpha + \beta) = 1$, then $\alpha + \beta = \tfrac{\pi}{2}$ and thus α and β are complementary angles.

16. (a)   The domain of $y = \sec^{-1} x$ is $(-\infty, -1] \cup [1, \infty)$, and the range is $\left[0, \frac{\pi}{2}\right) \cup \left(\frac{\pi}{2}, \pi\right]$.

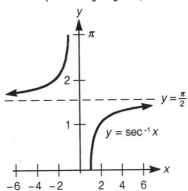

(b)   The domain of $y = \tan^{-1} x$ is $(-\infty, \infty)$, and the range is $\left(-\frac{\pi}{2}, \frac{\pi}{2}\right)$.

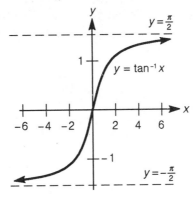

Chapter Five
Triangles and Vectors

5.1 The Law of Sines

1.   We draw the triangle:

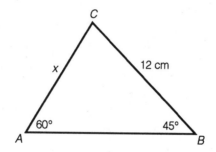

Using the law of sines:
$$\frac{\sin 45°}{x} = \frac{\sin 60°}{12}$$

Thus:
$$x = 12 \bullet \frac{\sin 45°}{\sin 60°} = 12 \bullet \frac{\sqrt{2}}{2} \bullet \frac{2}{\sqrt{3}} = \frac{12\sqrt{2}}{\sqrt{3}} = 4\sqrt{6} \text{ cm}$$

3.   We draw the triangle:

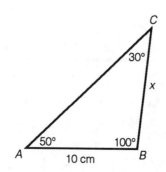

Using the law of sines:
$$\frac{\sin 50°}{x} = \frac{\sin 30°}{10}$$
Thus:
$$x = \frac{10\sin 50°}{\sin 30°} = \frac{10\sin 50°}{\frac{1}{2}} = 20\sin 50° \text{ cm}$$

5. We draw a triangle:

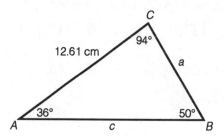

Using the law of sines:
$$\frac{\sin 36°}{a} = \frac{\sin 50°}{12.61}$$
Thus:
$$a = \frac{12.61\sin 36°}{\sin 50°} \approx 9.7 \text{ cm}$$
Using the law of sines:
$$\frac{\sin 94°}{c} = \frac{\sin 50°}{12.61}$$
Thus:
$$c = \frac{12.61\sin 94°}{\sin 50°} \approx 16.4 \text{ cm}$$

7. We draw a triangle:

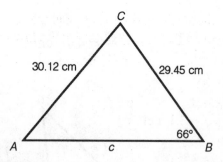

Using the law of sines:
$$\frac{\sin A}{29.45 \text{ cm}} = \frac{\sin 66°}{30.12 \text{ cm}}$$
Thus:
$$\sin A = \frac{29.45\sin 66°}{30.12} \approx 0.8932, \text{ so } A \approx 63.3°$$

Then $C = 180° - 66° - 63.3° \approx 50.7°$. Using the law of sines:
$$\frac{\sin 50.7°}{c} = \frac{\sin 66°}{30.12}$$
Thus:
$$c = \frac{30.12 \sin 50.7°}{\sin 66°} \approx 25.5 \text{ cm}$$

9. (a) Since $\sin B = \frac{\sqrt{2}}{2}$, then $\angle B = 45°$ or $\angle B = 135°$.

(b) Since $\cos E = \frac{\sqrt{2}}{2}$, then $\angle E = 45°$. Note that $\angle E \neq 135°$.

(c) Since $\sin H = \frac{1}{4}$, then $\angle H \approx 14.5°$ or $\angle H \approx 165.5°$.

(d) Since $\cos K = -\frac{2}{3}$, then $\angle K \approx 131.8°$.

11. (a) Using the law of sines, we have:
$$\frac{\sin 23.1°}{2.0} = \frac{\sin B}{6.0}$$
Thus:
$$\sin B = \frac{6.0 \sin 23.1°}{2.0} \approx 1.18$$
But $\sin B \leq 1$, so no such triangle exists.

(b) We first find $\angle B$:
$$\frac{\sin 23.1°}{2.0} = \frac{\sin B}{3.0}$$
Thus:
$$\sin B = \frac{3.0 \sin 23.1°}{2.0} \approx 0.5885, \text{ so } B \approx 36.05° \text{ or } B \approx 143.95°$$
Since $\angle B$ is obtuse, then $\angle B \approx 143.95°$. Therefore
$\angle C = 180° - 23.1° - 143.95° \approx 12.95°$, so by the law of sines:
$$\frac{\sin 12.95°}{c} = \frac{\sin 23.1°}{2.0}$$
Thus:
$$c = \frac{2.0 \sin 12.95°}{\sin 23.1°} \approx 1.1 \text{ feet}$$

13. (a) Using the law of sines, we have:
$$\frac{\sin A}{\sqrt{2}} = \frac{\sin 30°}{1}$$
Thus:
$$\sin A = \sqrt{2} \sin 30° = \sqrt{2} \cdot \frac{1}{2} = \frac{\sqrt{2}}{2}$$
Therefore $\angle A = 45°$ or $\angle A = 135°$.

(b)   If $\angle A = 45°$, then $\angle C = 180° - 45° - 30° = 105°$. Using the law of sines, we have:
$$\frac{\sin 105°}{c} = \frac{\sin 30°}{1}$$
Thus:
$$c = \frac{\sin 105°}{\sin 30°} \approx 1.93$$

(c)   If $\angle A = 135°$, then $\angle C = 180° - 135° - 30° = 15°$. Using the law of sines, we have:
$$\frac{\sin 15°}{c} = \frac{\sin 30°}{1}$$
$$c = \frac{\sin 15°}{\sin 30°} \approx 0.52$$

(d)   We find the area of each triangle:
$$A = \tfrac{1}{2}ab\sin C = \tfrac{1}{2} \bullet \sqrt{2} \bullet 1\sin 105° \approx 0.68$$
$$A = \tfrac{1}{2}ab\sin C = \tfrac{1}{2} \bullet \sqrt{2}\sin 15° \approx 0.18$$

15.   Using the law of sines, we have:
$$\frac{\sin 20°}{2} = \frac{\sin 100°}{a} = \frac{\sin 50°}{b}$$
$$\frac{\sin 70°}{c} = \frac{\sin 95°}{b} = \frac{\sin 15°}{d}$$
Hence:
$$a = \frac{2\sin 110°}{\sin 20°} = \frac{2\sin 70°}{\sin 20°} \text{ cm}$$
$$b = \frac{2\sin 50°}{\sin 20°} \text{ cm}$$
$$c = \frac{b\sin 70°}{\sin 95°} = \frac{2\sin 50°\sin 70°}{\sin 20°\sin 85°} \text{ cm}$$
$$d = \frac{b\sin 15°}{\sin 95°} = \frac{2\sin 50°\sin 15°}{\sin 20°\sin 85°} \text{ cm}$$

17.   (a)   The diameter of the circle is given by:
$$\frac{b}{\sin B} = \frac{12 \text{ cm}}{\sin 109°}$$

Since the radius is half of the diameter, then $r = \frac{6 \text{ cm}}{\sin 109°}$. Therefore the area of the circle is:
$$A = \pi r^2 = \pi \left(\frac{6}{\sin 109°}\right)^2 = \frac{36\pi}{\sin^2 109°} \approx 126.5 \text{ cm}^2$$

(b) First we find $\angle C = 180° - 38° - 109° = 33°$. Using the law of sines, we find side a:
$$\frac{\sin 109°}{12} = \frac{\sin 38°}{a}$$
Thus:
$$a = \frac{12\sin 38°}{\sin 109°}$$
So the area of the triangle is:
$$\tfrac{1}{2}ab\sin C = \tfrac{1}{2}\cdot\frac{12\sin 38°}{\sin 109°}\cdot 12\sin 33° = \frac{72\sin 38°\sin 33°}{\sin 109°}$$
So the area of the region within the circumscribed circle but outside of $\triangle ABC$ is given by:
$$A = \frac{36\pi}{\sin^2 109°} - \frac{72\sin 38°\sin 33°}{\sin 109°} \approx 101.0 \text{ cm}^2$$

19. (a) The diameter of the circle is given by:
$$\frac{b}{\sin B} = \frac{19 \text{ cm}}{\sin 130°}$$

Since the radius is half of the diameter, then $r = \frac{19 \text{ cm}}{2\sin 130°}$. Therefore the area of the circle is:
$$A = \pi r^2 = \pi\left(\frac{19}{2\sin 130°}\right)^2 = \frac{361\pi}{4\sin^2 130°} \approx 483.2 \text{ cm}^2$$

(b) We first use the law of sines to find $\angle C$:
$$\frac{\sin 130°}{19} = \frac{\sin C}{14}$$
Thus:
$$\sin C = \frac{14\sin 30°}{19} \approx 0.5645, \text{ so } C \approx 34.36°$$
Then $\angle A = 180° - 130° - 34.36° \approx 15.64°$. So the area of the triangle is:
$$\tfrac{1}{2}bc\sin A = \tfrac{1}{2}(19)(14)\bullet\sin 15.64° = 133\sin 15.64°$$
So the area of the region within the circumscribed circle but outside of $\triangle ABC$ is given by:
$$A = \frac{361\pi}{4\sin^2 130°} - 133\sin 15.64° \approx 447.3 \text{ cm}^2$$

21. We sketch the triangle:

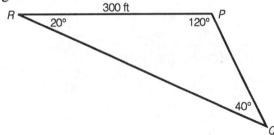

Applying the law of sines, we have:
$$\frac{\sin 40°}{300} = \frac{\sin 20°}{PQ}$$

So $PQ = \dfrac{300\sin 20°}{\sin 40°} \approx 160$ ft.

23. We draw the figure:

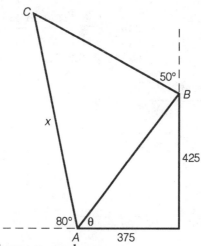

Using the Pythagorean theorem, we have:
$$AB = \sqrt{(375)^2 + (425)^2} \approx 566.8 \text{ ft}$$

Since $\tan\theta = \frac{425}{375}$, then $\theta \approx 48.6°$. We compute the following angles:

$\angle CAB = 180° - 80° - 48.6° = 51.4°$

$\angle ABC = 180° - 50° - (90° - 48.6°) = 88.6°$

$\angle ACB = 180° - 51.4° - 88.6° = 40°$

Using the law of sines, we have:
$$\frac{\sin 40°}{566.8} = \frac{\sin 88.6°}{x}$$

Thus:
$$x = \frac{566.8\sin 88.6°}{\sin 40°} \approx 881.5 \text{ ft}$$

25. Applying the law of sines, we have:
$$\frac{\sin(\angle BAD)}{n} = \frac{\sin\theta}{c} \quad \text{while} \quad \frac{\sin(\angle DAC)}{m} = \frac{\sin(180° - \theta)}{b}$$

But $\angle BAD = \angle DAC$ and $\sin\theta = \sin(180° - \theta)$, so:
$$\frac{n}{c} = \frac{\sin(\angle BAD)}{\sin\theta} = \frac{\sin(\angle DAC)}{\sin(180° - \theta)} = \frac{m}{b}$$

Hence:
$$\frac{n}{m} = \frac{c}{b}$$

27. **(a)** From the geometry theorem $\angle C = \frac{1}{2}\angle AOB$. Since $\angle OAD = \angle OBD$ (base angles of an isosceles triangle are congruent), then $\angle AOD = \angle BOD = \frac{1}{2}\angle AOB$, and thus $\angle C = \angle AOD$.

(b) Since $AD = \frac{1}{2}AB = \frac{c}{2}$, then:

$$\frac{\sin C}{c} = \frac{\sin \angle AOD}{2AD}$$

From the law of sines:

$$\frac{\sin \angle AOD}{AD} = \frac{\sin 90°}{r}$$

Hence:

$$\frac{1}{r} = \frac{2\sin C}{c}$$

$$r = \frac{c}{2\sin C}$$

(c) We draw similar figures to that given in the problem, except that D is now on $\overline{BC}$ and on $\overline{AC}$:

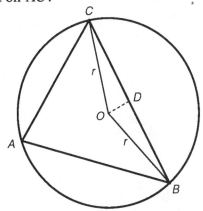 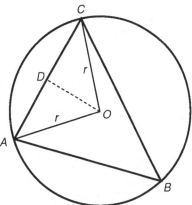

Using the same reasoning as (a) and (b), we have:

$$r = \frac{a}{2\sin A} \quad \text{and} \quad r = \frac{b}{2\sin B}$$

So:

$$\tfrac{1}{2}r = \frac{\sin C}{c}, \quad \tfrac{1}{2}r = \frac{\sin A}{a}, \quad \tfrac{1}{2}r = \frac{\sin B}{b}$$

Hence:

$$\frac{\sin A}{a} = \frac{\sin B}{b} = \frac{\sin C}{c}$$

(d) We draw the figure:

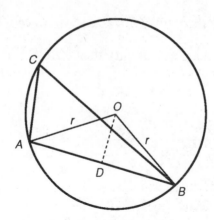

Using this figure and the reasoning from (a) and (b), we have:

$$r = \frac{c}{2\sin C}$$

29. Since K is the area of $\triangle ABC$, then:

$$K = \tfrac{1}{2}ab\sin C$$

But since $\mathcal{D} = \frac{a}{\sin A}$, then $a = \mathcal{D}\sin A$, and since $\mathcal{D} = \frac{b}{\sin B}$, then $b = \mathcal{D}\sin B$.
Therefore:

$$K = \tfrac{1}{2}ab\sin C = \tfrac{1}{2}\bullet\mathcal{D}\sin A\bullet\mathcal{D}\sin B\bullet\sin C = \frac{\mathcal{D}^2 \sin A\sin B\sin C}{2}$$

31. (a) Since $\tfrac{1}{2}A + \tfrac{1}{2}B + \tfrac{1}{2}C = 90°$, then $\tfrac{1}{2}(A+B) = 90°-\tfrac{1}{2}C$, so:

$$\sin\tfrac{1}{2}(A+B) = \sin\left(90°-\tfrac{1}{2}C\right) = \cos\tfrac{1}{2}C$$

(b) By the law of sines $\frac{a}{\sin A} = \frac{b}{\sin B} = \frac{c}{\sin C}$, so:

$$\frac{a}{c} = \frac{\sin A}{\sin C} \quad\text{and}\quad \frac{b}{c} = \frac{\sin B}{\sin C}$$

Therefore:

$$\frac{a+b}{c} = \frac{a}{c}+\frac{b}{c} = \frac{\sin A}{\sin C}+\frac{\sin B}{\sin C} = \frac{\sin A + \sin B}{\sin C}$$

(c) Using the sum-to-product formula:

$$\sin A + \sin B = 2\sin\tfrac{A+B}{2}\cos\tfrac{A-B}{2}$$

Using the double-angle formula for sine:

$$\sin C = 2\sin\tfrac{C}{2}\cos\tfrac{C}{2}$$

Therefore:

$$\frac{a+b}{c} = \frac{\sin A + \sin B}{\sin C} = \frac{2\sin\tfrac{A+B}{2}\cos\tfrac{A-B}{2}}{2\sin\tfrac{C}{2}\cos\tfrac{C}{2}} = \frac{\sin\tfrac{1}{2}(A+B)\cos\tfrac{1}{2}(A-B)}{\sin\tfrac{1}{2}C\cos\tfrac{1}{2}C}$$

(d)   Since $\sin\frac{1}{2}(A+B) = \cos\frac{1}{2}C$ from part (a), we have:

$$\frac{a+b}{c} = \frac{\sin\frac{1}{2}(A+B)\cos\frac{1}{2}(A-B)}{\sin\frac{1}{2}C\cos\frac{1}{2}C} = \frac{\cos\frac{1}{2}C\cos\frac{1}{2}(A-B)}{\sin\frac{1}{2}C\cos\frac{1}{2}C} = \frac{\cos\frac{1}{2}(A-B)}{\sin\frac{1}{2}C}$$

33.   (a)   Using $a = \mathcal{D}\sin A$ and $b = \mathcal{D}\sin B$, we have:

$$\frac{a-b}{a+b} = \frac{\mathcal{D}\sin A - \mathcal{D}\sin B}{\mathcal{D}\sin A + \mathcal{D}\sin B} = \frac{\mathcal{D}(\sin A - \sin B)}{\mathcal{D}(\sin A + \sin B)} = \frac{\sin A - \sin B}{\sin A + \sin B}$$

(b)   Using the sum-to-product formulas, we have:

$$\sin A - \sin B = 2\cos\frac{A+B}{2}\sin\frac{A-B}{2}$$

$$\sin A + \sin B = 2\sin\frac{A+B}{2}\cos\frac{A-B}{2}$$

Therefore our result from part (a) becomes:

$$\frac{a-b}{a+b} = \frac{2\cos\frac{A+B}{2}\sin\frac{A-B}{2}}{2\sin\frac{A+B}{2}\cos\frac{A-B}{2}} = \frac{\sin\frac{A-B}{2}}{\cos\frac{A-B}{2}} \div \frac{\sin\frac{A+B}{2}}{\cos\frac{A+B}{2}} = \frac{\tan\frac{1}{2}(A-B)}{\tan\frac{1}{2}(A+B)}$$

5.2 The Law of Cosines

1.   (a)   Using the law of cosines:

$$x^2 = 5^2 + 8^2 - 2(5)(8)\cos 60° = 25 + 64 - 80 \bullet \tfrac{1}{2} = 49$$

So $x = \sqrt{49} = 7$ cm.

(b)   Using the law of cosines:

$$x^2 = 5^2 + 8^2 - 2(5)(8)\cos 120° = 25 + 64 - 80 \bullet \left(-\tfrac{1}{2}\right) = 129$$

So $x = \sqrt{129}$ cm.

3.   (a)   Using the law of cosines:

$$x^2 = (7.3)^2 + (11.5)^2 - 2(7.3)(11.5)\cos 40°$$
$$= 53.29 + 132.25 - 167.9\cos 40°$$
$$= 185.54 - 167.9\cos 40°$$
$$\approx 56.92$$

So $x \approx \sqrt{56.92} \approx 7.5$ cm.

(b)   Using the law of cosines:

$$x^2 = (7.3)^2 + (11.5)^2 - 2(7.3)(11.5)\cos 140°$$
$$= 53.29 + 132.25 - 167.9\cos 140°$$
$$= 185.54 - 167.9\cos 140°$$
$$\approx 314.16$$

So $x \approx \sqrt{314.16} \approx 17.7$ cm.

5. This is incorrect because x is not the side opposite the 130° angle. The correct equation is:

$$6^2 = x^2 + 3^2 - 2(x)(3)\cos 130°$$

7. Using $a = 6$, $b = 7$, $c = 10$ and the law of cosines, we have:

$$6^2 = 7^2 + 10^2 - 2(7)(10)\cos A, \text{ so } \cos A = \tfrac{113}{140}$$
$$7^2 = 6^2 + 10^2 - 2(6)(10)\cos B, \text{ so } \cos B = \tfrac{87}{120} = \tfrac{29}{40}$$
$$10^2 = 6^2 + 7^2 - 2(6)(7)\cos C, \text{ so } \cos C = -\tfrac{15}{84} = -\tfrac{5}{28}$$

9. We use the law of cosines to find angle A:

$$7^2 = 8^2 + 13^2 - 2(8)(13)\cos A$$
$$49 = 64 + 169 - 208\cos A$$
$$-184 = -208\cos A$$
$$\cos A = \tfrac{184}{208}$$
$$A \approx 27.8°$$

Now use the law of cosines to find angle B:

$$8^2 = 7^2 + 13^2 - 2(7)(13)\cos B$$
$$64 = 49 + 169 - 182\cos B$$
$$-154 = -182\cos B$$
$$\cos B = \tfrac{154}{182}$$
$$B \approx 32.2°$$

Since $A + B + C = 180°$, we have:
$$C = 180° - A - B \approx 180° - 27.8° - 32.2° \approx 120°$$

11. We use the law of cosines to find angle A:

$$\left(\tfrac{2}{\sqrt{3}}\right)^2 = \left(\tfrac{2}{\sqrt{3}}\right)^2 + 2^2 - 2 \cdot \tfrac{2}{\sqrt{3}} \cdot 2\cos A$$
$$\tfrac{4}{3} = \tfrac{4}{3} + 4 - \tfrac{8}{\sqrt{3}}\cos A$$
$$-4 = -\tfrac{8}{\sqrt{3}}\cos A$$
$$\cos A = \tfrac{\sqrt{3}}{2}$$
$$A = 30°$$

Since $a = b$, then $B = 30°$. Since $A + B + C = 180°$, we have:
$$C = 180° - A - B = 180° - 30° - 30° = 120°$$

13. We first draw a figure, noting that the central angle is $\frac{360°}{5} = 72°$:

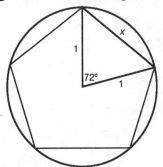

We apply the law of cosines to find x:
$$x^2 = 1^2 + 1^2 - 2(1)(1)\cos 72° = 2 - 2\cos 72° \approx 1.382$$
So $x \approx \sqrt{1.382} \approx 1.18$, and thus the perimeter is $5(1.18) \approx 5.9$ units.

15. (a) Using the law of cosines, we have:
$$\begin{aligned} a^2 &= (6.1)^2 + (3.2)^2 - 2(6.1)(3.2)\cos 40° \\ &= 37.21 + 10.24 - 39.04\cos 40° \\ &= 47.45 - 39.04\cos 40° \\ &\approx 17.54 \end{aligned}$$
So $a \approx \sqrt{17.54} \approx 4.2$ cm.

(b) Using the law of sines, we have:
$$\begin{aligned} \frac{\sin C}{3.2} &= \frac{\sin 40°}{4.2} \\ \sin C &= \frac{3.2\sin 40°}{4.2} \\ \sin C &\approx 0.49 \\ C &\approx 29.3° \end{aligned}$$

(c) Since $A + B + C = 180°$, we have:
$$B = 180° - A - C \approx 180° - 40° - 29.3° \approx 110.7°$$

17. We first draw a figure indicating the relationship between Town A, Town B and Town C, where Town A is centered at the origin:

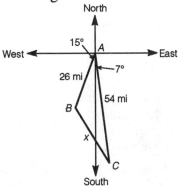

Using $\triangle ABC$ and the law of cosines, we have:
$$x^2 = 26^2 + 54^2 - 2(26)(54)\cos 22°$$
$$= 676 + 2916 - 2808\cos 22°$$
$$= 3592 - 2808\cos 22°$$
$$\approx 988.5$$
So $x \approx \sqrt{988.5} \approx 31$. The distance between Towns B and C is approximately 31 miles.

19. Using P to denote the plane, we have the following figure:

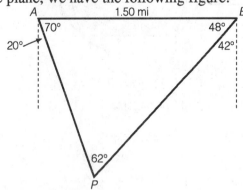

We use the law of sines to find AP:
$$\frac{AP}{\sin 48°} = \frac{1.5}{\sin 62°}$$
$$AP = \frac{1.5\sin 48°}{\sin 62°} \approx 1.26 \text{ mi}$$
We use the law of sines to find BP:
$$\frac{BP}{\sin 70°} = \frac{1.5}{\sin 62°}$$
$$BP = \frac{1.5\sin 70°}{\sin 62°} \approx 1.60 \text{ mi}$$
So the distance from the plane to lighthouse A is 1.26 miles and to lighthouse B is 1.60 miles.

21. Using the law of cosines, we have:
$$D^2 = d^2 + d^2 - 2(d)(d)\cos\tfrac{32°}{60}$$
$$= 2d^2 - 2d^2\cos\tfrac{32°}{60}$$
$$= 2d^2\left(1 - \cos\tfrac{32°}{60}\right)$$
$$= 2(92,690,000)^2\left(1 - \cos\tfrac{32°}{60}\right)$$
$$\approx 744,414,483,000$$
So $D \approx \sqrt{744,414,483,000} \approx 860,000$ miles.
Note: That is about 100 times the diameter of Earth!

23. (a) Using the law of cosines, we have:
$$\left(m^2 + n^2 + mn\right)^2 = \left(2mn + n^2\right)^2 + \left(m^2 - n^2\right)^2 - 2\left(2mn + n^2\right)\left(m^2 - n^2\right)\cos C$$

After carrying out the indicated squaring operations, and then combining like terms, the equation becomes:

$$2m^3n - 2mn^3 + m^2n^2 - n^4 = -2(2mn + n^2)(m^2 - n^2)\cos C$$

$$2mn(m^2 - n^2) + n^2(m^2 - n^2) = -2(2mn + n^2)(m^2 - n^2)\cos C$$

$$(m^2 - n^2)(2mn + n^2) = -2(2mn + n^2)(m^2 - n^2)\cos C$$

Therefore $\cos C = -\frac{1}{2}$, and consequently $C = 120°$.

(b)   Let $m = 2$ and $n = 1$. Then by means of the expressions in part (a), we obtain $a = 5$, $b = 3$, $c = 7$.

25. (a)   We use the law of cosines:

$$a^2 = b^2 + c^2 - 2bc\cos A$$

$$= \left(\frac{1}{\sqrt{6}-\sqrt{2}}\right)^2 + \left(\frac{1}{\sqrt{6}+\sqrt{2}}\right)^2 - 2\left(\frac{1}{\sqrt{6}-\sqrt{2}}\right)\left(\frac{1}{\sqrt{6}+\sqrt{2}}\right)\cos 60°$$

$$= \frac{1}{6 - 2\sqrt{12} + 2} + \frac{1}{6 + 2\sqrt{12} + 2} - \frac{2}{6 - 2} \cdot \frac{1}{2}$$

$$= \frac{1}{8 - 4\sqrt{3}} + \frac{1}{8 + 4\sqrt{3}} - \frac{1}{4}$$

$$= \frac{1}{4(2 - \sqrt{3})} + \frac{1}{4(2 + \sqrt{3})} - \frac{1}{4}$$

$$= \frac{(2 + \sqrt{3}) + (2 - \sqrt{3}) - (2 + \sqrt{3})(2 - \sqrt{3})}{4(2 + \sqrt{3})(2 - \sqrt{3})}$$

$$= \frac{4 - (4 - 3)}{4(4 - 3)}$$

$$= \frac{3}{4}$$

So $a = \sqrt{\frac{3}{4}} = \frac{\sqrt{3}}{2}$.

(b)   Using the law of sines to find $\sin B$:

$$\frac{\sin B}{b} = \frac{\sin A}{a}$$

$$\frac{\sin B}{\frac{1}{\sqrt{6}-\sqrt{2}}} = \frac{\sin 60°}{\frac{\sqrt{3}}{2}}$$

$$(\sqrt{6} - \sqrt{2})\sin B = \frac{2}{\sqrt{3}} \cdot \frac{\sqrt{3}}{2}$$

$$(\sqrt{6} - \sqrt{2})\sin B = 1$$

$$\sin B = \frac{1}{\sqrt{6} - \sqrt{2}}$$

Using the law of sines to find $\sin C$:

$$\frac{\sin C}{c} = \frac{\sin A}{a}$$

$$\frac{\sin C}{\frac{1}{\sqrt{6}+\sqrt{2}}} = \frac{\sin 60°}{\frac{\sqrt{3}}{2}}$$

$$\left(\sqrt{6}+\sqrt{2}\right)\sin C = \frac{2}{\sqrt{3}}\cdot\frac{\sqrt{3}}{2}$$

$$\left(\sqrt{6}+\sqrt{2}\right)\sin C = 1$$

$$\sin C = \frac{1}{\sqrt{6}+\sqrt{2}}$$

27. We draw the figure:

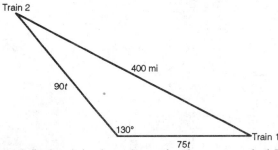

If t is the time of travel (in hours), then the trains have traveled $75t$ mi and $90t$ mi, as indicated in the figure. We use the law of cosines:

$$(400)^2 = (90t)^2 + (75t)^2 - 2(90t)(75t)\cos 130°$$

$$160000 = 8100t^2 + 5625t^2 - 13500\cos 130° t^2$$

$$t^2 = \frac{160000}{13725 - 13500\cos 130°} \approx 7.142$$

$$t \approx 2.672 \text{ hr} \approx 160 \text{ minutes}$$

The trains are 400 mi apart after 160 minutes, which occurs at 2:40 P.M.

29. Using the expressions:

$$\cos A = \frac{b^2 + c^2 - a^2}{2bc} \quad \text{and} \quad \cos B = \frac{c^2 + a^2 - b^2}{2ca}$$

We now substitute for $\cos A$ and $\cos B$ to obtain:

$$\frac{b^2 + c^2 - a^2}{2b^2c} = \frac{c^2 + a^2 - b^2}{2ca^2}$$

$$2ca^2\left(b^2 + c^2 - a^2\right) = 2b^2c\left(c^2 + a^2 - b^2\right)$$

$$a^2\left(b^2 + c^2 - a^2\right) = b^2\left(c^2 + a^2 - b^2\right) \quad \text{(since } 2c \neq 0\text{)}$$

$$a^2b^2 + a^2c^2 - a^4 = b^2c^2 + a^2b^2 - b^4$$

$$a^2c^2 - a^4 = b^2c^2 - b^4$$

$$a^2c^2 - b^2c^2 = a^4 - b^4$$

$$c^2\left(a^2 - b^2\right) = \left(a^2 + b^2\right)\left(a^2 - b^2\right)$$

$$\left(a^2 - b^2\right)\left(c^2 - a^2 - b^2\right) = 0$$

Setting each factor equal to 0, either $a^2 - b^2 = 0$ or $c^2 - a^2 - b^2 = 0$. If $a^2 - b^2 = 0$, then $a^2 = b^2$ and thus $a = b$. Therefore $\triangle ABC$ is isosceles. If $c^2 - a^2 - b^2 = 0$, then $a^2 + b^2 = c^2$, which is the Pythagorean theorem. Thus $\triangle ABC$ is a right triangle.

31. Following the hint that is given, we obtain the two equations:

$$\lambda^2 = a^2 + d^2 - 2ad\cos(180° - \theta) \qquad (1)$$
$$\lambda^2 = b^2 + c^2 - 2bc\cos\theta \qquad (2)$$

Equation (1) can be rewritten:

$$\lambda^2 = a^2 + d^2 + 2ad\cos\theta$$

Upon solving this last equation for $\cos\theta$, we obtain:

$$\cos\theta = \frac{\lambda^2 - a^2 - d^2}{2ad}$$

Now we use this expression for $\cos\theta$ in equation (2) to obtain:

$$\lambda^2 = b^2 + c^2 - 2bc\left(\frac{\lambda^2 - a^2 - d^2}{2ad}\right)$$

$$\lambda^2 ad = b^2 ad + c^2 ad - bc\lambda^2 + a^2 bc + bcd^2$$

$$\lambda^2(ad + bc) = b^2 ad + c^2 ad + a^2 bc + d^2 bc$$

$$\lambda^2(ad + bc) = \left(c^2 ad + a^2 bc\right) + \left(b^2 ad + d^2 bc\right)$$

$$\lambda^2(ad + bc) = ac(cd + ab) + bd(ab + cd)$$

$$\lambda^2(ad + bc) = (ab + cd)(ac + bd)$$

$$\lambda^2 = \frac{(ab + cd)(ac + bd)}{ad + bc}$$

33. Using the suggestion, we make the substitutions:

$$a^2 = b^2 + c^2 - 2bc\cos A$$
$$b^2 = a^2 + c^2 - 2ac\cos B$$
$$c^2 = a^2 + b^2 - 2ab\cos C$$

Substituting into the right-hand side of the identity, we have:

$$\frac{a^2 + b^2 + c^2}{2abc} = \frac{b^2 + c^2 - 2bc\cos A + a^2 + c^2 - 2ac\cos B + a^2 + b^2 - 2ab\cos C}{2abc}$$

$$= \frac{2(a^2 + b^2 + c^2) - 2bc\cos A - 2ac\cos B - 2ab\cos C}{2abc}$$

$$= \frac{a^2 + b^2 + c^2}{abc} - \frac{\cos A}{a} - \frac{\cos B}{b} - \frac{\cos C}{c}$$

Subtracting the first term from each side of this equation yields:

$$-\frac{a^2 + b^2 + c^2}{2abc} = -\frac{\cos A}{a} - \frac{\cos B}{b} - \frac{\cos C}{c}$$

$$\frac{a^2 + b^2 + c^2}{2abc} = \frac{\cos A}{a} + \frac{\cos B}{b} + \frac{\cos C}{c}$$

35. Using the hint, we have:
$$a^4 + b^4 + c^4 = 2(a^2 + b^2)c^2$$
$$c^4 - 2(a^2 + b^2)c^2 + (a^4 + b^4) = 0$$
Using the Pythagorean theorem, we have:
$$c^2 = \frac{2(a^2 + b^2) \pm \sqrt{4(a^2 + b^2)^2 - 4(a^4 + b^4)}}{2}$$
$$= a^2 + b^2 \pm \sqrt{a^4 + 2a^2b^2 + b^4 - a^4 - b^4}$$
$$= a^2 + b^2 \pm \sqrt{2a^2b^2}$$
$$= a^2 + b^2 \pm ab\sqrt{2}$$
By the law of cosines, we have:
$$c^2 = a^2 + b^2 - 2ab\cos C$$
Setting these expressions equal:
$$a^2 + b^2 - 2ab\cos C = a^2 + b^2 \pm ab\sqrt{2}$$
$$-2ab\cos C = \pm ab\sqrt{2}$$
$$\cos C = \pm\tfrac{\sqrt{2}}{2}$$
$$C = 45° \text{ or } 135°$$

37. (a) Simplifying each of the fractions:
$$\frac{\sin A}{a} = \frac{\frac{\sqrt{T}}{2bc}}{a} = \frac{\sqrt{T}}{2abc}$$
$$\frac{\sin B}{b} = \frac{\frac{\sqrt{T}}{2ac}}{b} = \frac{\sqrt{T}}{2abc}$$
$$\frac{\sin C}{c} = \frac{\frac{\sqrt{T}}{2ab}}{c} = \frac{\sqrt{T}}{2abc}$$

(b) Since each of the fractions is equal to $\frac{\sqrt{T}}{2abc}$, then:
$$\frac{\sin A}{a} = \frac{\sin B}{b} = \frac{\sin C}{c}$$

39. (a) Computing areas of each triangle:
$$\text{Area}_{\text{left}} = \tfrac{1}{2}(\text{base})(\text{height}) = \tfrac{1}{2}af\sin\tfrac{C}{2}$$
$$\text{Area}_{\text{right}} = \tfrac{1}{2}(\text{base})(\text{height}) = \tfrac{1}{2}bf\sin\tfrac{C}{2}$$
$$\text{Area}_{\text{entire}} = \tfrac{1}{2}(\text{base})(\text{height}) = \tfrac{1}{2}ab\sin C$$
Adding the left and right triangles, we have:
$$\tfrac{1}{2}af\sin\tfrac{C}{2} + \tfrac{1}{2}bf\sin\tfrac{C}{2} = \tfrac{1}{2}ab\sin C$$

(b) Since $\sin C = 2\sin\tfrac{C}{2}\cos\tfrac{C}{2}$, we have:
$$\tfrac{1}{2}af\sin\tfrac{C}{2} + \tfrac{1}{2}bf\sin\tfrac{C}{2} = ab\sin\tfrac{C}{2}\cos\tfrac{C}{2}$$

Multiplying by $\frac{2}{\sin(C/2)}$, we have:

$$af + bf = 2ab\cos\frac{C}{2}$$

$$f(a+b) = 2ab\cos\frac{C}{2}$$

$$f = \frac{2ab\cos\frac{C}{2}}{a+b}$$

(c)   Using the half-angle formula, we have:

$$\cos\frac{C}{2} = \sqrt{\frac{1+\cos C}{2}}$$

$$= \sqrt{\frac{1+\frac{a^2+b^2-c^2}{2ab}}{2}}$$

$$= \sqrt{\frac{a^2+2ab+b^2-c^2}{4ab}}$$

$$= \frac{1}{2}\sqrt{\frac{(a+b)^2-c^2}{ab}}$$

$$= \frac{1}{2}\sqrt{\frac{(a+b-c)(a+b+c)}{ab}}$$

(d)   Combining our results from (b) and (c), we have:

$$f = \frac{2ab}{a+b}\cos\frac{C}{2}$$

$$= \frac{ab}{a+b}\sqrt{\frac{(a+b-c)(a+b+c)}{ab}}$$

$$= \frac{\sqrt{ab}}{a+b}\sqrt{(a+b-c)(a+b+c)}$$

5.3 Vectors in the Plane, A Geometric Approach

1.   We graph the vector:

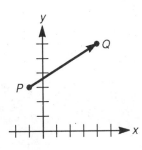

The magnitude is given by:

$$|\vec{PQ}| = \sqrt{(4-(-1))^2+(6-3)^2} = \sqrt{25+9} = \sqrt{34}$$

3.   We graph the vector:

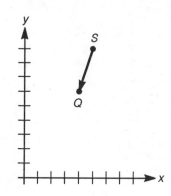

The magnitude is given by:

$$|\vec{SQ}| = \sqrt{(4-5)^2 + (6-9)^2} = \sqrt{1+9} = \sqrt{10}$$

5.   We graph the vector:

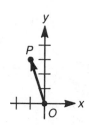

The magnitude is given by:

$$|\vec{OP}| = \sqrt{(-1-0)^2 + (3-0)^2} = \sqrt{1+9} = \sqrt{10}$$

7.   We graph the vector sum:

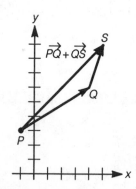

The magnitude is given by:

$$|\vec{PQ} + \vec{QS}| = |\vec{PS}| = \sqrt{(5-(-1))^2 + (9-3)^2} = \sqrt{36+36} = 6\sqrt{2}$$

9. We graph the vector sum:

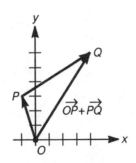

The magnitude is given by:

$$|\overrightarrow{OP} + \overrightarrow{PQ}| = |\overrightarrow{OQ}| = \sqrt{(4-0)^2 + (6-0)^2} = \sqrt{16+36} = 2\sqrt{13}$$

11. We graph the vector sum:

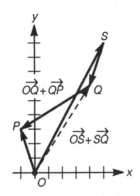

The magnitude is given by:

$$|\overrightarrow{OS} + \overrightarrow{SQ} + \overrightarrow{QP}| = |\overrightarrow{OQ} + \overrightarrow{QP}| = |\overrightarrow{OP}| = \sqrt{(-1-0)^2 + (3-0)^2} = \sqrt{1+9} = \sqrt{10}$$

13. We graph the vector sum:

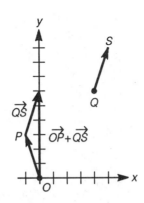

The magnitude is given by:

$$|\overrightarrow{OP} + \overrightarrow{QS}| = \sqrt{(0-0)^2 + (6-0)^2} = \sqrt{0+36} = 6$$

15. We graph the vector sum:

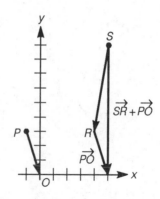

The magnitude is given by:

$$|\vec{SR}+\vec{PO}| = \sqrt{(5-5)^2+(9-0)^2} = \sqrt{0+81} = 9$$

17. We graph the vector sum:

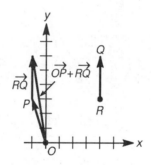

The magnitude is given by:

$$|\vec{OP}+\vec{RQ}| = \sqrt{(-1-0)^2+(6-0)^2} = \sqrt{1+36} = \sqrt{37}$$

19. We graph the vector sum:

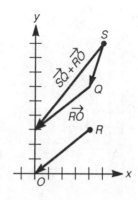

The magnitude is given by:

$$|\vec{SQ}+\vec{RO}| = \sqrt{(5-0)^2+(9-3)^2} = \sqrt{25+36} = \sqrt{61}$$

21. We graph the vector sum:

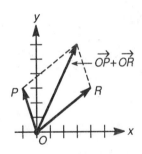

The magnitude is given by:

$$\left|\overrightarrow{OP}+\overrightarrow{OR}\right|=\sqrt{(3-0)^2+(6-0)^2}=\sqrt{9+36}=3\sqrt{5}$$

23. We graph the vector sum:

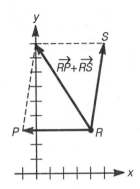

The magnitude is given by:

$$\left|\overrightarrow{RP}+\overrightarrow{RS}\right|=\sqrt{(0-4)^2+(9-3)^2}=\sqrt{16+36}=2\sqrt{13}$$

25. We graph the vector sum:

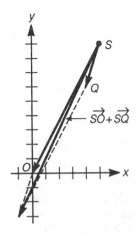

The magnitude is given by:

$$\left|\overrightarrow{SO}+\overrightarrow{SQ}\right|=\sqrt{(-1-5)^2+(-3-9)^2}=\sqrt{36+144}=6\sqrt{5}$$

27. We draw the figure:

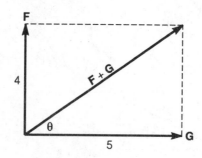

Now compute the magnitude and direction of $\mathbf{F} + \mathbf{G}$:

$$|\mathbf{F} + \mathbf{G}| = \sqrt{4^2 + 5^2} = \sqrt{16 + 25} = \sqrt{41} \text{ N}$$

$$\theta = \tan^{-1}\left(\tfrac{4}{5}\right) \approx 38.7°$$

29. We draw the figure:

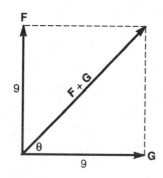

Now compute the magnitude and direction of $\mathbf{F} + \mathbf{G}$:

$$|\mathbf{F} + \mathbf{G}| = \sqrt{9^2 + 9^2} = 9\sqrt{2} \text{ N}$$

$$\theta = \tan^{-1}\left(\tfrac{9}{9}\right) = \tan^{-1}1 = 45°$$

31. We draw the figure:

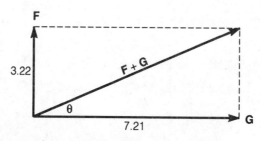

Now compute the magnitude and direction of $\mathbf{F} + \mathbf{G}$:

$$|\mathbf{F} + \mathbf{G}| = \sqrt{3.22^2 + 7.21^2} = \sqrt{62.3525} \approx 7.90 \text{ N}$$

$$\theta = \tan^{-1}\left(\tfrac{3.22}{7.21}\right) \approx 24.1°$$

33. We draw the parallelogram:

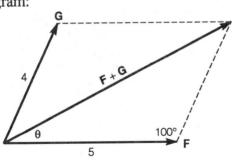

Let $d = |\mathbf{F} + \mathbf{G}|$. Then using the law of cosines:
$$d^2 = 5^2 + 4^2 - 2(5)(4)\cos 100°$$
$$d^2 = 41 - 40\cos 100°$$
$$d = \sqrt{41 - 40\cos 100°} \approx 6.92 \text{ N}$$
We find θ by the law of sines:
$$\frac{\sin\theta}{4} = \frac{\sin 100°}{d}$$
$$\sin\theta = \frac{4\sin 100°}{\sqrt{41 - 40\cos 100°}} \approx 0.5689$$
$$\theta \approx 34.67°$$

35. We draw the parallelogram:

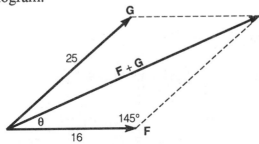

Let $d = |\mathbf{F} + \mathbf{G}|$. Then using the law of cosines:
$$d^2 = 16^2 + 25^2 - 2(16)(25)\cos 145°$$
$$d^2 = 881 - 800\cos 145°$$
$$d = \sqrt{881 - 800\cos 145°} \approx 39.20 \text{ N}$$
We find θ by the law of sines:
$$\frac{\sin\theta}{25} = \frac{\sin 145°}{d}$$
$$\sin\theta = \frac{25\sin 145°}{\sqrt{881 - 800\cos 145°}} \approx 0.3658$$
$$\theta \approx 21.46°$$

37. We draw the parallelogram:

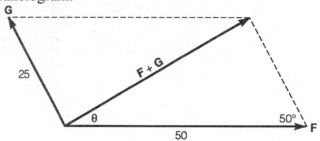

Let $d = |\mathbf{F} + \mathbf{G}|$. Then using the law of cosines:
$$d^2 = 50^2 + 25^2 - 2(50)(25)\cos 50°$$
$$d^2 = 3125 - 2500\cos 50°$$
$$d = \sqrt{3125 - 2500\cos 50°} \approx 38.96 \text{ N}$$

We find θ by the law of sines:
$$\frac{\sin\theta}{25} = \frac{\sin 50°}{d}$$
$$\sin\theta = \frac{25\sin 50°}{\sqrt{3125 - 2500\cos 50°}} \approx 0.4915$$
$$\theta \approx 29.44°$$

39. We compute the horizontal and vertical components:
$$V_x = 16\cos 30° \approx 13.86 \text{ cm/sec}$$
$$V_y = 16\sin 30° = 8 \text{ cm/sec}$$

41. We compute the horizontal and vertical components:
$$F_x = 14\cos 75° \approx 3.62 \text{ N}$$
$$F_y = 14\sin 75° \approx 13.52 \text{ N}$$

43. We compute the horizontal and vertical components:
$$V_x = 1\cos 135° \approx -0.71 \text{ cm/sec}$$
$$V_y = 1\sin 135° \approx 0.71 \text{ cm/sec}$$

45. We compute the horizontal and vertical components:
$$F_x = 1.25\cos 145° \approx -1.02 \text{ N}$$
$$F_y = 1.25\sin 145° \approx 0.72 \text{ N}$$

47. We draw the vectors:

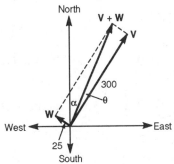

Let θ be the drift angle. Then:
$$\tan\theta = \frac{25}{300} = \frac{1}{12}$$
$$\theta = \tan^{-1}\left(\frac{1}{12}\right) \approx 4.76°$$
The ground speed is given by:
$$|V+W| = \sqrt{25^2 + 300^2} = \sqrt{90625} \approx 301.04 \text{ mph}$$
Let α be the bearing. Then:
$$\alpha = 30° - \theta \approx 30° - 4.76° \approx 25.24°$$

49. We draw the vectors:

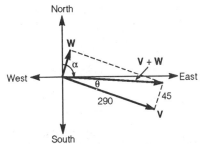

Let θ be the drift angle. Then:
$$\tan\theta = \frac{45}{290} = \frac{9}{58}$$
$$\theta = \tan^{-1}\left(\frac{9}{58}\right) \approx 8.82°$$
The ground speed is given by:
$$|V+W| = \sqrt{290^2 + 45^2} = \sqrt{86125} \approx 293.47 \text{ mph}$$
Let α be the bearing. Then:
$$\alpha = 100° - \theta \approx 100° - 8.82° \approx 91.18°$$

51. We draw a figure:

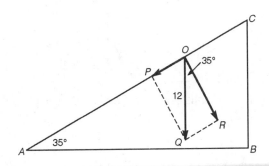

Notice that $\angle QOR = 35°$ since it is complementary to $\angle POQ$, but $\angle POQ = \angle ACB$. So the desired components are given by:

$$|\overrightarrow{OR}| = 12\cos 35° \approx 9.83 \text{ lb}$$

$$|\overrightarrow{OP}| = 12\sin 35° \approx 6.88 \text{ lb}$$

53. Using the same approach as in Exercise 51, we have:
perpendicular: $12\cos 10° \approx 11.82 \text{ lb}$
parallel: $12\sin 10° \approx 2.08 \text{ lb}$

55. (a) We draw the vector sum $(\mathbf{A} + \mathbf{B}) + \mathbf{C}$:

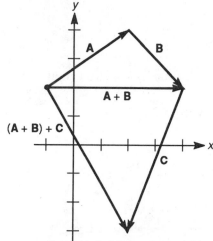

From the diagram, we see that the initial point of $(\mathbf{A} + \mathbf{B}) + \mathbf{C}$ is $(-1, 2)$ and the terminal point is $(2, -3)$.

(b) We draw the vector sum $\mathbf{A} + (\mathbf{B} + \mathbf{C})$:

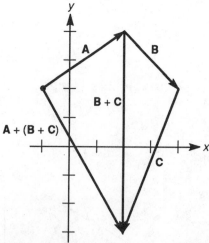

From the diagram, we see that the initial point of $\mathbf{A} + (\mathbf{B} + \mathbf{C})$ is $(-1, 2)$ and the terminal point is $(2, -3)$.

5.4 Vectors in the Plane, An Algebraic Approach

1. We compute the length of the vector:

$$|\langle 4,3\rangle| = \sqrt{4^2 + 3^2} = \sqrt{25} = 5$$

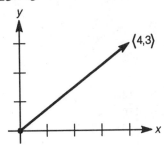

3. We compute the length of the vector:

$$|\langle -4,2\rangle| = \sqrt{(-4)^2 + 2^2} = \sqrt{20} = 2\sqrt{5}$$

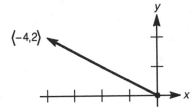

5. We compute the length of the vector:

$$\left|\left\langle \tfrac{3}{4}, -\tfrac{1}{2}\right\rangle\right| = \sqrt{\left(\tfrac{3}{4}\right)^2 + \left(-\tfrac{1}{2}\right)^2} = \sqrt{\tfrac{9}{16} + \tfrac{1}{4}} = \tfrac{\sqrt{13}}{4}$$

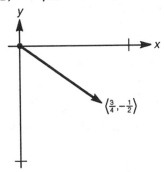

7. Subtracting components, we have:

$$\overrightarrow{PQ} = \langle 3-2, 7-3\rangle = \langle 1,4\rangle$$

9. Subtracting components, we have:

$$\overrightarrow{PQ} = \langle -3-(-2), -2-(-3)\rangle = \langle -3+2, -2+3\rangle = \langle -1,1\rangle$$

11. Subtracting components, we have:

$$\overrightarrow{PQ} = \langle 3-(-5), -4-1\rangle = \langle 3+5, -5\rangle = \langle 8,-5\rangle$$

13. $\mathbf{a} + \mathbf{b} = \langle 2 + 5, 3 + 4 \rangle = \langle 7, 7 \rangle$

15. $2\mathbf{a} + 4\mathbf{b} = \langle 4, 6 \rangle + \langle 20, 16 \rangle = \langle 24, 22 \rangle$

17. Since $\mathbf{b} + \mathbf{c} = \langle 5 + 6, 4 - 1 \rangle = \langle 11, 3 \rangle$, then:
$$|\mathbf{b} + \mathbf{c}| = \sqrt{11^2 + 3^2} = \sqrt{130}$$

19. Since $\mathbf{a} + \mathbf{c} = \langle 2 + 6, 3 - 1 \rangle = \langle 8, 2 \rangle$, then:
$$|\mathbf{a} + \mathbf{c}| = \sqrt{8^2 + 2^2} = \sqrt{68} = 2\sqrt{17}$$
$$|\mathbf{a}| = \sqrt{2^2 + 3^2} = \sqrt{13}$$
$$|\mathbf{c}| = \sqrt{6^2 + (-1)^2} = \sqrt{37}$$
So $|\mathbf{a} + \mathbf{c}| - |\mathbf{a}| - |\mathbf{c}| = 2\sqrt{17} - \sqrt{13} - \sqrt{37}$.

21. Since $\mathbf{b} + \mathbf{c} = \langle 5 + 6, 4 - 1 \rangle = \langle 11, 3 \rangle$, then:
$$\mathbf{a} + (\mathbf{b} + \mathbf{c}) = \langle 2, 3 \rangle + \langle 11, 3 \rangle = \langle 13, 6 \rangle$$

23. $3\mathbf{a} + 4\mathbf{a} = \langle 6, 9 \rangle + \langle 8, 12 \rangle = \langle 14, 21 \rangle$

25. $\mathbf{a} - \mathbf{b} = \langle 2, 3 \rangle - \langle 5, 4 \rangle = \langle -3, -1 \rangle$

27. $3\mathbf{b} - 4\mathbf{d} = \langle 15, 12 \rangle - \langle -8, 0 \rangle = \langle 15 + 8, 12 - 0 \rangle = \langle 23, 12 \rangle$

29. Since $\mathbf{b} + \mathbf{c} = \langle 11, 3 \rangle$, then:
$$\mathbf{a} - (\mathbf{b} + \mathbf{c}) = \langle 2, 3 \rangle - \langle 11, 3 \rangle = \langle -9, 0 \rangle$$

31. Since $\mathbf{c} + \mathbf{d} = \langle 4, -1 \rangle$ and $\mathbf{c} - \mathbf{d} = \langle 8, -1 \rangle$, then:
$$|\mathbf{c} + \mathbf{d}| = \sqrt{16 + 1} = \sqrt{17} \quad \text{and} \quad |\mathbf{c} - \mathbf{d}| = \sqrt{64 + 1} = \sqrt{65}$$
Therefore:
$$|\mathbf{c} + \mathbf{d}|^2 - |\mathbf{c} - \mathbf{d}|^2 = 17 - 65 = -48$$

33. Separating individual components, we have:
$$\langle 3, 8 \rangle = \langle 3, 0 \rangle + \langle 0, 8 \rangle = 3\mathbf{i} + 8\mathbf{j}$$

35. Separating individual components, we have:
$$\langle -8, -6 \rangle = \langle -8, 0 \rangle + \langle 0, -6 \rangle = -8\mathbf{i} - 6\mathbf{j}$$

37. Separating individual components, we have:
$$3\langle 5,3\rangle + 2\langle 2,7\rangle = 3(5\mathbf{i} + 3\mathbf{j}) + 2(2\mathbf{i} + 7\mathbf{j})$$
$$= 15\mathbf{i} + 9\mathbf{j} + 4\mathbf{i} + 14\mathbf{j}$$
$$= 19\mathbf{i} + 23\mathbf{j}$$

39. $\mathbf{i} + \mathbf{j} = \langle 1,1\rangle$

41. $5\mathbf{i} - 4\mathbf{j} = \langle 5,-4\rangle$

43. We first compute the length of the vector:
$$|\langle 4,8\rangle| = \sqrt{4^2 + 8^2} = \sqrt{80} = 4\sqrt{5}$$
So a unit vector would be given by:
$$\tfrac{1}{4\sqrt{5}}\langle 4,8\rangle = \left\langle \tfrac{1}{\sqrt{5}}, \tfrac{2}{\sqrt{5}}\right\rangle = \left\langle \tfrac{\sqrt{5}}{5}, \tfrac{2\sqrt{5}}{5}\right\rangle$$

45. We first compute the length of the vector:
$$|\langle 6,-3\rangle| = \sqrt{6^2 + (-3)^2} = \sqrt{45} = 3\sqrt{5}$$
So a unit vector would be given by:
$$\tfrac{1}{3\sqrt{5}}\langle 6,-3\rangle = \left\langle \tfrac{2}{\sqrt{5}}, -\tfrac{1}{\sqrt{5}}\right\rangle = \left\langle \tfrac{2\sqrt{5}}{5}, -\tfrac{\sqrt{5}}{5}\right\rangle$$

47. We first compute the length of the vector:
$$|8\mathbf{i} - 9\mathbf{j}| = \sqrt{8^2 + (-9)^2} = \sqrt{145}$$
So a unit vector would be given by:
$$\tfrac{1}{\sqrt{145}}(8\mathbf{i} - 9\mathbf{j}) = \tfrac{8}{\sqrt{145}}\mathbf{i} - \tfrac{9}{\sqrt{145}}\mathbf{j} = \tfrac{8\sqrt{145}}{145}\mathbf{i} - \tfrac{9\sqrt{145}}{145}\mathbf{j}$$

49. We compute the components u_1 and u_2:
$$u_1 = \cos\tfrac{\pi}{6} = \tfrac{\sqrt{3}}{2}$$
$$u_2 = \sin\tfrac{\pi}{6} = \tfrac{1}{2}$$

51. We compute the components u_1 and u_2:
$$u_1 = \cos\tfrac{2\pi}{3} = -\tfrac{1}{2}$$
$$u_2 = \sin\tfrac{2\pi}{3} = \tfrac{\sqrt{3}}{2}$$

53. We compute the components u_1 and u_2:
$$u_1 = \cos\tfrac{5\pi}{6} = -\tfrac{\sqrt{3}}{2}$$
$$u_2 = \sin\tfrac{5\pi}{6} = \tfrac{1}{2}$$

55. We verify property 1:

$$\mathbf{u} + (\mathbf{v} + \mathbf{w}) = \langle u_1, u_2 \rangle + \left(\langle v_1, v_2 \rangle + \langle w_1, w_2 \rangle \right)$$
$$= \langle u_1, u_2 \rangle + \langle v_1 + w_1, v_2 + w_2 \rangle$$
$$= \langle u_1 + v_1 + w_1, u_2 + v_2 + w_2 \rangle$$
$$= \langle u_1 + v_1, u_2 + v_2 \rangle + \langle w_1, w_2 \rangle$$
$$= \left(\langle u_1, u_2 \rangle + \langle v_1, v_2 \rangle \right) + \langle w_1, w_2 \rangle$$
$$= (\mathbf{u} + \mathbf{v}) + \mathbf{w}$$

We verify property 2:

$$\mathbf{0} + \mathbf{v} = \langle 0, 0 \rangle + \langle v_1, v_2 \rangle$$
$$= \langle 0 + v_1, 0 + v_2 \rangle$$
$$= \langle v_1 + 0, v_2 + 0 \rangle$$
$$= \langle v_1, v_2 \rangle + \langle 0, 0 \rangle$$
$$= \mathbf{v} + \mathbf{0}$$

$$\mathbf{v} + \mathbf{0} = \langle v_1, v_2 \rangle + \langle 0, 0 \rangle$$
$$= \langle v_1 + 0, v_2 + 0 \rangle$$
$$= \langle v_1, v_2 \rangle$$
$$= \mathbf{v}$$

57. We verify property 5:

$$a(\mathbf{u} + \mathbf{v}) = a \left(\langle u_1, u_2 \rangle + \langle v_1, v_2 \rangle \right)$$
$$= a \langle u_1 + v_1, u_2 + v_2 \rangle$$
$$= \left\langle a(u_1 + v_1), a(u_2 + v_2) \right\rangle$$
$$= \langle au_1 + av_1, au_2 + av_2 \rangle$$
$$= \langle au_1, au_2 \rangle + \langle av_1, av_2 \rangle$$
$$= a \langle u_1, u_2 \rangle + a \langle v_1, v_2 \rangle$$
$$= a\mathbf{u} + a\mathbf{v}$$

We verify property 6:

$$(a + b)\mathbf{v} = (a + b)\langle v_1, v_2 \rangle$$
$$= \left\langle (a + b)v_1, (a + b)v_2 \right\rangle$$
$$= \langle av_1 + bv_1, av_2 + bv_2 \rangle$$
$$= \langle av_1, av_2 \rangle + \langle bv_1, bv_2 \rangle$$
$$= a \langle v_1, v_2 \rangle + b \langle v_1, v_2 \rangle$$
$$= a\mathbf{v} + b\mathbf{v}$$

59. (a)   We compute the dot products:

$$\mathbf{u} \bullet \mathbf{v} = \langle -4, 5 \rangle \bullet \langle 3, 4 \rangle = (-4)(3) + (5)(4) = -12 + 20 = 8$$
$$\mathbf{v} \bullet \mathbf{u} = \langle 3, 4 \rangle \bullet \langle -4, 5 \rangle = (3)(-4) + (4)(5) = -12 + 20 = 8$$

(b)   We compute the dot products:
$$\mathbf{v} \bullet \mathbf{w} = \langle 3,4 \rangle \bullet \langle 2,-5 \rangle = (3)(2)+(4)(-5) = 6-20 = -14$$
$$\mathbf{w} \bullet \mathbf{v} = \langle 2,-5 \rangle \bullet \langle 3,4 \rangle = (2)(3)+(-5)(4) = 6-20 = -14$$

(c)   Let $A = \langle x_1, y_1 \rangle$ and $B = \langle x_2, y_2 \rangle$. We compute each dot product:
$$A \bullet B = \langle x_1, y_1 \rangle \bullet \langle x_2, y_2 \rangle = x_1 x_2 + y_1 y_2$$
$$B \bullet A = \langle x_2, y_2 \rangle \bullet \langle x_1, y_1 \rangle = x_2 x_1 + y_2 y_1 = x_1 x_2 + y_1 y_2$$
Thus $A \bullet B = B \bullet A$.

61. (a)   We compute each quantity:
$$\mathbf{v} \bullet \mathbf{v} = \langle 3,4 \rangle \bullet \langle 3,4 \rangle = (3)(3)+(4)(4) = 9+16 = 25$$
$$|\mathbf{v}| = \sqrt{3^2+4^2} = \sqrt{9+16} = \sqrt{25} = 5, \text{ so } |\mathbf{v}|^2 = 25$$

(b)   We compute each quantity:
$$\mathbf{w} \bullet \mathbf{w} = \langle 2,-5 \rangle \bullet \langle 2,-5 \rangle = (2)(2)+(-5)(-5) = 4+25 = 29$$
$$|\mathbf{w}| = \sqrt{2^2+(-5)^2} = \sqrt{4+25} = \sqrt{29}, \text{ so } |\mathbf{w}|^2 = 29$$

63.   We first do the computations:
$$|A| = \sqrt{16+1} = \sqrt{17}$$
$$|B| = \sqrt{4+36} = \sqrt{40} = 2\sqrt{10}$$
$$A \bullet B = 8+6 = 14$$
Thus:
$$\cos\theta = \frac{14}{\sqrt{17} \bullet 2\sqrt{10}} = \frac{7}{\sqrt{170}}$$
So $\theta \approx 57.53°$ or $\theta \approx 1.00$ radian.

65.   We first do the computations:
$$|A| = \sqrt{25+36} = \sqrt{61}$$
$$|B| = \sqrt{9+49} = \sqrt{58}$$
$$A \bullet B = -15-42 = -57$$
Thus:
$$\cos\theta = \frac{-57}{\sqrt{61} \bullet \sqrt{58}} = \frac{-57}{\sqrt{3538}}$$
So $\theta \approx 163.39°$ or $\theta \approx 2.85$ radians.

67. (a)   We first do the computations:
$$|A| = \sqrt{64+4} = \sqrt{68} = 2\sqrt{17}$$
$$|B| = \sqrt{1+9} = \sqrt{10}$$
$$A \bullet B = -8-6 = -14$$
Thus:
$$\cos\theta = \frac{-14}{2\sqrt{17} \bullet \sqrt{10}} = \frac{-7}{\sqrt{170}}$$
So $\theta \approx 122.47°$ or $\theta \approx 2.14$ radians.

(b) Again, we do the computations:
$$|A| = \sqrt{64+4} = \sqrt{68} = 2\sqrt{17}$$
$$|B| = \sqrt{1+9} = \sqrt{10}$$
$$A \bullet B = 8+6 = 14$$
Thus:
$$\cos\theta = \frac{14}{2\sqrt{17} \bullet \sqrt{10}} = \frac{7}{\sqrt{170}}$$
So $\theta \approx 57.53°$ or $\theta \approx 1.00$ radian.

69. (a) We first compute:
$$|\langle 2,5 \rangle| = \sqrt{4+25} = \sqrt{29}$$
$$|\langle -5,2 \rangle| = \sqrt{25+4} = \sqrt{29}$$
$$\langle 2,5 \rangle \bullet \langle -5,2 \rangle = -10+10 = 0$$

So $\cos\theta = \frac{0}{29} = 0$.

(b) Since the angle between the vectors is 90°, the vectors must be perpendicular.

(c) We draw the sketch:

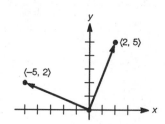

71. Since $\cos\theta = \frac{A \bullet B}{|A||B|}$, then $A \bullet B = 0$ implies $\cos\theta = 0$, and thus $\theta = 90°$. So the vectors are perpendicular.

73. Call such a vector $\langle x,y \rangle$, so $x^2 + y^2 = 1$. Now we know $\langle x,y \rangle \bullet \langle -12,5 \rangle = 0$, so:
$$-12x + 5y = 0$$
$$5y = 12x$$
$$y = \frac{12}{5}x$$
Substituting into $x^2 + y^2 = 1$, we have:
$$x^2 + \frac{144}{25}x^2 = 1$$
$$\frac{169}{25}x^2 = 1$$
$$x^2 = \frac{25}{169}$$
$$x = \pm\frac{5}{13}$$
$$y = \pm\frac{12}{13}$$

So the two unit vectors are $\left\langle \frac{5}{13}, \frac{12}{13} \right\rangle$ and $\left\langle -\frac{5}{13}, -\frac{12}{13} \right\rangle$.

75. (a)   We first find:
$$C = B - A = \langle x_2, y_2 \rangle - \langle x_1, y_1 \rangle = \langle x_2 - x_1, y_2 - y_1 \rangle$$

So:
$$|C| = \sqrt{(x_2 - x_1)^2 + (y_2 - y_1)^2} = \sqrt{x_1^2 + y_1^2 + x_2^2 + y_2^2 - 2x_1 x_2 - 2y_1 y_2}$$

(b)   Working from the left-hand side, we have:
$$|C|^2 = x_1^2 + y_1^2 + x_2^2 + y_2^2 - 2x_1 x_2 - 2y_1 y_2 = |A|^2 + |B|^2 - 2(A \bullet B)$$

(c)   Using the suggestion given, we have:
$$|A|^2 + |B|^2 - 2|A||B|\cos\theta = |A|^2 + |B|^2 - 2(A \bullet B)$$
$$-2|A||B|\cos\theta = -2(A \bullet B)$$
$$|A||B|\cos\theta = A \bullet B$$
$$\cos\theta = \frac{A \bullet B}{|A||B|}$$

Chapter Five Review Exercises

1.   We draw the triangle:

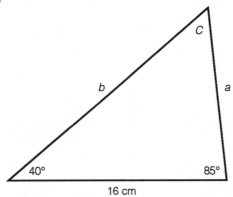

First note that $\angle C = 180° - 40° - 85° = 55°$. Using the law of sines:
$$\frac{\sin 40°}{a} = \frac{\sin 55°}{16}, \text{ so } a = \frac{16\sin 40°}{\sin 55°} \approx 12.6 \text{ cm}$$

Using the law of sines:
$$\frac{\sin 85°}{b} = \frac{\sin 55°}{16}, \text{ so } b = \frac{16\sin 85°}{\sin 55°} \approx 19.5 \text{ cm}$$

3.   (a)   We draw the triangle:

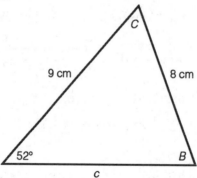

Using the law of sines:
$$\frac{\sin B}{9} = \frac{\sin 52°}{8}, \text{ so } \sin B = \frac{9\sin 52°}{8} \approx 0.8865$$
Thus $\angle B \approx 62.4°$. Now note that $\angle C \approx 180° - 52° - 62.4° \approx 65.6°$. Using the law of sines:
$$\frac{\sin 65.6°}{c} = \frac{\sin 52°}{8}, \text{ so } c = \frac{8\sin 65.6°}{\sin 52°} \approx 9.2 \text{ cm}$$

(b)   We have $\sin B \approx 0.8865$, thus $\angle B \approx 117.6°$. Now note that $\angle C = 180° - 52° - 117.6° \approx 10.4°$. Using the law of sines:
$$\frac{\sin 10.4°}{c} = \frac{\sin 52°}{8}, \text{ so } c = \frac{8\sin 10.4°}{\sin 52°} \approx 1.8 \text{ cm}$$

5.   We draw the triangle:

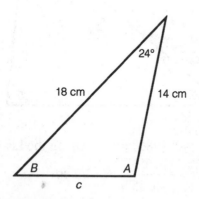

Using the law of cosines:
$$c^2 = 18^2 + 14^2 - 2(18)(14)\cos 24°, \text{ so } c \approx 7.7 \text{ cm}$$
Using the law of sines:
$$\frac{\sin A}{18} = \frac{\sin 24°}{7.7}, \text{ so } \sin A = \frac{18\sin 24°}{7.7} \approx 0.9486$$
Then either $\angle A \approx 71.5°$ or $\angle A \approx 108.5°$. If $\angle A \approx 71.5°$, then $\angle B \approx 180° - 24° - 71.5° \approx 84.5°$. But this is impossible since $\angle B < \angle A$. If $\angle A \approx 108.5°$, then $\angle B \approx 180° - 24° - 108.5° \approx 47.5°$.

7.   We draw the triangle:

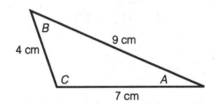

Using the law of cosines:
$$9^2 = 4^2 + 7^2 - 2(4)(7)\cos C, \text{ so } \cos C \approx -0.2857, \text{ thus } C \approx 106.6°$$
Using the law of sines:
$$\frac{\sin 106.6°}{9} = \frac{\sin B}{7}, \text{ so } \sin B = \frac{7\sin 106.6°}{9} \approx 0.7454$$
Thus $B \approx 48.2°$. Therefore $\angle A \approx 180° - 106.6° - 48.2° \approx 25.2°$.

9.   Note that $\angle AEB = 86°$. Using the law of sines, we have:
$$\frac{\sin 50°}{BE} = \frac{\sin 86°}{12}, \text{ so } BE = \frac{12\sin 50°}{\sin 86°} \approx 9.21 \text{ cm}$$

11.   Using the area formula, we have:
$$A = \tfrac{1}{2}(BC)(BE)\sin 36° = \tfrac{1}{2}(12 \text{ cm})(9.21 \text{ cm})(\sin 36°) \approx 32.48 \text{ cm}^2$$

13.   Using the area formula, we have:
$$A = \tfrac{1}{2}(12 \text{ cm})(BD\sin 44°) = \tfrac{1}{2}(12 \text{ cm})(13.25 \text{ cm})(\sin 44°) \approx 55.23 \text{ cm}^2$$

15.   Using the law of cosines, we have:
$$(CD)^2 = (BC)^2 + (BD)^2 - 2(BC)(BD)\cos 36°$$
$$\approx (12)^2 + (13.25)^2 - 2(12)(13.25)\cos 36°$$
$$\approx 319.56 - 318\cos 36°$$
$$\approx 62.29$$
So $CD \approx \sqrt{62.29} \approx 7.89$ cm.

17.   Using the law of sines, we have:
$$\frac{\sin 80°}{AC} = \frac{\sin 50°}{12}, \text{ so } AC = \frac{12\sin 80°}{\sin 50°} \approx 15.43 \text{ cm}$$

19.   We re-draw the figure:

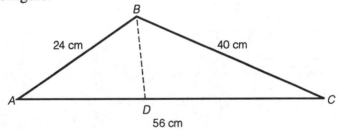

We use the law of cosines to find $\angle ABC$:
$$56^2 = 40^2 + 24^2 - 2(40)(24)\cos(\angle ABC)$$
$$-0.5 = \cos(\angle ABC)$$
$$\angle ABC = 120°$$
We now use the law of sines to find $\angle A$:
$$\frac{\sin A}{40} = \frac{\sin 120°}{56}, \text{ so } \sin A \approx 0.6186, \text{ thus } A \approx 38.21°$$
Then $\angle ADB \approx 180° - 60° - 38.21° \approx 81.79°$. We find BD by the law of sines:
$$\frac{\sin 81.79°}{24} = \frac{\sin 38.21°}{BD}, \text{ so } BD = \frac{24\sin 38.21°}{\sin 81.79°} \approx 15 \text{ cm}$$

21. We re-draw the figure:

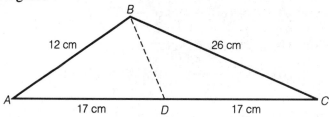

Using the law of cosines, we find $\angle A$:
$$26^2 = 12^2 + 34^2 - 2(12)(34)\cos A$$
$$\cos A \approx 0.7647$$
$$A \approx 40.12°$$
We can now find BD by using the law of cosines (on $\triangle ADB$):
$$(BD)^2 = 12^2 + 17^2 - 2(12)(17)\cos 40.12°$$
$$(BD)^2 = 121$$
$$BD = 11 \text{ cm}$$

23. (a) Using the law of cosines, we have:
$$a^2 = b^2 + c^2 - 2bc\cos A$$
$$4^2 = 5^2 + 6^2 - 2(5)(6)\cos A$$
$$16 = 25 + 36 - 60\cos A$$
$$-45 = -60\cos A$$
$$\tfrac{3}{4} = \cos A$$
Also using the law of cosines, we have:
$$c^2 = a^2 + b^2 - 2ab\cos C$$
$$6^2 = 4^2 + 5^2 - 2(4)(5)\cos C$$
$$36 = 16 + 25 - 40\cos C$$
$$-5 = -40\cos C$$
$$\tfrac{1}{8} = \cos C$$

(b)   Since $\cos A = \frac{3}{4}$, then:

$$\sin A = \sqrt{1 - \cos^2 A} = \sqrt{1 - \frac{9}{16}} = \sqrt{\frac{7}{16}} = \frac{\sqrt{7}}{4}$$

Therefore:

$$\cos 2A = \cos^2 A - \sin^2 A = \left(\frac{3}{4}\right)^2 - \left(\frac{\sqrt{7}}{4}\right)^2 = \frac{9}{16} - \frac{7}{16} = \frac{1}{8}$$

But since $\cos C = \frac{1}{8}$, then $C = 2A$.

25.   Since $BE = BD$, then $\triangle BDE$ is isosceles and thus its base angles are congruent. Therefore, calling $\angle BED = \theta$ we have:

$$\theta + \theta + 112° = 180°$$
$$2\theta = 68°$$
$$\theta = 34°$$

So $\angle BED = 34°$.

27.   Note that from Exercise 26 $\angle ABE = 34°$. Using the law of sines on $\triangle ABE$, we have:

$$\frac{\sin \angle AEB}{16} = \frac{\sin 34°}{11}$$
$$\sin \angle AEB = \frac{16 \sin 34°}{11} \approx 0.8134$$
$$\angle AEB \approx 54.43°$$

29.   Note that from Exercise 28 $\angle BAE \approx 91.57°$. Using the law of sines on $\triangle ABE$, we have:

$$\frac{\sin 91.57°}{BE} = \frac{\sin 34°}{11}$$
$$BE = \frac{11 \sin 91.57°}{\sin 34°} \approx 19.66 \text{ m}$$

31.   (a)   Let O be the center of the circle, so that $OA = OB = 10$ cm. Since

$\angle AOB = \frac{360°}{5} = 72°$, then we can find AB using the law of cosines on $\triangle AOB$:

$$(AB)^2 = (OA)^2 + (OB)^2 - 2(OA)(OB)\cos 72°$$
$$= (10)^2 + (10)^2 - 2(10)(10)\cos 72°$$
$$= 200 - 200 \cos 72°$$

So $AB = \sqrt{200 - 200 \cos 72°} \approx 11.76$ cm.

(b)   The interior angles of the pentagon are $\frac{180°(5-2)}{5} = 108°$, so $\angle ABC = 108°$. Using the law of cosines on $\triangle ABC$, we have (using the exact value from (a)):

$$(AC)^2 = (AB)^2 + (BC)^2 - 2(AB)(BC)\cos 108°$$
$$\approx (11.76)^2 + (11.76)^2 - 2(11.76)(11.76)\cos 108°$$
$$\approx 361.80$$

So $AC \approx \sqrt{361.80} \approx 19.02$ cm.

33. For the triangle, note the following figure:

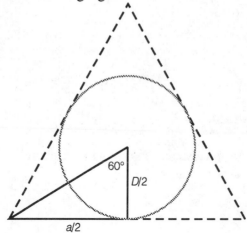

So $\tan 60° = \frac{a/2}{D/2}$, or $\sqrt{3} = \frac{a}{D}$, thus $a = D\sqrt{3}$. For the hexagon, note the following figure:

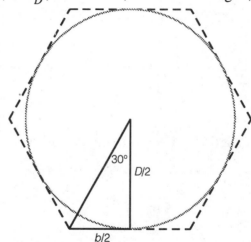

So $\tan 30° = \frac{b/2}{D/2}$, or $\frac{1}{\sqrt{3}} = \frac{b}{D}$, thus $b = \frac{D}{\sqrt{3}}$. Thus we have:

$$ab = \left(D\sqrt{3}\right)\left(\frac{D}{\sqrt{3}}\right) = D^2$$

35. The area of the triangle is given by:

$$A = \tfrac{1}{2}(\text{base})(\text{height}) = \tfrac{1}{2}(a)(b\sin 60°) = \frac{\sqrt{3}}{4}ab$$

Since the area is $10\sqrt{3}$ cm^2, we have:

$$\frac{\sqrt{3}}{4}ab = 10\sqrt{3}$$
$$ab = 40$$

We can find the third side d in terms of a and b using the law of cosines:

$$d^2 = a^2 + b^2 - 2ab\cos 60°$$
$$d^2 = a^2 + b^2 - 2(40)\left(\tfrac{1}{2}\right)$$
$$d = \sqrt{a^2 + b^2 - 40}$$

Since the perimeter of the triangle is 20 cm, we have:
$$a + b + \sqrt{a^2 + b^2 - 40} = 20$$
$$\sqrt{a^2 + b^2 - 40} = 20 - (a + b)$$
$$a^2 + b^2 - 40 = 400 - 40(a + b) + (a + b)^2$$
$$a^2 + b^2 - 40 = 400 - 40a - 40b + a^2 + 2ab + b^2$$
$$-440 = -40a - 40b + 2(40)$$
$$-520 = -40a - 40b$$
$$13 = a + b$$
So, we have the system of equations:
$$a + b = 13$$
$$ab = 40$$
Solving the first equation for b yields $b = 13 - a$, now substitute:
$$a(13 - a) = 40$$
$$13a - a^2 = 40$$
$$a^2 - 13a + 40 = 0$$
$$(a - 8)(a - 5) = 0$$
$$a = 8, 5$$
Since a is the smaller of the two numbers, we have $a = 5$ and $b = 8$.

37. We draw a figure:

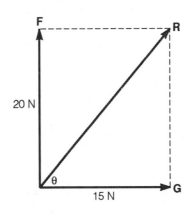

The resultant has a magnitude and direction given by:
$$|\mathbf{R}| = \sqrt{15^2 + 20^2} = \sqrt{625} = 25 \text{ N}$$
$$\theta = \tan^{-1}\left(\tfrac{20}{15}\right) \approx 53.1°$$

39. The components are given by:
$$v_x = 50\cos 35° \approx 41.0 \text{ cm/sec}$$
$$v_y = 50\sin 35° \approx 28.7 \text{ cm/sec}$$

41. We draw the figure:

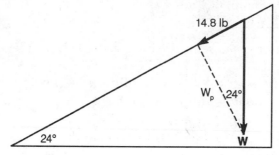

So the desired weights are given by:

$$\tan 24° = \frac{14.8}{\left|\mathbf{W}_p\right|}, \text{ so } \left|\mathbf{W}_p\right| = \frac{14.8}{\tan 24°} \approx 33.2 \text{ lb}$$

$$\sin 24° = \frac{14.8}{\left|\mathbf{W}\right|}, \text{ so } \left|\mathbf{W}\right| = \frac{14.8}{\sin 24°} \approx 36.4 \text{ lb}$$

43. We first find the lengths of $\langle 2,6 \rangle$ and $\langle -5,b \rangle$:

$$\left|\langle 2,6 \rangle\right| = \sqrt{2^2 + 6^2} = \sqrt{40}$$
$$\left|\langle -5,b \rangle\right| = \sqrt{(-5)^2 + b^2} = \sqrt{b^2 + 25}$$

We find b by solving the equation:

$$\sqrt{b^2 + 25} = \sqrt{40}$$
$$b^2 + 25 = 40$$
$$b^2 = 15$$
$$b = \pm\sqrt{15}$$

45. $\mathbf{a} + \mathbf{b} = \langle 3,5 \rangle + \langle 7,4 \rangle = \langle 10,9 \rangle$

47. We compute the quantity:

$$3\mathbf{c} + 2\mathbf{a} = 3\langle 2,-1 \rangle + 2\langle 3,5 \rangle = \langle 6,-3 \rangle + \langle 6,10 \rangle = \langle 12,7 \rangle$$

49. Since $\mathbf{b} + \mathbf{d} = \langle 7,7 \rangle$ and $\mathbf{b} - \mathbf{d} = \langle 7,1 \rangle$, then:

$$\left|\mathbf{b} + \mathbf{d}\right|^2 - \left|\mathbf{b} - \mathbf{d}\right|^2 = \left(7^2 + 7^2\right) - \left(7^2 + 1^2\right) = 98 - 50 = 48$$

51. $(\mathbf{a} + \mathbf{b}) + \mathbf{c} = \langle 10,9 \rangle + \langle 2,-1 \rangle = \langle 12,8 \rangle$

53. $(\mathbf{a} - \mathbf{b}) - \mathbf{c} = \langle -4,1 \rangle - \langle 2,-1 \rangle = \langle -6,2 \rangle$

55. $4\mathbf{c} + 2\mathbf{a} - 3\mathbf{b} = \langle 8,-4 \rangle + \langle 6,10 \rangle - \langle 21,12 \rangle = \langle -7,-6 \rangle$

57. $\langle 7,-6 \rangle = 7\mathbf{i} - 6\mathbf{j}$

59. We first compute the length of $\langle 6,4 \rangle$:
$$|\langle 6,4 \rangle| = \sqrt{36+16} = \sqrt{52} = 2\sqrt{13}$$
Therefore a unit vector in the same direction as $\langle 6,4 \rangle$ would be:
$$\tfrac{1}{2\sqrt{13}} \langle 6,4 \rangle = \left\langle \tfrac{3}{\sqrt{13}}, \tfrac{2}{\sqrt{13}} \right\rangle = \left\langle \tfrac{3\sqrt{13}}{13}, \tfrac{2\sqrt{13}}{13} \right\rangle$$

Chapter Five Test

1. We draw the triangle:

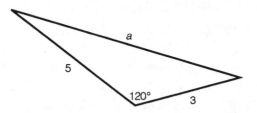

Using the law of cosines:
$$a^2 = 3^2 + 5^2 - 2(3)(5)\cos 120° = 9 + 25 - 30\left(-\tfrac{1}{2}\right) = 49$$
So $a = 7$ cm.

2. We draw the triangle:

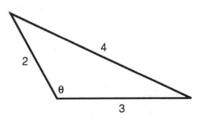

We use the law of cosines:
$$4^2 = 2^2 + 3^2 - 2(2)(3)\cos\theta$$
$$16 = 4 + 9 - 12\cos\theta$$
$$3 = -12\cos\theta$$
$$\cos\theta = -\tfrac{1}{4}$$
The angle opposite the 4 cm side must be obtuse (not acute), since its cosine is negative.

3. We use the law of sines:
$$\frac{\sin 45°}{20\sqrt{2}} = \frac{\sin 30°}{x}$$
$$x = \frac{20\sqrt{2}\sin 30°}{\sin 45°} = \frac{20\sqrt{2} \cdot \tfrac{1}{2}}{\tfrac{\sqrt{2}}{2}} = 20 \text{ cm}$$

4.   We draw the figure:

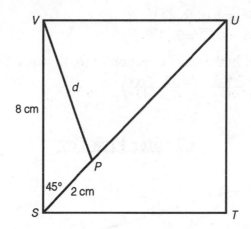

Note that the diagonals of a square bisect the vertex angles. We use the law of cosines with $\triangle SPV$:

$$d^2 = 8^2 + 2^2 - 2(8)(2)\cos 45° = 64 + 4 - 32\left(\tfrac{\sqrt{2}}{2}\right) = 68 - 16\sqrt{2}$$

$$d = \sqrt{68 - 16\sqrt{2}} = 2\sqrt{17 - 4\sqrt{2}}$$

So $PV = 2\sqrt{17 - 4\sqrt{2}}$ cm.

5.   We use the law of cosines to find a:

$$a^2 = (5.8)^2 + (3.2)^2 - 2(5.8)(3.2)\cos 27° = 43.88 - 37.12\cos 27° \approx 10.81$$

So $a \approx \sqrt{10.81} \approx 3.3$ cm. Now we use the law of sines to find $\angle C$:

$$\frac{\sin C}{3.2} = \frac{\sin 27°}{3.3}$$

$$\sin C = \frac{3.2\sin 27°}{3.3} \approx 0.442$$

$$C \approx 26.2°$$

Finally, we find $B \approx 180° - 27° - 26.2° \approx 126.8°$. The diameter of the circumscribed circle is given by:

$$\mathcal{D} = \frac{a}{\sin A} \approx \frac{3.3}{\sin 27°} \approx 7.2 \text{ cm}$$

6.   We draw the figure:

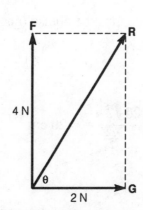

(a) We find the magnitude of the resultant:
$$|\mathbf{R}| = \sqrt{2^2 + 4^2} = \sqrt{20} = 2\sqrt{5} \text{ N}$$

(b) $\tan\theta = \frac{4}{2} = 2$

7. We draw the figure:

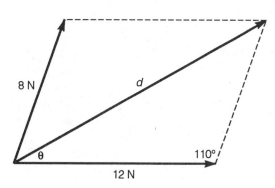

8 N d

θ 110°

12 N

(a) Using the law of cosines, we have:
$$d^2 = 12^2 + 8^2 - 2(12)(8)\cos110° = 208 - 192\cos110°$$
So $d = \sqrt{208 - 192\cos110°} = 4\sqrt{13 - 12\cos110°}$ N.

(b) Using the law of sines:
$$\frac{\sin\theta}{8} = \frac{\sin110°}{4\sqrt{13 - 12\cos110°}}$$
$$\sin\theta = \frac{2\sin110°}{\sqrt{13 - 12\cos110°}}$$

8. We draw the figure:

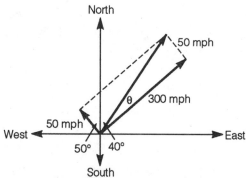

North

50 mph

θ 300 mph

50 mph

West   East

50°   40°

South

The heading vector and the wind vector are perpendicular to each other, so:
$$\text{ground speed} = \sqrt{300^2 + 50^2} = 50\sqrt{37} \text{ mph}$$
$$\tan\theta = \frac{50}{300} = \frac{1}{6}$$

9.   (a) $2\mathbf{A} + 3\mathbf{B} = 2\langle 2,4\rangle + 3\langle 3,-1\rangle = \langle 13,5\rangle$

   (b)   We compute the length:
$$|2\mathbf{A} + 3\mathbf{B}| = |\langle 13,5\rangle| = \sqrt{194}$$

   (c) $\mathbf{C} - \mathbf{B} = \langle 4,-4\rangle - \langle 3,-1\rangle = \langle 1,-3\rangle = \mathbf{i} - 3\mathbf{j}$

10. Subtracting components, we have:
$$\overrightarrow{PQ} = \langle -7,2\rangle - \langle 4,5\rangle = \langle -11,-3\rangle$$
The required unit vector is:

$$\frac{\overrightarrow{PQ}}{|\overrightarrow{PQ}|} = \frac{1}{\sqrt{130}}\langle -11,-3\rangle = \left\langle \frac{-11\sqrt{130}}{130}, \frac{-3\sqrt{130}}{130}\right\rangle$$

Chapter Six
Complex Numbers

6.1 The Complex Number System

1. We complete the table:

i^2	i^3	i^4	i^5	i^6	i^7	i^8
-1	$-i$	1	i	-1	$-i$	1

3. (a) The real part is 4 and the imaginary part is 5.

(b) The real part is 4 and the imaginary part is -5.

(c) The real part is $\frac{1}{2}$ and the imaginary part is -1.

(d) The real part is 0 and the imaginary part is 16.

5. Equating the real parts gives $2c = 8$, and therefore $c = 4$. Similarly, equating the imaginary parts yields $d = -3$.

7. (a) $(5 - 6i) + (9+2i) = (5 + 9) + (-6 + 2)i = 14 - 4i$

(b) $(5 - 6i) - (9+2i) = (5 - 9) + (-6 - 2)i = -4 - 8i$

9. (a) $(3 - 4i)(5 + i) = 15 - 17i - 4i^2 = 19 - 17i$

(b) $(5 + i)(3 - 4i) = 19 - 17i$, from part (a)

(c) We compute the quotient:
$$\frac{3 - 4i}{5 + i} \bullet \frac{5 - i}{5 - i} = \frac{15 - 23i + 4i^2}{25 - i^2} = \frac{11 - 23i}{26} = \frac{11}{26} - \frac{23}{26}i$$

(d)   We compute the quotient:
$$\frac{5+i}{3-4i} \bullet \frac{3+4i}{3+4i} = \frac{15+23i+4i^2}{9-16i^2} = \frac{11+23i}{25} = \frac{11}{25} + \frac{23}{25}i$$

11. (a) $z+w = (2+3i)+(9-4i) = 11-i$

     (b) $\bar{z}+w = (2-3i)+(9-4i) = 11-7i$

     (c) $z+\bar{z} = (2+3i)+(2-3i) = 4$

13. We evaluate the expression:
$$(z+w)+w_1 = \left[(2+3i)+(9-4i)\right]+(-7-i) = (11-i)+(-7-i) = 4-2i$$

15. We compute the product:
$$zw = (2+3i)(9-4i) = 18+19i-12i^2 = 30+19i$$

17. $z\bar{z} = (2+3i)(2-3i) = 4-9i^2 = 13$

19. We first compute the product:
$$ww_1 = (9-4i)(-7-i) = -63+19i+4i^2 = -67+19i$$
Therefore:
$$z(ww_1) = (2+3i)(-67+19i) = -134-163i+57i^2 = -191-163i$$

21. We first compute the sum:
$$w+w_1 = (9-4i)+(-7-i) = 2-5i$$
Therefore:
$$z(w+w_1) = (2+3i)(2-5i) = 4-4i-15i^2 = 19-4i$$

23. We first compute the powers:
$$z^2 = (2+3i)(2+3i) = 4+12i+9i^2 = -5+12i$$
$$w^2 = (9-4i)(9-4i) = 81-72i+16i^2 = 65-72i$$
Therefore:
$$z^2-w^2 = (-5+12i)-(65-72i) = -70+84i$$

25. Since $zw = 30 + 19i$ (from Exercise 15), then:
$$(zw)^2 = (30+19i)(30+19i) = 900+1140i+361i^2 = 539+1140i$$

27. Since $z^2 = -5 + 12i$ (from Exercise 23), then:
$$z^3 = z \bullet z^2 = (2+3i)(-5+12i) = -10+9i+36i^2 = -46+9i$$

29. We compute the quotient:
$$\frac{z}{w} = \frac{2+3i}{9-4i} \cdot \frac{9+4i}{9+4i} = \frac{18+35i+12i^2}{81-16i^2} = \frac{6+35i}{97} = \frac{6}{97} + \frac{35}{97}i$$

31. Using the values of $\bar{z}$ and $\bar{w}$, we have:
$$\frac{\bar{z}}{\bar{w}} = \frac{2-3i}{9+4i} \cdot \frac{9-4i}{9-4i} = \frac{18-35i+12i^2}{81-16i^2} = \frac{6-35i}{97} = \frac{6}{97} - \frac{35}{97}i$$

33. Using the value of $\bar{z}$, we have:
$$\frac{z}{\bar{z}} = \frac{2+3i}{2-3i} \cdot \frac{2+3i}{2+3i} = \frac{4+12i+9i^2}{4-9i^2} = \frac{-5+12i}{13} = -\frac{5}{13} + \frac{12}{13}i$$

35. Since $w - \bar{w} = (9-4i) - (9+4i) = -8i$, then:
$$\frac{w-\bar{w}}{2i} = \frac{-8i}{2i} = -4$$

37. We compute the quotient:
$$\frac{i}{5+i} \cdot \frac{5-i}{5-i} = \frac{5i-i^2}{25-i^2} = \frac{1+5i}{26} = \frac{1}{26} + \frac{5}{26}i$$

39. We compute the quotient:
$$\frac{1}{i} \cdot \frac{i}{i} = \frac{i}{i^2} = \frac{i}{-1} = -i$$

41. By writing the numbers in complex form, we have:
$$\sqrt{-49} + \sqrt{-9} + \sqrt{-4} = 7i + 3i + 2i = 12i$$

43. By writing the numbers in complex form, we have:
$$\sqrt{-20} - 3\sqrt{-45} + \sqrt{-80} = \sqrt{4}\sqrt{-5} - 3\sqrt{9}\sqrt{-5} + \sqrt{16}\sqrt{-5}$$
$$= 2\sqrt{5}i - 9\sqrt{5}i + 4\sqrt{5}i$$
$$= -3\sqrt{5}i$$

45. By writing the numbers in complex form, we have:
$$1 + \sqrt{-36}\sqrt{-36} = 1 + (6i)(6i) = 1 + 36i^2 = -35$$

47. By writing the numbers in complex form, we have:
$$3\sqrt{-128} - 4\sqrt{-18} = 3\sqrt{-64}\sqrt{2} - 4\sqrt{-9}\sqrt{2}$$
$$= 3(8i)\sqrt{2} - 4(3i)\sqrt{2}$$
$$= 24\sqrt{2}i - 12\sqrt{2}i$$
$$= 12\sqrt{2}i$$

49. (a) Notice that the result does agree with the definition:
$$z + w = (a+bi) + (c+di) = (a+c) + (b+d)i$$

(b)   Notice that the result does agree with the definition:
$$z - w = (a + bi) - (c + di) = (a - c) + (b - d)i$$

(c)   Notice that the result does agree with the definition:
$$zw = (a + bi)(c + di) = ac + bci + adi + bdi^2 = (ac - bd) + (bc + ad)i$$

(d)   Notice that the result does agree with the definition:
$$\frac{z}{w} = \frac{a + bi}{c + di} \cdot \frac{c - di}{c - di} = \frac{ac + bci - adi - bdi^2}{c^2 - d^2 i^2} = \frac{ac + bd}{c^2 + d^2} + \frac{bc - ad}{c^2 + d^2} i$$

51. (a)   We compute each power:

$$z^3 = \left(\frac{-1 + \sqrt{3}i}{2}\right)^3 \qquad\qquad w^3 = \left(\frac{-1 - \sqrt{3}i}{2}\right)^3$$

$$= \frac{(-1 + \sqrt{3}i)^3}{8} \qquad\qquad\quad = \frac{(-1 - \sqrt{3}i)^3}{8}$$

$$= \frac{(-1 + \sqrt{3}i)(-1 + \sqrt{3}i)^2}{8} \qquad = \frac{(-1 - \sqrt{3}i)(-1 - \sqrt{3}i)^2}{8}$$

$$= \frac{(-1 + \sqrt{3}i)(1 - 2\sqrt{3}i + 3i^2)}{8} \qquad = \frac{(-1 - \sqrt{3}i)(1 + 2\sqrt{3}i + 3i^2)}{8}$$

$$= \frac{(-1 + \sqrt{3}i)(-2 - 2\sqrt{3}i)}{8} \qquad = \frac{(-1 - \sqrt{3}i)(-2 + 2\sqrt{3}i)}{8}$$

$$= \frac{2 - 2\sqrt{3}i + 2\sqrt{3}i - 6i^2}{8} \qquad = \frac{2 + 2\sqrt{3}i - 2\sqrt{3}i - 6i^2}{8}$$

$$= \frac{8}{8} \qquad\qquad\qquad\qquad\quad = \frac{8}{8}$$

$$= 1 \qquad\qquad\qquad\qquad\qquad = 1$$

(b)   We compute the product:
$$zw = \left(\frac{-1 + \sqrt{3}i}{2}\right)\left(\frac{-1 - \sqrt{3}i}{2}\right) = \frac{1 - \sqrt{3}i + \sqrt{3}i - 3i^2}{4} = \frac{4}{4} = 1$$

(c)   We compute each power:

$$w^2 = \left(\frac{-1 - \sqrt{3}i}{2}\right)^2 \qquad\qquad z^2 = \left(\frac{-1 + \sqrt{3}i}{2}\right)^2$$

$$= \frac{1 + 2\sqrt{3}i + 3i^2}{4} \qquad\qquad = \frac{1 - 2\sqrt{3}i + 3i^2}{4}$$

$$= \frac{-2 + 2\sqrt{3}i}{4} \qquad\qquad\quad = \frac{-2 - 2\sqrt{3}i}{4}$$

$$= \frac{-1 + \sqrt{3}i}{2} \qquad\qquad\quad = \frac{-1 - \sqrt{3}i}{2}$$

$$= z \qquad\qquad\qquad\qquad\quad = w$$

(d) We compute the product:
$$\left(1-z+z^2\right)\left(1+z-z^2\right) = (1-z+w)(1+z-w)$$
$$= \left(\frac{2}{2} - \frac{-1+\sqrt{3}i}{2} + \frac{-1-\sqrt{3}i}{2}\right)\left(\frac{2}{2} + \frac{-1+\sqrt{3}i}{2} - \frac{-1-\sqrt{3}i}{2}\right)$$
$$= \left(\frac{2-2\sqrt{3}i}{2}\right)\left(\frac{2+2\sqrt{3}i}{2}\right)$$
$$= \left(1-\sqrt{3}i\right)\left(1+\sqrt{3}i\right)$$
$$= 1 - 3i^2$$
$$= 4$$

53. (a) Let $z = a + bi$, then:
$$0 + z = (0+0i) + (a+bi) = a + bi = z$$
$$z + 0 = (a+bi) + (0+0i) = a + bi = z$$

(b) Let $z = a + bi$, then:
$$0 \bullet z = (0+0i)(a+bi) = 0 + 0i = 0$$
$$z \bullet 0 = (a+bi)(0+0i) = 0$$

55. (a) We compute each sum:
$$z + w = (a+bi) + (c+di) = (a+c) + (b+d)i$$
$$w + z = (c+di) + (a+bi) = (c+a) + (d+b)i = z + w$$

(b) We compute each product:
$$zw = (a+bi)(c+di) = ac + bci + adi + bdi^2 = (ac - bd) + (bc + ad)i$$
$$wz = (c+di)(a+bi) = ac + adi + bci + bdi^2 = (ac - bd) + (bc + ad)i = zw$$

57. We compute each quotient then add the complex numbers:
$$\frac{a+bi}{a-bi} + \frac{a-bi}{a+bi} = \frac{a+bi}{a-bi} \bullet \frac{a+bi}{a+bi} + \frac{a-bi}{a+bi} \bullet \frac{a-bi}{a-bi}$$
$$= \frac{(a+bi)^2}{a^2+b^2} + \frac{(a-bi)^2}{a^2+b^2}$$
$$= \frac{a^2 + 2abi - b^2 + a^2 - 2abi - b^2}{a^2+b^2}$$
$$= \frac{2a^2 - 2b^2}{a^2+b^2}$$

Thus the real part is $\frac{2a^2-2b^2}{a^2+b^2}$, and the imaginary part is 0.

59. We compute each quotient then add the complex numbers:

$$\frac{(a+bi)^2}{a-bi} - \frac{(a-bi)^2}{a+bi} = \frac{(a+bi)^3 - (a-bi)^3}{(a-bi)(a+bi)}$$

$$= \frac{\left(a^3 + 3a^2bi + 3ab^2i^2 + b^3i^3\right) - \left(a^3 - 3a^2bi + 3ab^2i^2 - b^3i^3\right)}{a^2 - b^2i^2}$$

$$= \frac{\left(a^3 - 3ab^2\right) + \left(3a^2b - b^3\right)i - \left(a^3 - 3ab^2\right) + \left(3a^2b - b^3\right)}{a^2 + b^2}$$

$$= \frac{6a^2b - 2b^3}{a^2 + b^2}i$$

Thus the real part is 0.

6.2 Trigonometric Form for Complex Numbers

1. The complex number $4 + 2i$ is identified with the point $(4, 2)$:

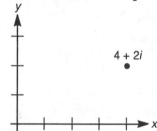

3. The complex number $-5 + i$ is identified with the point $(-5, 1)$:

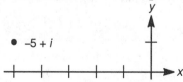

5. The complex number $1 - 4i$ is identified with the point $(1, -4)$:

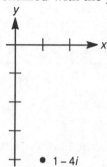

7. The complex number $-i$, or $0-1i$, is identified with the point $(0,-1)$:

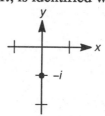

9. We convert the complex number to rectangular form:
$$2\left[\cos\tfrac{\pi}{4}+i\sin\tfrac{\pi}{4}\right]=2\left[\tfrac{\sqrt{2}}{2}+\tfrac{\sqrt{2}}{2}i\right]=\sqrt{2}+\sqrt{2}i$$

11. We convert the complex number to rectangular form:
$$4\left[\cos\tfrac{5\pi}{6}+i\sin\tfrac{5\pi}{6}\right]=4\left[-\tfrac{\sqrt{3}}{2}+\tfrac{1}{2}i\right]=-2\sqrt{3}+2i$$

13. We convert the complex number to rectangular form:
$$\sqrt{2}\left[\cos225°+i\sin225°\right]=\sqrt{2}\left[-\tfrac{\sqrt{2}}{2}-\tfrac{\sqrt{2}}{2}i\right]=-1-i$$

15. We convert the complex number to rectangular form:
$$\sqrt{3}\left[\cos\tfrac{\pi}{2}+i\sin\tfrac{\pi}{2}\right]=\sqrt{3}(0+1i)=\sqrt{3}i$$

17. We use the hint to find $\cos75°$ and $\sin75°$:
$$\cos75°=\cos(30°+45°)=\cos30°\cos45°-\sin30°\sin45°=\tfrac{\sqrt{3}}{2}\cdot\tfrac{\sqrt{2}}{2}-\tfrac{1}{2}\cdot\tfrac{\sqrt{2}}{2}=\tfrac{\sqrt{6}-\sqrt{2}}{4}$$
$$\sin75°=\sin(30°+45°)=\sin30°\cos45°+\cos30°\sin45°=\tfrac{1}{2}\cdot\tfrac{\sqrt{2}}{2}+\tfrac{\sqrt{3}}{2}\cdot\tfrac{\sqrt{2}}{2}=\tfrac{\sqrt{6}+\sqrt{2}}{4}$$
Therefore:
$$4\left[\cos75°+i\sin75°\right]=4\left[\tfrac{\sqrt{6}-\sqrt{2}}{4}+\tfrac{\sqrt{6}+\sqrt{2}}{4}i\right]=\left(\sqrt{6}-\sqrt{2}\right)+\left(\sqrt{6}+\sqrt{2}\right)i$$

19. Here $a=\tfrac{\sqrt{3}}{2}$ and $b=\tfrac{1}{2}$, so:
$$r=\sqrt{a^2+b^2}=\sqrt{\tfrac{3}{4}+\tfrac{1}{4}}=1$$

We now find θ such that $\cos\theta=\tfrac{a}{r}=\tfrac{\sqrt{3}}{2}$ and $\sin\theta=\tfrac{b}{r}=\tfrac{1}{2}$. Such a θ is $\theta=\tfrac{\pi}{6}$. Thus:
$$\tfrac{\sqrt{3}}{2}+\tfrac{1}{2}i=\cos\tfrac{\pi}{6}+i\sin\tfrac{\pi}{6}$$

21. Here $a=-1$ and $b=\sqrt{3}$, so:
$$r=\sqrt{a^2+b^2}=\sqrt{1+3}=\sqrt{4}=2$$

We now find θ such that $\cos\theta=\tfrac{a}{r}=-\tfrac{1}{2}$ and $\sin\theta=\tfrac{b}{r}=\tfrac{\sqrt{3}}{2}$. Such a θ is $\theta=\tfrac{2\pi}{3}$. Thus:
$$-1+\sqrt{3}i=2\left[\cos\tfrac{2\pi}{3}+i\sin\tfrac{2\pi}{3}\right]$$

23. Here $a = -2\sqrt{3}$ and $b = -2$, so:
$$r = \sqrt{a^2 + b^2} = \sqrt{12 + 4} = \sqrt{16} = 4$$

We now find θ such that $\cos\theta = \frac{a}{r} = -\frac{2\sqrt{3}}{4} = -\frac{\sqrt{3}}{2}$ and $\sin\theta = \frac{b}{r} = -\frac{2}{4} = -\frac{1}{2}$. Such a θ is

$\theta = \frac{7\pi}{6}$. Thus:
$$-2\sqrt{3} - 2i = 4\left[\cos\frac{7\pi}{6} + i\sin\frac{7\pi}{6}\right]$$

25. Here $a = 0$ and $b = -6$, so:
$$r = \sqrt{a^2 + b^2} = \sqrt{0 + 36} = 6$$

We now find θ such that $\cos\theta = \frac{a}{r} = \frac{0}{6} = 0$ and $\sin\theta = \frac{b}{r} = -\frac{6}{6} = -1$. Such a θ is $\theta = \frac{3\pi}{2}$.
Thus:
$$-6i = 6\left[\cos\frac{3\pi}{2} + i\sin\frac{3\pi}{2}\right]$$

27. Here $a = \frac{\sqrt{3}}{4}$ and $b = -\frac{1}{4}$, so:
$$r = \sqrt{a^2 + b^2} = \sqrt{\frac{3}{16} + \frac{1}{16}} = \sqrt{\frac{1}{4}} = \frac{1}{2}$$

We now find θ such that $\cos\theta = \frac{a}{r} = \frac{\sqrt{3}/4}{1/2} = \frac{\sqrt{3}}{2}$ and $\sin\theta = \frac{b}{r} = \frac{-1/4}{1/2} = -\frac{1}{2}$. Such a θ is

$\theta = \frac{11\pi}{6}$. Thus:
$$\frac{\sqrt{3}}{4} - \frac{1}{4}i = \frac{1}{2}\left[\cos\frac{11\pi}{6} + i\sin\frac{11\pi}{6}\right]$$

29. Performing the multiplication and adding angles, we have:
$$2\left[\cos 22° + i\sin 22°\right] \bullet 3\left[\cos 38° + i\sin 38°\right] = 6\left[\cos 60° + i\sin 60°\right]$$
$$= 6\left(\frac{1}{2} + \frac{\sqrt{3}}{2}i\right)$$
$$= 3 + 3\sqrt{3}i$$

31. Performing the multiplication and adding angles, we have:
$$\sqrt{2}\left[\cos\frac{\pi}{3} + i\sin\frac{\pi}{3}\right] \bullet \sqrt{2}\left[\cos\frac{4\pi}{3} + i\sin\frac{4\pi}{3}\right] = 2\left[\cos\frac{5\pi}{3} + i\sin\frac{5\pi}{3}\right]$$
$$= 2\left(\frac{1}{2} - \frac{\sqrt{3}}{2}i\right)$$
$$= 1 - \sqrt{3}i$$

33. Performing the multiplication and adding angles, we have:
$$3\left[\cos\frac{\pi}{7} + i\sin\frac{\pi}{7}\right] \bullet \sqrt{2}\left[\cos\frac{\pi}{7} + i\sin\frac{\pi}{7}\right] = 3\sqrt{2}\left[\cos\frac{2\pi}{7} + i\sin\frac{2\pi}{7}\right]$$
$$= 3\sqrt{2}\cos\frac{2\pi}{7} + \left(3\sqrt{2}\sin\frac{2\pi}{7}\right)i$$

35. Performing the division and subtracting angles, we have:
$$6[\cos 50° + i\sin 50°] \div 2[\cos 5° + i\sin 5°] = 3[\cos 45° + i\sin 45°]$$
$$= 3\left(\frac{\sqrt{2}}{2} + \frac{\sqrt{2}}{2}i\right)$$
$$= \frac{3\sqrt{2}}{2} + \frac{3\sqrt{2}}{2}i$$

37. Performing the division and subtracting angles, we have:
$$2^{4/3}\left[\cos \frac{5\pi}{12} + i\sin \frac{5\pi}{12}\right] \div 2^{1/3}\left[\cos \frac{\pi}{4} + i\sin \frac{\pi}{4}\right] = 2\left[\cos \frac{\pi}{6} + i\sin \frac{\pi}{6}\right] = 2\left(\frac{\sqrt{3}}{2} + \frac{1}{2}i\right) = \sqrt{3} + i$$

39. Performing the division and subtracting angles, we have:
$$\left[\cos \frac{2\pi}{5} + i\sin \frac{2\pi}{5}\right] \div \left[\cos \frac{2\pi}{5} + i\sin \frac{2\pi}{5}\right] = \cos 0 + i\sin 0 = 1 + 0i = 1$$

41. Performing the exponent (multiplication) and adding angles, we have:
$$[3(\cos 15° + i\sin 15°)]^2 = 9(\cos 30° + i\sin 30°) = 9\left(\frac{\sqrt{3}}{2} + \frac{1}{2}i\right) = \frac{9\sqrt{3}}{2} + \frac{9}{2}i$$

43. Performing the exponent (multiplication) and adding angles, we have:
$$\left[\sqrt{3}\left(\cos \frac{\pi}{7} + i\sin \frac{\pi}{7}\right)\right]^2 = 3\left(\cos \frac{2\pi}{7} + i\sin \frac{2\pi}{7}\right) = 3\cos \frac{2\pi}{7} + \left(3\sin \frac{2\pi}{7}\right)i$$

45. Since the multiplication results in a difference of squares, we have:
$$(\cos \theta + i\sin \theta)(\cos \theta - i\sin \theta) = \cos^2 \theta - i^2 \sin^2 \theta$$
$$= \cos^2 \theta + \sin^2 \theta \quad (\text{since } i^2 = -1)$$
$$= 1$$

47. Looking at the corresponding real and imaginary parts on each side of the equation, we must show that:

(1) $1 + \cos \theta = 2\cos^2 \frac{\theta}{2}$

(2) $\sin \theta = 2\cos \frac{\theta}{2}\sin \frac{\theta}{2}$

But (1) is just a restatement of the half-angle identity for cosine, since:
$$\cos^2 \frac{\theta}{2} = \frac{1 + \cos \theta}{2}, \text{ so } 1 + \cos \theta = 2\cos^2 \frac{\theta}{2}$$

And (2) is just a restatement of the double-angle identity for sine, since

$\sin 2x = 2\sin x \cos x$, so when $x = \frac{\theta}{2}$, we have $\sin \theta = 2\sin \frac{\theta}{2}\cos \frac{\theta}{2}$. Thus the result is true since the corresponding real and imaginary parts are indeed equal.

49. For convenience, denote $\sin \theta$ and $\cos \theta$ by S and C, respectively. Then, following the hint in the text, we have:

$$\frac{(1+S)+iC}{(1+S)-iC} \bullet \frac{(1+S)+iC}{(1+S)+iC} = \frac{(1+S)^2 + 2iC(1+S)-C^2}{(1+S)^2+C^2}$$

$$= \frac{(1+S)^2 + 2iC(1+S)-(1-S^2)}{(1+S)^2+(1-S^2)}$$

$$= \frac{(1+S)^2 + 2iC(1+S)-(1-S)(1+S)}{(1+S)^2+(1-S)(1+S)}$$

Notice that every term in the numerator and in the denominator of this last expression contains the common factor $(1 + S)$, and this factor is nonzero since $\theta \neq \frac{3\pi}{2}k$ and thus $\sin \theta \neq -1$. Dividing out this factor, we see that the expression becomes:

$$\frac{(1+S)+2iC-(1-S)}{(1+S)+(1-S)} = \frac{2S+2iC}{2} = S+iC$$

Since this expression represents $\sin \theta + i \cos \theta$, we have proven the result.

6.3 De Moivre's Theorem

1. Using De Moivre's theorem, we have:

$$\left[3\left(\cos\tfrac{\pi}{4}+i\sin\tfrac{\pi}{4}\right)\right]^2 = 3^2\left(\cos\tfrac{\pi}{2}+i\sin\tfrac{\pi}{2}\right) = 9(0+i) = 9i$$

3. Using De Moivre's theorem, we have:

$$\left[2(\cos 10°+i\sin 10°)\right]^3 = 2^3\left(\cos 30°+i\sin 30°\right) = 8\left(\tfrac{\sqrt{3}}{2}+\tfrac{1}{2}i\right) = 4\sqrt{3}+4i$$

5. Using De Moivre's theorem, we have:

$$\left[3\left(\cos\tfrac{\pi}{3}+i\sin\tfrac{\pi}{3}\right)\right]^5 = 3^5\left(\cos\tfrac{5\pi}{3}+i\sin\tfrac{5\pi}{3}\right) = 243\left(\tfrac{1}{2}-\tfrac{\sqrt{3}}{2}i\right) = \tfrac{243}{2} - \tfrac{243\sqrt{3}}{2}i$$

7. Using De Moivre's theorem, we have:

$$\left[\tfrac{1}{2}\left(\cos\tfrac{\pi}{24}+i\sin\tfrac{\pi}{24}\right)\right]^6 = \left(\tfrac{1}{2}\right)^6\left(\cos\tfrac{\pi}{4}+i\sin\tfrac{\pi}{4}\right) = \tfrac{1}{64}\left(\tfrac{\sqrt{2}}{2}+\tfrac{\sqrt{2}}{2}i\right) = \tfrac{\sqrt{2}}{128}+\tfrac{\sqrt{2}}{128}i$$

9. Using De Moivre's theorem, we have:

$$\left[2^{1/5}(\cos 63°+i\sin 63°)\right]^{10} = \left(2^{1/5}\right)^{10}(\cos 630°+i\sin 630°) = 4(0-1i) = -4i$$

11. First computing the product by adding angles, then using De Moivre's theorem, we have:

$$\left[(\cos 5°+i\sin 5°)(\cos 3°+i\sin 3°)\right]^{15} = (\cos 8°+i\sin 8°)^{15}$$

$$= \cos 120°+i\sin 120°$$

$$= -\tfrac{1}{2}+\tfrac{\sqrt{3}}{2}i$$

13. We first write $\frac{1}{2} - \frac{\sqrt{3}}{2}i$ in trigonometric form:

$$\frac{1}{2} - \frac{\sqrt{3}}{2}i = \cos\frac{5\pi}{3} + i\sin\frac{5\pi}{3}$$

Now using De Moivre's theorem, we have:

$$\left(\frac{1}{2} - \frac{\sqrt{3}}{2}i\right)^5 = \cos\frac{25\pi}{3} + i\sin\frac{25\pi}{3} = \frac{1}{2} + \frac{\sqrt{3}}{2}i$$

15. We first write $-2 - 2i$ in trigonometric form:

$$-2 - 2i = 2\sqrt{2}\left(-\frac{\sqrt{2}}{2} - \frac{\sqrt{2}}{2}i\right) = 2\sqrt{2}\left(\cos\frac{5\pi}{4} + i\sin\frac{5\pi}{4}\right)$$

Now using De Moivre's theorem, we have:

$$(-2 - 2i)^5 = \left(2\sqrt{2}\right)^5\left(\cos\frac{25\pi}{4} + i\sin\frac{25\pi}{4}\right) = 128\sqrt{2}\left(\frac{\sqrt{2}}{2} + \frac{\sqrt{2}}{2}i\right) = 128 + 128i$$

17. We first write $-2\sqrt{3} - 2i$ in trigonometric form:

$$-2\sqrt{3} - 2i = 4\left(-\frac{\sqrt{3}}{2} - \frac{1}{2}i\right) = 4\left(\cos\frac{7\pi}{6} + i\sin\frac{7\pi}{6}\right)$$

Now using De Moivre's theorem, we have:

$$(-2\sqrt{3} - 2i)^4 = 4^4\left[\cos\frac{14\pi}{3} + i\sin\frac{14\pi}{3}\right] = 256\left(-\frac{1}{2} + \frac{\sqrt{3}}{2}i\right) = -128 + 128\sqrt{3}i$$

19. We begin by writing $-27i$ in trigonometric form:

$$-27i = 27(0 - 1i) = 27\left[\cos\frac{3\pi}{2} + i\sin\frac{3\pi}{2}\right]$$

Now let $z = r(\cos\theta + i\sin\theta)$ denote a cube root of $-27i$. Then:

$$z^3 = r^3(\cos3\theta + i\sin3\theta) = 27\left[\cos\frac{3\pi}{2} + i\sin\frac{3\pi}{2}\right]$$

Then $r^3 = 27$ so $r = 3$, and $3\theta = \frac{3\pi}{2} + 2\pi k$, so $\theta = \frac{\pi}{2} + \frac{2\pi}{3}k$.

When $k = 0$, we have:

$$z_1 = 3\left[\cos\frac{\pi}{2} + i\sin\frac{\pi}{2}\right] = 3(0 + i) = 3i$$

When $k = 1$, we have:

$$z_2 = 3\left[\cos\frac{7\pi}{6} + i\sin\frac{7\pi}{6}\right] = 3\left(-\frac{\sqrt{3}}{2} - \frac{1}{2}i\right) = -\frac{3\sqrt{3}}{2} - \frac{3}{2}i$$

When $k = 2$, we have:

$$z_3 = 3\left[\cos\frac{11\pi}{6} + i\sin\frac{11\pi}{6}\right] = 3\left(\frac{\sqrt{3}}{2} - \frac{1}{2}i\right) = \frac{3\sqrt{3}}{2} - \frac{3}{2}i$$

So the cube roots of $-27i$ are $3i$, $-\frac{3\sqrt{3}}{2} - \frac{3}{2}i$ and $\frac{3\sqrt{3}}{2} - \frac{3}{2}i$.

21. We begin by writing i in trigonometric form:

$$i = 1(0 + 1i) = 1\left[\cos\frac{\pi}{2} + i\sin\frac{\pi}{2}\right]$$

Now let $z = r(\cos\theta + i\sin\theta)$ denote a square root of i. Then:

$$z^2 = r^2(\cos2\theta + i\sin2\theta) = 1\left[\cos\frac{\pi}{2} + i\sin\frac{\pi}{2}\right]$$

Then $r^2 = 1$ so $r = 1$, and $2\theta = \frac{\pi}{2} + 2\pi k$, so $\theta = \frac{\pi}{4} + \pi k$.

When $k = 0$, we have:
$$z_1 = 1\left[\cos\tfrac{\pi}{4} + i\sin\tfrac{\pi}{4}\right] = \tfrac{\sqrt{2}}{2} + \tfrac{\sqrt{2}}{2}i$$
When $k = 1$, we have:
$$z_2 = 1\left[\cos\tfrac{5\pi}{4} + i\sin\tfrac{5\pi}{4}\right] = -\tfrac{\sqrt{2}}{2} - \tfrac{\sqrt{2}}{2}i$$

So the square roots of i are $\tfrac{\sqrt{2}}{2} + \tfrac{\sqrt{2}}{2}i$ and $-\tfrac{\sqrt{2}}{2} - \tfrac{\sqrt{2}}{2}i$.

23. We begin by writing 2 in trigonometric form:
$$2 = 2(1 + 0i) = 2[\cos 0 + i\sin 0]$$
Now let $z = r(\cos\theta + i\sin\theta)$ denote a cube root of 2. Then:
$$z^3 = r^3(\cos 3\theta + i\sin 3\theta) = 2[\cos 0 + i\sin 0]$$

Then $r^3 = 2$ so $r = 2^{1/3}$, and $3\theta = 0 + 2\pi k$, so $\theta = 0 + \tfrac{2\pi}{3}k$.
When $k = 0$, we have:
$$z_1 = 2^{1/3}[\cos 0 + i\sin 0] = 2^{1/3}(1 + 0i) = 2^{1/3}$$
When $k = 1$, we have:
$$z_2 = 2^{1/3}\left[\cos\tfrac{2\pi}{3} + i\sin\tfrac{2\pi}{3}\right] = 2^{1/3}\left(-\tfrac{1}{2} + \tfrac{\sqrt{3}}{2}i\right) = -\frac{1}{2^{2/3}} + \frac{\sqrt{3}}{2^{2/3}}i$$
When $k = 2$, we have:
$$z_3 = 2^{1/3}\left[\cos\tfrac{4\pi}{3} + i\sin\tfrac{4\pi}{3}\right] = 2^{1/3}\left(-\tfrac{1}{2} - \tfrac{\sqrt{3}}{2}i\right) = -\frac{1}{2^{2/3}} - \frac{\sqrt{3}}{2^{2/3}}i$$

So the cube roots of 2 are $2^{1/3}$, $-\tfrac{1}{2^{2/3}} + \tfrac{\sqrt{3}}{2^{2/3}}i$ and $-\tfrac{1}{2^{2/3}} - \tfrac{\sqrt{3}}{2^{2/3}}i$.

25. We begin by writing -1 in trigonometric form:
$$-1 = 1(-1 + 0i) = 1[\cos\pi + i\sin\pi]$$
Now let $z = r(\cos\theta + i\sin\theta)$ denote an eighth root of -1. Then:
$$z^8 = r^8(\cos 8\theta + i\sin 8\theta) = 1[\cos\pi + i\sin\pi]$$

Then $r^8 = 1$ so $r = 1$, and $8\theta = \pi + 2\pi k$, so $\theta = \tfrac{\pi}{8} + \tfrac{\pi}{4}k$.
When $k = 0$, we have:
$$z_1 = 1\left[\cos\tfrac{\pi}{8} + i\sin\tfrac{\pi}{8}\right] = \frac{\sqrt{2+\sqrt{2}}}{2} + \frac{\sqrt{2-\sqrt{2}}}{2}i$$
When $k = 1$, we have:
$$z_2 = 1\left[\cos\tfrac{3\pi}{8} + i\sin\tfrac{3\pi}{8}\right] = \frac{\sqrt{2-\sqrt{2}}}{2} + \frac{\sqrt{2+\sqrt{2}}}{2}i$$
When $k = 2$, we have:
$$z_3 = 1\left[\cos\tfrac{5\pi}{8} + i\sin\tfrac{5\pi}{8}\right] = -\frac{\sqrt{2-\sqrt{2}}}{2} + \frac{\sqrt{2+\sqrt{2}}}{2}i$$
When $k = 3$, we have:
$$z_4 = 1\left[\cos\tfrac{7\pi}{8} + i\sin\tfrac{7\pi}{8}\right] = -\frac{\sqrt{2+\sqrt{2}}}{2} + \frac{\sqrt{2-\sqrt{2}}}{2}i$$
When $k = 4$, we have:
$$z_5 = 1\left[\cos\tfrac{9\pi}{8} + i\sin\tfrac{9\pi}{8}\right] = -\frac{\sqrt{2+\sqrt{2}}}{2} - \frac{\sqrt{2-\sqrt{2}}}{2}i$$

When $k = 5$, we have:
$$z_6 = 1\left[\cos\tfrac{11\pi}{8} + i\sin\tfrac{11\pi}{8}\right] = -\tfrac{\sqrt{2-\sqrt{2}}}{2} - \tfrac{\sqrt{2+\sqrt{2}}}{2}i$$
When $k = 6$, we have:
$$z_7 = 1\left[\cos\tfrac{13\pi}{8} + i\sin\tfrac{13\pi}{8}\right] = \tfrac{\sqrt{2-\sqrt{2}}}{2} - \tfrac{\sqrt{2+\sqrt{2}}}{2}i$$
When $k = 7$, we have:
$$z_8 = 1\left[\cos\tfrac{15\pi}{8} + i\sin\tfrac{15\pi}{8}\right] = \tfrac{\sqrt{2+\sqrt{2}}}{2} - \tfrac{\sqrt{2-\sqrt{2}}}{2}i$$

So the eighth roots of -1 are $\tfrac{\sqrt{2+\sqrt{2}}}{2} + \tfrac{\sqrt{2-\sqrt{2}}}{2}i$, $\tfrac{\sqrt{2-\sqrt{2}}}{2} + \tfrac{\sqrt{2+\sqrt{2}}}{2}i$, $-\tfrac{\sqrt{2-\sqrt{2}}}{2} + \tfrac{\sqrt{2+\sqrt{2}}}{2}i$,

$-\tfrac{\sqrt{2+\sqrt{2}}}{2} + \tfrac{\sqrt{2-\sqrt{2}}}{2}i$, $-\tfrac{\sqrt{2+\sqrt{2}}}{2} - \tfrac{\sqrt{2-\sqrt{2}}}{2}i$, $-\tfrac{\sqrt{2-\sqrt{2}}}{2} - \tfrac{\sqrt{2+\sqrt{2}}}{2}i$, $\tfrac{\sqrt{2-\sqrt{2}}}{2} - \tfrac{\sqrt{2+\sqrt{2}}}{2}i$ and

$\tfrac{\sqrt{2+\sqrt{2}}}{2} - \tfrac{\sqrt{2-\sqrt{2}}}{2}i$.

These final results were obtained using the half-angle formulas for sine and cosine.

27. We begin by writing $4\sqrt{2} - 4\sqrt{2}i$ in trigonometric form:
$$4\sqrt{2} - 4\sqrt{2}i = 8\left(\tfrac{\sqrt{2}}{2} - \tfrac{\sqrt{2}}{2}i\right) = 8\left[\cos\tfrac{7\pi}{4} + i\sin\tfrac{7\pi}{4}\right]$$
Now let $z = r(\cos\theta + i\sin\theta)$ denote a cube root of $4\sqrt{2} - 4\sqrt{2}i$. Then:
$$z^3 = r^3(\cos 3\theta + i\sin 3\theta) = 8\left[\cos\tfrac{7\pi}{4} + i\sin\tfrac{7\pi}{4}\right]$$

Then $r^3 = 8$ so $r = 2$, and $3\theta = \tfrac{7\pi}{4} + 2\pi k$, so $\theta = \tfrac{7\pi}{12} + \tfrac{2\pi}{3}k$.
When $k = 0$, we have:
$$z_1 = 2\left[\cos\tfrac{7\pi}{12} + i\sin\tfrac{7\pi}{12}\right] = \tfrac{\sqrt{2}-\sqrt{6}}{2} + \tfrac{\sqrt{2}+\sqrt{6}}{2}i$$
When $k = 1$, we have:
$$z_2 = 2\left[\cos\tfrac{5\pi}{4} + i\sin\tfrac{5\pi}{4}\right] = -\sqrt{2} - \sqrt{2}i$$
When $k = 2$, we have:
$$z_3 = 2\left[\cos\tfrac{23\pi}{12} + i\sin\tfrac{23\pi}{12}\right] = \tfrac{\sqrt{2}+\sqrt{6}}{2} + \tfrac{\sqrt{2}-\sqrt{6}}{2}i$$

So the cube roots of $4\sqrt{2} - 4\sqrt{2}i$ are $\tfrac{\sqrt{2}-\sqrt{6}}{2} + \tfrac{\sqrt{2}+\sqrt{6}}{2}i$, $-\sqrt{2} - \sqrt{2}i$ and $\tfrac{\sqrt{2}+\sqrt{6}}{2} + \tfrac{\sqrt{2}-\sqrt{6}}{2}i$.
These final results were obtained using the addition formulas for sine and cosine.

29. We first write $9 + 9i$ in trigonometric form:
$$9 + 9i = 9\sqrt{2}\left(\tfrac{1}{\sqrt{2}} + \tfrac{1}{\sqrt{2}}i\right) = 9\sqrt{2}\left[\cos\tfrac{\pi}{4} + i\sin\tfrac{\pi}{4}\right]$$
Using De Moivre's theorem, we have:
$$(9 + 9i)^6 = (9\sqrt{2})^6\left[\cos\tfrac{3\pi}{2} + i\sin\tfrac{3\pi}{2}\right] = (9^6)(2^3)(0 - i) = -4,251,528i$$

31. We begin by writing $1 + 2i$ in trigonometric form:
$$1 + 2i = \sqrt{5}\left(\tfrac{1}{\sqrt{5}} + \tfrac{2}{\sqrt{5}}i\right) = \sqrt{5}[\cos 63.4349° + i\sin 63.4349°]$$
Now let z be a cube root of $1 + 2i$. Then $z = r(\cos\theta + i\sin\theta)$, and:
$$z^3 = r^3(\cos 3\theta + i\sin 3\theta) = \sqrt{5}[\cos 63.4349° + i\sin 63.4349°]$$

So $r^3 = \sqrt{5}$ and $r = 5^{1/6}$, and $3\theta = 63.4349° + 360°k$, so $\theta = 21.1450° + 120°k$.
When $k = 0$, we have:
$$z_1 = 1.3077[\cos 21.1450° + i \sin 21.1450°] = 1.22 + 0.47i$$
When $k = 1$, we have:
$$z_2 = 1.3077[\cos 141.1450° + i \sin 141.1450°] = -1.02 + 0.82i$$
When $k = 2$, we have:
$$z_3 = 1.3077[\cos 261.1450° + i \sin 261.1450°] = -0.20 - 1.29i$$
So the cube roots of $1 + 2i$ (to two decimal places) are $1.22 + 0.47i$, $-1.02 + 0.82i$ and $-0.20 - 1.29i$.

33. We begin by writing i in trigonometric form:
$$i = 1(0 + 1i) = 1\left[\cos\tfrac{\pi}{2} + i \sin\tfrac{\pi}{2}\right]$$
Now let $z = r(\cos\theta + i \sin\theta)$ denote a fourth root of i. Then:
$$z^4 = r^4(\cos 4\theta + i \sin 4\theta) = 1\left[\cos\tfrac{\pi}{2} + i \sin\tfrac{\pi}{2}\right]$$

Then $r^4 = 1$ so $r = 1$, and $4\theta = \tfrac{\pi}{2} + 2\pi k$, so $\theta = \tfrac{\pi}{8} + \tfrac{\pi}{2}k$.
When $k = 0$, we have:
$$z_1 = 1\left[\cos\tfrac{\pi}{8} + i \sin\tfrac{\pi}{8}\right] = \frac{\sqrt{2+\sqrt{2}}}{2} + \frac{\sqrt{2-\sqrt{2}}}{2}i$$
When $k = 1$, we have:
$$z_2 = 1\left[\cos\tfrac{5\pi}{8} + i \sin\tfrac{5\pi}{8}\right] = -\frac{\sqrt{2-\sqrt{2}}}{2} + \frac{\sqrt{2+\sqrt{2}}}{2}i$$
When $k = 2$, we have:
$$z_3 = 1\left[\cos\tfrac{9\pi}{8} + i \sin\tfrac{9\pi}{8}\right] = -\frac{\sqrt{2+\sqrt{2}}}{2} - \frac{\sqrt{2-\sqrt{2}}}{2}i$$
When $k = 3$, we have:
$$z_4 = 1\left[\cos\tfrac{13\pi}{8} + i \sin\tfrac{13\pi}{8}\right] = \frac{\sqrt{2-\sqrt{2}}}{2} - \frac{\sqrt{2+\sqrt{2}}}{2}i$$

So the fourth roots of i are $\frac{\sqrt{2+\sqrt{2}}}{2} + \frac{\sqrt{2-\sqrt{2}}}{2}i$, $-\frac{\sqrt{2-\sqrt{2}}}{2} + \frac{\sqrt{2+\sqrt{2}}}{2}i$, $-\frac{\sqrt{2+\sqrt{2}}}{2} - \frac{\sqrt{2-\sqrt{2}}}{2}i$ and

$\frac{\sqrt{2-\sqrt{2}}}{2} - \frac{\sqrt{2+\sqrt{2}}}{2}i$. These final results were obtained using the half-angle formulas for sine and cosine.

35. We write $7 + 24i = 25\left(\tfrac{7}{25} + \tfrac{24}{25}i\right) = 25(\cos w + i \sin w)$ for some angle w which satisfies the

conditions $\cos w = \tfrac{7}{25}$ and $0 < w < \tfrac{\pi}{2}$. Now let $z = r(\cos\theta + i \sin\theta)$ denote a square root of $7 + 24i$. Then:
$$z^2 = r^2(\cos 2\theta + i \sin 2\theta) = 25(\cos w + i \sin w)$$

Then $r^2 = 25$ so $r = 5$, and $2\theta = w + 2\pi k$, so $\theta = \tfrac{w}{2} + \pi k$.

When $k = 0$, we have $z_1 = 5\left[\cos\frac{w}{2} + i\sin\frac{w}{2}\right]$. Now using the half-angle formulas, we have:

$$\cos\frac{w}{2} = \sqrt{\frac{1+\cos w}{2}} = \sqrt{\frac{1+\frac{7}{25}}{2}} = \sqrt{\frac{32}{50}} = \sqrt{\frac{16}{25}} = \frac{4}{5}$$

$$\sin\frac{w}{2} = \sqrt{\frac{1-\cos w}{2}} = \sqrt{\frac{1-\frac{7}{25}}{2}} = \sqrt{\frac{18}{50}} = \sqrt{\frac{9}{25}} = \frac{3}{5}$$

So $z_1 = 5\left(\frac{4}{5} + \frac{3}{5}i\right) = 4 + 3i$. When $k = 1$, we have $z_2 = 5\left[\cos\left(\frac{w}{2} + \pi\right) + i\sin\left(\frac{w}{2} + \pi\right)\right]$. Now $\cos\left(\frac{w}{2} + \pi\right) = -\cos\frac{w}{2} = -\frac{4}{5}$ and $\sin\left(\frac{w}{2} + \pi\right) = -\sin\frac{w}{2} = -\frac{3}{5}$. So $z_2 = 5\left(-\frac{4}{5} - \frac{3}{5}i\right) = -4 - 3i$. So the square roots of $7 + 24i$ are $4 + 3i$ and $-4 - 3i$.

37. (a) Let $z = r(\cos\theta + i\sin\theta)$, so $z^4 = r^4(\cos 4\theta + i\sin 4\theta)$. Since $1 = 1[\cos 0 + i\sin 0]$, we have $r^4 = 1$ so $r = 1$ and $4\theta = 0 + 2\pi k$, so $\theta = 0 + \frac{\pi}{2}k$.

When $k = 0$: $\quad z_1 = 1[\cos 0 + i\sin 0] = 1(1 + 0i) = 1$

When $k = 1$: $\quad z_2 = 1\left[\cos\frac{\pi}{2} + i\sin\frac{\pi}{2}\right] = 1(0 + 1i) = i$

When $k = 2$: $\quad z_3 = 1[\cos\pi + i\sin\pi] = 1(-1 + 0i) = -1$

When $k = 3$: $\quad z_4 = 1\left[\cos\frac{3\pi}{2} + i\sin\frac{3\pi}{2}\right] = 1(0 - 1i) = -i$

So the fourth roots of 1 are 1, i, -1 and $-i$.

(b) We compute the sum of the four roots:
$$z_1 + z_2 + z_3 + z_4 = 1 + i - 1 - i = 0$$

39. Using trigonometric forms and De Moivre's theorem, we compute the powers:

$$\left[\frac{-1+\sqrt{3}i}{2}\right]^6 = \left[-\frac{1}{2} + \frac{\sqrt{3}}{2}i\right]^6 = \left[\cos\frac{2\pi}{3} + i\sin\frac{2\pi}{3}\right]^6 = \cos 4\pi + i\sin 4\pi = 1$$

$$\left[\frac{-1-\sqrt{3}i}{2}\right]^6 = \left[-\frac{1}{2} - \frac{\sqrt{3}}{2}i\right]^6 = \left[\cos\frac{4\pi}{3} + i\sin\frac{4\pi}{3}\right]^6 = \cos 8\pi + i\sin 8\pi = 1$$

Now compute the sum:

$$\left[\frac{-1+\sqrt{3}i}{2}\right]^6 + \left[\frac{-1-\sqrt{3}i}{2}\right]^6 = 1 + 1 = 2$$

41. Using De Moivre's theorem, we have:

$$\left[\sqrt{2}\left(\cos\frac{\pi}{5} + i\sin\frac{\pi}{5}\right)\right]^5 = \left(\sqrt{2}\right)^5[\cos\pi + i\sin\pi] = 4\sqrt{2}(-1 + 0i) = -4\sqrt{2} + 0i$$

Therefore $Re\,(z) = -4\sqrt{2}$ and $Im\,(z) = 0$.

43. We first compute the product:
$$(1+i\tan\theta)^2 = 1+(2\tan\theta)i + \tan^2\theta i^2 = \left(1-\tan^2\theta\right)+2\tan\theta i$$
Therefore:
$$\frac{Im\left[(1+i\tan\theta)^2\right]}{Re\left[(1+i\tan\theta)^2\right]} = \frac{2\tan\theta}{1-\tan^2\theta} = \tan 2\theta$$

45. You may have learned mathematical induction in your previous algebra course. It proceeds as follows. Let P_n denote the statement:
$$\left[r(\cos\theta+i\sin\theta)\right]^n = r^n(\cos n\theta + i\sin n\theta)$$
For $n = 1$, we have:
$$\left[r(\cos\theta+i\sin\theta)\right] = r(\cos\theta+i\sin\theta), \text{ clearly a true statement.}$$
Assume P_k is true:
$$\left[r(\cos\theta+i\sin\theta)\right]^k = r^k(\cos k\theta + i\sin k\theta)$$

For P_{k+1}:
$$\left[r(\cos\theta+i\sin\theta)\right]^{k+1}$$
$$= r(\cos\theta+i\sin\theta) \bullet \left[r(\cos\theta+i\sin\theta)\right]^k$$
$$= r(\cos\theta+i\sin\theta) \bullet r^k(\cos k\theta + i\sin k\theta)$$
$$= r^{k+1}\left[(\cos\theta\cos k\theta - \sin\theta\sin k\theta) + i(\sin\theta\cos k\theta + \cos\theta\sin k\theta)\right]$$
$$= r^{k+1}(\cos(k+1)\theta + i\sin(k+1)\theta)$$
Therefore P_{k+1} is true. Now, by mathematical induction we conclude that P_n is true for all positive integers n.

Chapter Six Review Exercises

1. $5i^2 - 4i = 5(-1) - 4i = -5 - 4i$

3. $(6-2i)+(4+3i) = 10+i$

5. We square the quantity:
$$(3+5i)^2 = 9+30i+25i^2 = 9+30i-25 = -16+30i$$

7. We perform the computations:
$$(8-i)(8+i)+(1-i)^2 = 16-i^2+1-2i+i^2 = 16+1+1-2i-1 = 17-2i$$

9. We perform the computations:
$$(1+i\sqrt{3})(1-i\sqrt{3})+(\sqrt{3}+i)(\sqrt{3}-i) = 1-3i^2+3-i^2 = 1+3+3+1 = 8$$

11. We compute the quotient:
$$\frac{2-3i}{1+2i}\cdot\frac{1-2i}{1-2i}=\frac{2-7i+6i^2}{1-4i^2}=\frac{2-7i-6}{1+4}=-\frac{4}{5}-\frac{7}{5}i$$

13. We compute the quotient:
$$\frac{2+3i}{1+i}\cdot\frac{1-i}{1-i}=\frac{2+i-3i^2}{1-i^2}=\frac{5+i}{2}=\frac{5}{2}+\frac{1}{2}i$$

15. We compute the quantity:
$$\frac{1+i}{1-i}+\frac{1-i}{1+i}=\frac{(1+i)^2+(1-i)^2}{(1-i)(1+i)}$$
$$=\frac{1+2i+i^2+1-2i+i^2}{1-i^2}$$
$$=\frac{1+2i-1+1-2i-1}{1+1}$$
$$=\frac{0}{2}$$
$$=0$$

17. We write the quantity in complex form, then simplify:
$$\frac{\sqrt{-4}-\sqrt{-3}\sqrt{-3}}{\sqrt{-100}}=\frac{2i-(i\sqrt{3})(i\sqrt{3})}{10i}=\frac{2i-3i^2}{10i}=\frac{i(2-3i)}{10i}=\frac{1}{5}-\frac{3}{10}i$$

19. (a) $\bar{z}=3+4i$

(b)   We compute the quantity:
$$\tfrac{1}{2}(z+\bar{z})=\tfrac{1}{2}(3-4i+3+4i)=\tfrac{1}{2}(6)=3$$

(c)   We compute the quantity:
$$\tfrac{1}{2i}(z-\bar{z})=\tfrac{1}{2i}(3-4i-3-4i)=\tfrac{1}{2i}(-8i)=-4$$

21. We first simplify z:
$$z=(5+6i)^2=25+60i+36i^2=25+60i-36=-11+60i$$

(a) $\bar{z}=-11-60i$

(b)   We compute the quantity:
$$\tfrac{1}{2}(z+\bar{z})=\tfrac{1}{2}(-11+60i-11-60i)=\tfrac{1}{2}(-22)=-11$$

(c)   We compute the quantity:
$$\tfrac{1}{2i}(z-\bar{z})=\tfrac{1}{2i}(-11+60i+11+60i)=\tfrac{1}{2i}(120i)=60$$

23. We convert to rectangular form:
$$3\left[\cos\tfrac{\pi}{3}+i\sin\tfrac{\pi}{3}\right]=3\left(\tfrac{1}{2}+\tfrac{\sqrt{3}}{2}i\right)=\tfrac{3}{2}+\tfrac{3\sqrt{3}}{2}i$$

25. We convert to rectangular form:
$$2^{1/4}\left[\cos\tfrac{7\pi}{4}+i\sin\tfrac{7\pi}{4}\right]=2^{1/4}\left(\tfrac{\sqrt{2}}{2}-\tfrac{\sqrt{2}}{2}i\right)=2^{-1/4}-2^{-1/4}i$$

27. We convert to trigonometric form:
$$\tfrac{1}{2}+\tfrac{\sqrt{3}}{2}i=1\left[\cos\tfrac{\pi}{3}+i\sin\tfrac{\pi}{3}\right]$$

29. We convert to trigonometric form:
$$-3\sqrt{2}-3\sqrt{2}i=6\left(-\tfrac{\sqrt{2}}{2}-\tfrac{\sqrt{2}}{2}i\right)=6\left[\cos\tfrac{5\pi}{4}+i\sin\tfrac{5\pi}{4}\right]$$

31. We compute the product:
$$5\left[\cos\tfrac{\pi}{7}+i\sin\tfrac{\pi}{7}\right]\bullet 2\left[\cos\tfrac{3\pi}{28}+i\sin\tfrac{3\pi}{28}\right]=10\left[\cos\tfrac{7\pi}{28}+i\sin\tfrac{7\pi}{28}\right]$$
$$=10\left[\cos\tfrac{\pi}{4}+i\sin\tfrac{\pi}{4}\right]$$
$$=10\left(\tfrac{\sqrt{2}}{2}+\tfrac{\sqrt{2}}{2}i\right)$$
$$=5\sqrt{2}+5\sqrt{2}i$$

33. We compute the quotient:
$$8\left[\cos\tfrac{\pi}{12}+i\sin\tfrac{\pi}{12}\right]\div 4\left[\cos\tfrac{\pi}{3}+i\sin\tfrac{\pi}{3}\right]=2\left[\cos\left(-\tfrac{3\pi}{12}\right)+i\sin\left(-\tfrac{3\pi}{12}\right)\right]$$
$$=2\left[\cos\left(-\tfrac{\pi}{4}\right)+i\sin\left(-\tfrac{\pi}{4}\right)\right]$$
$$=2\left(\tfrac{\sqrt{2}}{2}-\tfrac{\sqrt{2}}{2}i\right)$$
$$=\sqrt{2}-\sqrt{2}i$$

35. We compute the product:
$$\left[\cos\tfrac{\pi}{9}+i\sin\tfrac{\pi}{9}\right]\bullet 3\left[\cos\tfrac{4\pi}{9}+i\sin\tfrac{4\pi}{9}\right]=3\left[\cos\tfrac{5\pi}{9}+i\sin\tfrac{5\pi}{9}\right]$$

37. We compute the power using De Moivre's theorem:
$$\left[3^{1/4}\left(\cos\tfrac{\pi}{36}+i\sin\tfrac{\pi}{36}\right)\right]^{12}=3^3\left[\cos\tfrac{12\pi}{36}+i\sin\tfrac{12\pi}{36}\right]$$
$$=27\left[\cos\tfrac{\pi}{3}+i\sin\tfrac{\pi}{3}\right]$$
$$=27\left(\tfrac{1}{2}+\tfrac{\sqrt{3}}{2}i\right)$$
$$=\tfrac{27}{2}+\tfrac{27\sqrt{3}}{2}i$$

39. We first write $\sqrt{3} + i$ in trigonometric form:
$$\sqrt{3} + i = 2\left(\tfrac{\sqrt{3}}{2} + \tfrac{1}{2}i\right) = 2\left[\cos\tfrac{\pi}{6} + i\sin\tfrac{\pi}{6}\right]$$
Now compute the power using De Moivre's theorem:
$$(\sqrt{3} + i)^{10} = \left[2\left(\cos\tfrac{\pi}{6} + i\sin\tfrac{\pi}{6}\right)\right]^{10}$$
$$= 2^{10}\left[\cos\tfrac{5\pi}{3} + i\sin\tfrac{5\pi}{3}\right]$$
$$= 2^{10}\left(\tfrac{1}{2} - \tfrac{\sqrt{3}}{2}i\right)$$
$$= 2^9 - 2^9\sqrt{3}i$$
$$= 512 - 512\sqrt{3}i$$

41. Let $z = r(\cos\theta + i\sin\theta)$ and $1 = 1[\cos 0 + i\sin 0]$, so:
$$z^6 = r^6(\cos 6\theta + i\sin 6\theta) = 1(\cos 0 + i\sin 0)$$

Thus $r = 1$ and $6\theta = 0 + 2\pi k$, so $\theta = 0 + \tfrac{\pi}{3}k$.
When $k = 0$, we have:
$$z_1 = 1[\cos 0 + i\sin 0] = 1(1 + 0i) = 1$$
When $k = 1$, we have:
$$z_2 = 1\left[\cos\tfrac{\pi}{3} + i\sin\tfrac{\pi}{3}\right] = \tfrac{1}{2} + \tfrac{\sqrt{3}}{2}i$$
When $k = 2$, we have:
$$z_3 = 1\left[\cos\tfrac{2\pi}{3} + i\sin\tfrac{2\pi}{3}\right] = -\tfrac{1}{2} + \tfrac{\sqrt{3}}{2}i$$
When $k = 3$, we have:
$$z_4 = 1[\cos\pi + i\sin\pi] = -1 + 0i = -1$$
When $k = 4$, we have:
$$z_5 = 1\left[\cos\tfrac{4\pi}{3} + i\sin\tfrac{4\pi}{3}\right] = -\tfrac{1}{2} - \tfrac{\sqrt{3}}{2}i$$
When $k = 5$, we have:
$$z_6 = 1\left[\cos\tfrac{5\pi}{3} + i\sin\tfrac{5\pi}{3}\right] = \tfrac{1}{2} - \tfrac{\sqrt{3}}{2}i$$
So the sixth roots of 1 are $1, \tfrac{1}{2} + \tfrac{\sqrt{3}}{2}i, -\tfrac{1}{2} + \tfrac{\sqrt{3}}{2}i, -1, -\tfrac{1}{2} - \tfrac{\sqrt{3}}{2}i$ and $\tfrac{1}{2} - \tfrac{\sqrt{3}}{2}i$.

43. Let $z = r(\cos\theta + i\sin\theta)$ and write $\sqrt{2} - \sqrt{2}i$ in trigonometric form:
$$\sqrt{2} - \sqrt{2}i = 2\left(\tfrac{\sqrt{2}}{2} - \tfrac{\sqrt{2}}{2}i\right) = 2\left[\cos\tfrac{7\pi}{4} + i\sin\tfrac{7\pi}{4}\right]$$

Then $z^2 = r^2(\cos 2\theta + i\sin 2\theta) = 2\left[\cos\tfrac{7\pi}{4} + i\sin\tfrac{7\pi}{4}\right]$.

So $r^2 = 2$ and $r = \sqrt{2}$, and also $2\theta = \tfrac{7\pi}{4} + 2\pi k$, so $\theta = \tfrac{7\pi}{8} + \pi k$.

When $k = 0$, we have:
$$z_1 = \sqrt{2}\left[\cos\frac{7\pi}{8} + i\sin\frac{7\pi}{8}\right]$$
$$= \sqrt{2}\left[-\frac{\sqrt{2+\sqrt{2}}}{2} + i\frac{\sqrt{2-\sqrt{2}}}{2}\right]$$
(by the half - angle identities)
$$= -\frac{\sqrt{4+2\sqrt{2}}}{2} + \frac{\sqrt{4-2\sqrt{2}}}{2}i$$
When $k = 1$, we have:
$$z_2 = \sqrt{2}\left[\cos\frac{15\pi}{8} + i\sin\frac{15\pi}{8}\right] = \sqrt{2}\left[\frac{\sqrt{2+\sqrt{2}}}{2} - i\frac{\sqrt{2-\sqrt{2}}}{2}\right] = \frac{\sqrt{4+2\sqrt{2}}}{2} - \frac{\sqrt{4-2\sqrt{2}}}{2}i$$

So the square roots of $\sqrt{2} - \sqrt{2}i$ are $-\frac{\sqrt{4+2\sqrt{2}}}{2} + \frac{\sqrt{4-2\sqrt{2}}}{2}i$ and $\frac{\sqrt{4+2\sqrt{2}}}{2} - \frac{\sqrt{4-2\sqrt{2}}}{2}i$.

45. Let $z = r(\cos\theta + i\sin\theta)$, and write $1 + i$ in trigonometric form:
$$1 + i = \sqrt{2}\left(\frac{\sqrt{2}}{2} + \frac{\sqrt{2}}{2}i\right) = \sqrt{2}\left[\cos\frac{\pi}{4} + i\sin\frac{\pi}{4}\right]$$
Then $z^5 = r^5[\cos 5\theta + i\sin 5\theta] = \sqrt{2}[\cos 45° + i\sin 45°]$.
So $r^5 = \sqrt{2}$ and $r = 2^{1/10} \approx 1.0718$, and $5\theta = 45° + 360°k$, so $\theta = 9° + 72°k$.

When $k = 0$: $z_1 = 1.0718[\cos 9° + i\sin 9°] \approx 1.06 + 0.17i$

When $k = 1$: $z_2 = 1.0718[\cos 81° + i\sin 81°] \approx 0.17 + 1.06i$

When $k = 2$: $z_3 = 1.0718[\cos 153° + i\sin 153°] \approx -0.95 + 0.49i$

When $k = 3$: $z_4 = 1.0718[\cos 225° + i\sin 225°] \approx -0.76 - 0.76i$

When $k = 4$: $z_5 = 1.0718[\cos 297° + i\sin 297°] \approx 0.49 - 0.95i$

So the fifth roots of $1 + i$ are $1.06 + 0.17i$, $0.17 + 1.06i$, $-0.95 + 0.49i$, $-0.76 - 0.76i$ and $0.49 - 0.95i$.

47. We expand the expression on the left-hand side:
$$(\cos\theta + i\sin\theta)^3 = (\cos\theta)^3 + 3(\cos\theta)^2(i\sin\theta) + 3(\cos\theta)(i\sin\theta)^2 + (i\sin\theta)^3$$
$$= \cos^3\theta + 3\cos^2\theta\sin\theta i - 3\cos\theta\sin^2\theta - \sin^3\theta i$$
$$= (\cos^3\theta - 3\cos\theta\sin^2\theta) + (3\cos^2\theta\sin\theta - \sin^3\theta)i$$
Since this is equal to $\cos 3\theta + i\sin 3\theta$, and the real and imaginary parts must be equal, we have the identities:
$$\cos 3\theta = \cos^3\theta - 3\cos\theta\sin^2\theta$$
$$\sin 3\theta = 3\cos^2\theta\sin\theta - \sin^3\theta$$

49. Working from the left-hand side, we have:
$$\frac{1}{a - bi} - \frac{1}{a + bi} = \frac{(a + bi) - (a - bi)}{(a - bi)(a + bi)} = \frac{2bi}{a^2 - b^2i^2} = \frac{2bi}{a^2 + b^2}$$

51. Working from the left-hand side, we have:
$$\frac{a + bi}{a - bi} - \frac{a - bi}{a + bi} = \frac{(a + bi)^2 - (a - bi)^2}{(a - bi)(a + bi)} = \frac{a^2 + 2abi - b^2 - a^2 + 2abi + b^2}{a^2 + b^2} = \frac{4abi}{a^2 + b^2}$$

Chapter Six Test

1. We simplify the expression:
$$i^3 - i^4 + \tfrac{1}{i} = -i - 1 + \tfrac{1}{i} \cdot \tfrac{i}{i} = -i - 1 - i = -1 - 2i$$

2. We evaluate the expression:
$$\begin{aligned}(w+z)(w-z) &= (1-4i+2+3i)(1-4i-2-3i) \\ &= (3-i)(-1-7i) \\ &= -3-20i+7i^2 \\ &= -10-20i\end{aligned}$$

3. We evaluate the expression:
$$\begin{aligned}z^2 + w^2 &= (2+3i)^2 + (1-4i)^2 \\ &= 4+12i+9i^2+1-8i+16i^2 \\ &= 4+12i-9+1-8i-16 \\ &= -20+4i\end{aligned}$$

4. We evaluate the quotient:
$$\frac{w}{z} = \frac{1-4i}{2+3i} \bullet \frac{2-3i}{2-3i} = \frac{2-11i+12i^2}{4-9i^2} = \frac{2-11i-12}{4+9} = -\frac{10}{13} - \frac{11}{13}i$$

5. We evaluate the expression:
$$\bar{z} + 3w = 2 - 3i + 3(1-4i) = 2 - 3i + 3 - 12i = 5 - 15i$$

6. We convert to rectangular form:
$$z = 2\left(\cos\tfrac{2\pi}{3} + i\sin\tfrac{2\pi}{3}\right) = 2\left(-\tfrac{1}{2} + \tfrac{\sqrt{3}}{2}i\right) = -1 + \sqrt{3}i$$

7. We convert to trigonometric form:
$$\sqrt{2} - \sqrt{2}i = 2\left(\tfrac{\sqrt{2}}{2} - \tfrac{\sqrt{2}}{2}i\right) = 2\left(\cos\tfrac{7\pi}{4} + i\sin\tfrac{7\pi}{4}\right)$$

8. $r^n(\cos n\theta + i\sin n\theta)$

9. We compute the product:
$$\begin{aligned}zw &= 3\left(\cos\tfrac{2\pi}{9} + i\sin\tfrac{2\pi}{9}\right) \bullet 5\left(\cos\tfrac{\pi}{9} + i\sin\tfrac{\pi}{9}\right) \\ &= 15\left(\cos\tfrac{\pi}{3} + i\sin\tfrac{\pi}{3}\right) \\ &= 15\left(\tfrac{1}{2} + \tfrac{\sqrt{3}}{2}i\right) \\ &= \tfrac{15}{2} + \tfrac{15\sqrt{3}}{2}i\end{aligned}$$

10. Let $z = r(\cos\theta + i\sin\theta)$ be a cube root of $64i$. We write $64i$ in trigonometric form:

$$64i = 64(0+1i) = 64\left(\cos\tfrac{\pi}{2} + i\sin\tfrac{\pi}{2}\right)$$

Thus:

$$z^3 = r^3(\cos 3\theta + i\sin 3\theta) = 64\left(\cos\tfrac{\pi}{2} + i\sin\tfrac{\pi}{2}\right)$$

So $r^3 = 64$ and thus $r = 4$, and $3\theta = \tfrac{\pi}{2} + 2\pi k$ thus $\theta = \tfrac{\pi}{6} + \tfrac{2\pi}{3}k$.
When $k = 0$, we have:

$$z_1 = 4\left(\cos\tfrac{\pi}{6} + i\sin\tfrac{\pi}{6}\right) = 4\left(\tfrac{\sqrt{3}}{2} + \tfrac{1}{2}i\right) = 2\sqrt{3} + 2i$$

When $k = 1$, we have:

$$z_2 = 4\left(\cos\tfrac{5\pi}{6} + i\sin\tfrac{5\pi}{6}\right) = 4\left(-\tfrac{\sqrt{3}}{2} + \tfrac{1}{2}i\right) = -2\sqrt{3} + 2i$$

When $k = 2$, we have:

$$z_3 = 4\left(\cos\tfrac{3\pi}{2} + i\sin\tfrac{3\pi}{2}\right) = 4(0 - i) = -4i$$

So the cube roots of $64i$ are $2\sqrt{3} + 2i$, $-2\sqrt{3} + 2i$ and $-4i$.

Chapter Seven
Analytic Geometry

7.1 The Basic Equations

1. Using the distance formula, we have:
$$d = \sqrt{(-5-3)^2 + (-6+1)^2} = \sqrt{64+25} = \sqrt{89}$$

3. We find the slope of the line:
$$4x - 5y - 20 = 0$$
$$-5y = -4x + 20$$
$$y = \tfrac{4}{5}x - 4$$

So the perpendicular slope is $m = -\tfrac{5}{4}$. We now find the y-intercept:
$$x - y + 1 = 0$$
$$-y = -x - 1$$
$$y = x + 1$$

So $b = 1$, and the equation is $y = -\tfrac{5}{4}x + 1$. Multiplying by 4, we have:
$$4y = -5x + 4, \text{ or } 5x + 4y - 4 = 0$$

5. We find the slope of the line segment:
$$m = \tfrac{7-1}{6-2} = \tfrac{6}{4} = \tfrac{3}{2}$$

So the perpendicular slope is $m = -\tfrac{2}{3}$. We find the midpoint:
$$M = \left(\tfrac{2+6}{2}, \tfrac{1+7}{2}\right) = \left(\tfrac{8}{2}, \tfrac{8}{2}\right) = (4,4)$$
Now use the point-slope formula:
$$y - 4 = -\tfrac{2}{3}(x-4)$$
$$y - 4 = -\tfrac{2}{3}x + \tfrac{8}{3}$$
$$y = -\tfrac{2}{3}x + \tfrac{20}{3}$$
Multiplying by 3, we have $3y = -2x + 20$, or $2x + 3y - 20 = 0$.

7. Since the center is $(1,0)$ and the radius is 5, the equation must be $(x-1)^2 + y^2 = 25$. To find the x-intercepts, let $y = 0$ and solve the resulting equation for x:
$$y = 0$$
$$(x-1)^2 = 25$$
$$x - 1 = \pm 5$$
$$x = 6, -4$$
To find the y-intercepts, let $x = 0$ and solve the resulting equation for y:
$$x = 0$$
$$(-1)^2 + y^2 = 25$$
$$y^2 = 24$$
$$y = \pm\sqrt{24} = \pm 2\sqrt{6}$$
The x-intercepts are 6 and -4, and the y-intercepts are $\pm 2\sqrt{6}$.

9. We first find the midpoint of $\overline{AB}$:
$$M = \left(\tfrac{1+6}{2}, \tfrac{2+1}{2}\right) = \left(\tfrac{7}{2}, \tfrac{3}{2}\right)$$
Now find the slope of the line:
$$m = \frac{8 - \frac{3}{2}}{7 - \frac{7}{2}} = \frac{\frac{13}{2}}{\frac{7}{2}} = \frac{13}{7}$$
Using the point $(7, 8)$ in the point-slope formula:
$$y - 8 = \tfrac{13}{7}(x - 7)$$
$$y - 8 = \tfrac{13}{7}x - 13$$
$$y = \tfrac{13}{7}x - 5$$
Multiply by 7 to get $7y = 13x - 35$, or $13x - 7y - 35 = 0$.

11. Since C is the x-intercept, we let $y = 0$ to obtain $\frac{x}{7} = 1$, or $x = 7$. So the coordinates of C are $(7,0)$. Similarly, since B is the y-intercept, we let $x = 0$ to obtain $\frac{y}{5} = 1$, or $y = 5$. So the coordinates of B are $(0,5)$. We find BC by the distance formula:
$$BC = \sqrt{(7-0)^2 + (0-5)^2} = \sqrt{49+25} = \sqrt{74}$$
So, the perimeter is given by:
$$P = AB + BC + AC = 5 + \sqrt{74} + 7 = 12 + \sqrt{74}$$

13. Since $\tan\theta = \sqrt{3}$, then $\theta = \frac{\pi}{3}$ or $60°$.

15. (a) Since $\tan\theta = 5$, then $\theta = 1.37$ or $78.69°$.
(b) Since $\tan\theta = -5$, then $\theta = 1.77$ or $101.31°$.

17. (a) Here $(x_0, y_0) = (1,4)$, $m = 1$, $b = -2$, so:
$$d = \frac{|1 - 2 - 4|}{\sqrt{1+1}} = \frac{5}{\sqrt{2}} = \frac{5\sqrt{2}}{2}$$

(b)   Using $x - y\ -2 = 0$, we have $A = 1, B = -1, C = -2$:
$$d = \frac{|1 - 4 - 2|}{\sqrt{1+1}} = \frac{5}{\sqrt{2}} = \frac{5\sqrt{2}}{2}$$

19.   (a)   Here $(x_0, y_0) = (-3, 5)$. We convert to slope-intercept form:
$$4x + 5y + 6 = 0$$
$$5y = -4x - 6$$
$$y = -\tfrac{4}{5}x - \tfrac{6}{5}$$

So $m = -\tfrac{4}{5}$ and $b = -\tfrac{6}{5}$:
$$d = \frac{\left|\frac{12}{5} - \frac{6}{5} - 5\right|}{\sqrt{1 + \frac{16}{25}}} = \frac{\frac{19}{5}}{\frac{\sqrt{41}}{5}} = \frac{19}{\sqrt{41}} = \frac{19\sqrt{41}}{41}$$

(b)   We have $A = 4, B = 5, C = 6$:
$$d = \frac{|-12 + 25 + 6|}{\sqrt{16 + 25}} = \frac{19}{\sqrt{41}} = \frac{19\sqrt{41}}{41}$$

21.   (a)   The radius of the circle is the distance from the point $(-2, -3)$ to the line $2x + 3y - 6 = 0$. Using $(x_0, y_0) = (-2, -3), A = 2, B = 3, C = -6$:
$$r = \frac{|-4 - 9 - 6|}{\sqrt{4 + 9}} = \frac{19}{\sqrt{13}}$$

So the equation of the circle is $(x + 2)^2 + (y + 3)^2 = \frac{361}{13}$.

(b)   Since the radius is the distance from the point $(1, 3)$ to the line $y = \tfrac{1}{2}x + 5$, we have $(x_0, y_0) = (1, 3), m = \tfrac{1}{2}, b = 5$:
$$r = \frac{\left|\frac{1}{2} + 5 - 3\right|}{\sqrt{1 + \frac{1}{4}}} = \frac{\frac{5}{2}}{\frac{\sqrt{5}}{2}} = \sqrt{5}$$

23.   Using the suggestion, we work with $\triangle ABC$ and $\triangle CDA$. For $\triangle ABC$, we find the base using the distance formula:
$$AB = \sqrt{(8 - 0)^2 + (2 - 0)^2} = \sqrt{64 + 4} = \sqrt{68} = 2\sqrt{17}$$
Now we find the equation of the line through A and B. We find the slope:
$$m = \tfrac{2-0}{8-0} = \tfrac{1}{4}$$

So the equation is $y = \tfrac{1}{4}x$. We find the distance from $C(4, 7)$ to this line:
$$h = \frac{|1 + 0 - 7|}{\sqrt{1 + \frac{1}{16}}} = \frac{6}{\frac{\sqrt{17}}{4}} = \frac{24}{\sqrt{17}}$$

So $\triangle ABC$ will have an area of:

$$\text{Area}_{\triangle ABC} = \tfrac{1}{2}(\text{base})(\text{height}) = \tfrac{1}{2}(2\sqrt{17})\left(\tfrac{24}{\sqrt{17}}\right) = 24$$

For $\triangle CDA$, we find the base using the distance formula:

$$CD = \sqrt{(1-4)^2 + (6-7)^2} = \sqrt{9+1} = \sqrt{10}$$

Now we find the equation of the line through C and D. We find the slope:

$$m = \tfrac{6-7}{1-4} = \tfrac{-1}{-3} = \tfrac{1}{3}$$

Using $C(4,7)$, in the point-slope formula:

$$y - 7 = \tfrac{1}{3}(x - 4)$$
$$y - 7 = \tfrac{1}{3}x - \tfrac{4}{3}$$
$$y = \tfrac{1}{3}x + \tfrac{17}{3}$$
$$3y = x + 17 \text{ or } x - 3y + 17 = 0$$

We find the distance from $A\,(0,0)$ to this line:

$$h = \frac{|0+0+17|}{\sqrt{1+9}} = \frac{17}{\sqrt{10}}$$

So $\triangle CDA$ will have an area of :

$$\text{Area}_{\triangle CDA} = \tfrac{1}{2}(\text{base})(\text{height}) = \tfrac{1}{2}\left(\sqrt{10}\right)\left(\tfrac{17}{\sqrt{10}}\right) = \tfrac{17}{2}$$

Thus, the total combined area of quadrilateral $ABCD$ is:

$$24 + \tfrac{17}{2} = \tfrac{65}{2}$$

25. Using the point $(0,-5)$ in the point-slope formula:

$$y + 5 = m(x - 0)$$
$$y = mx - 5$$

Now, since the distance from the center $(3,0)$ to this line is 2, we have:

$$2 = \frac{|3m - 5 - 0|}{\sqrt{1 + m^2}} = \frac{|3m - 5|}{\sqrt{1 + m^2}}$$

Squaring each side, we have:

$$4 = \frac{9m^2 - 30m + 25}{1 + m^2}$$
$$4 + 4m^2 = 9m^2 - 30m + 25$$
$$0 = 5m^2 - 30m + 21$$

Using the quadratic formula:

$$m = \frac{30 \pm \sqrt{900 - 420}}{10} = \frac{30 \pm \sqrt{480}}{10} = \frac{30 \pm 4\sqrt{30}}{10} = \frac{15 \pm 2\sqrt{30}}{5}$$

27. We use the same approach as in the preceeding exercise. Let d_1 and d_2 represent the distances from $(0,0)$ to the lines $3x + 4y - 12 = 0$ and $3x + 4y - 24 = 0$, respectively. For d_1, we have $(x_0, y_0) = (0,0)$, $A = 3$, $B = 4$, $C = -12$:

$$d_1 = \frac{|0+0-12|}{\sqrt{9+16}} = \frac{12}{\sqrt{25}} = \frac{12}{5}$$

For d_2, we have $(x_0, y_0) = (0, 0)$, $A = 3$, $B = 4$, $C = -24$:

$$d_2 = \frac{|0 + 0 - 24|}{\sqrt{9 + 16}} = \frac{24}{\sqrt{25}} = \frac{24}{5}$$

So the distance between the lines is:

$$d_2 - d_1 = \frac{24}{5} - \frac{12}{5} = \frac{12}{5}$$

29. For the described line segment to be bisected by the point $(2, 6)$, then $(2, 6)$ must be the midpoint of the x- and y-intercepts (as points) of the line. Using the point-slope formula:

$$y - 6 = m(x - 2)$$
$$y - 6 = mx - 2m$$
$$y = mx - 2m + 6$$

To find the x-intercept, let $y = 0$ and solve the resulting equation for x:

$$y = 0$$
$$mx - 2m + 6 = 0$$
$$mx = 2m - 6$$
$$x = \frac{2m - 6}{m}$$

To find the y-intercept, let $x = 0$ and solve the resulting equation for y:

$$x = 0$$
$$y = -2m + 6$$

Since $(2, 6)$ must be the midpoint of $\left(\frac{2m-6}{m}, 0\right)$ and $(0, -2m + 6)$, we have the two sets of equations:

$$\frac{2m - 6}{2m} = 2 \qquad\qquad \frac{-2m + 6}{2} = 6$$
$$2m - 6 = 4m \qquad\qquad -2m + 6 = 12$$
$$-6 = 2m \qquad\qquad -2m = 6$$
$$-3 = m \qquad\qquad m = -3$$

Now use the point-slope formula with the point $(2, 6)$:

$$y - 6 = -3(x - 2)$$
$$y - 6 = -3x + 6$$
$$y = -3x + 12$$

31. Using the hint, we let d_1 and d_2 represent the distances from (x, y) to $x - y + 1 = 0$ and $x + 7y - 49 = 0$, respectively. Then:

$$d_1 = \frac{|x - y + 1|}{\sqrt{1 + 1}} = \frac{|x - y + 1|}{\sqrt{2}}$$
$$d_2 = \frac{|x + 7y - 49|}{\sqrt{1 + 49}} = \frac{|x + 7y - 49|}{\sqrt{50}}$$

Since $d_1 = d_2$, we have:

$$\frac{|x - y + 1|}{\sqrt{2}} = \frac{|x + 7y - 49|}{\sqrt{50}}$$

Multiplying each side by $\sqrt{2}$, we have:

$$|x-y+1| = \frac{|x+7y-49|}{5}$$

$$5|x-y+1| = |x+7y-49|$$

Rather than squaring each side, we will note that one of the two equations must hold:

$$5(x-y+1) = x+7y-49$$
$$5x-5y+5 = x+7y-49$$
$$4x-12y+54 = 0$$
$$2x-6y+27 = 0$$
$$y = \tfrac{1}{3}x + \tfrac{9}{2}$$

$$5(x-y+1) = -x-7y+49$$
$$5x-5y+5 = -x-7y+49$$
$$6x+2y-44 = 0$$
$$3x+y-22 = 0$$
$$y = -3x+22$$

Note that the first of these lines, $y = \tfrac{1}{3}x + \tfrac{9}{2}$, is the solution as indicated in the figure. The second line we found is the second angle bisector passing through the same intersection point but bisecting the larger vertical angles.

33. (a)   The standard form for a circle is $(x-h)^2 + (y-k)^2 = r^2$. We substitute each point for (x, y):

$$(-12-h)^2 + (1-k)^2 = r^2$$
$$(2-h)^2 + (1-k)^2 = r^2$$
$$(0-h)^2 + (7-k)^2 = r^2$$

Setting the first two equations equal, we have:

$$(-12-h)^2 + (1-k)^2 = (2-h)^2 + (1-k)^2$$
$$144 + 24h + h^2 = 4 - 4h + h^2$$
$$28h = -140$$
$$h = -5$$

Setting the second two equations equal, we have:

$$(2-h)^2 + (1-k)^2 = (0-h)^2 + (7-k)^2$$
$$4 - 4h + h^2 + 1 - 2k + k^2 = h^2 + 49 - 14k + k^2$$
$$-4h + 12k = 44$$
$$h - 3k = -11$$

Substituting $h = -5$, we have:

$$-5 - 3k = -11$$
$$-3k = -6$$
$$k = 2$$

Now substitute into the second equation to find r:

$$(7)^2 + (-1)^2 = r^2$$
$$50 = r^2$$
$$5\sqrt{2} = r$$

So the center is $(-5, 2)$ and the radius is $5\sqrt{2}$.

(b) We first draw the sketch:

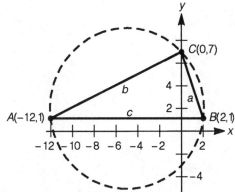

We find the lengths of three sides (using the distance formula):
$$a = \sqrt{(2-0)^2 + (1-7)^2} = \sqrt{4+36} = \sqrt{40} = 2\sqrt{10}$$
$$b = \sqrt{(-12-0)^2 + (1-7)^2} = \sqrt{144+36} = \sqrt{180} = 6\sqrt{5}$$
$$c = \sqrt{(-12-2)^2 + (1-1)^2} = \sqrt{196+0} = \sqrt{196} = 14$$
Using $R = 5\sqrt{2}$ from part (a), we have:
$$\frac{abc}{4R} = \frac{(2\sqrt{10})(6\sqrt{5})(14)}{4(5\sqrt{2})} = \frac{42\sqrt{50}}{5\sqrt{2}} = 42$$
For the area of the triangle, we have base = 14 and height = 6, so:
$$\text{Area} = \tfrac{1}{2}(\text{base})(\text{height}) = \tfrac{1}{2}(14)(6) = 42$$
This verifies the result.

35. We will first find points P and Q. Solving for x, we have $x = 7y - 44$. Now substitute:
$$(7y - 44)^2 - 4(7y - 44) + y^2 - 6y = 12$$
$$49y^2 - 616y + 1936 - 28y + 176 + y^2 - 6y = 12$$
$$50y^2 - 650y + 2100 = 0$$
$$50(y^2 - 13y + 42) = 0$$
$$50(y - 7)(y - 6) = 0$$
$$y = 7, 6$$
When $y = 7$, $x = 49 - 44 = 5$. When $y = 6$, $x = 42 - 44 = -2$. So the two points are $P(5,7)$ and $Q(-2,6)$. We now use the distance formula:
$$PQ = \sqrt{(-2-5)^2 + (6-7)^2} = \sqrt{49+1} = \sqrt{50} = 5\sqrt{2}$$

37. Let d_1 be the distance from $(0, c)$ to $ax + y = 0$:
$$d_1 = \frac{|0 + c + 0|}{\sqrt{a^2+1}} = \frac{|c|}{\sqrt{a^2+1}}$$
Let d_2 be the distance from $(0, c)$ to $x + by = 0$:
$$d_1 = \frac{|0 + bc + 0|}{\sqrt{1+b^2}} = \frac{|bc|}{\sqrt{1+b^2}}$$

So the product of these distances is given by:

$$d_1 d_2 = \frac{|c|}{\sqrt{a^2+1}} \cdot \frac{|bc|}{\sqrt{1+b^2}} = \frac{|bc^2|}{\sqrt{a^2+a^2b^2+b^2+1}}$$

39. (a)   We draw the triangle:

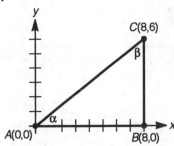

Since the triangle is a right triangle, note that $\sin\alpha = \frac{6}{10} = \frac{3}{5}$, $\cos\alpha = \frac{8}{10} = \frac{4}{5}$,

$\sin\beta = \frac{8}{10} = \frac{4}{5}$ and $\cos\beta = \frac{6}{10} = \frac{3}{5}$. To find the bisector for A, note that:

$$m = \tan\frac{\alpha}{2} = \frac{\sin\alpha}{1+\cos\alpha} = \frac{\frac{3}{5}}{1+\frac{4}{5}} = \frac{3}{9} = \frac{1}{3}$$

Since this line passes through the point $(0,0)$, its equation is $y = \frac{1}{3}x$. To find the
bisector for B, note that its slope is -1 and that it passes through $(8,0)$, so:

$$y - 0 = -1(x-8)$$
$$y = -x + 8$$

To find the bisector for C, first draw the figure:

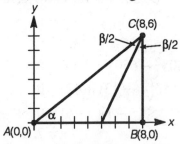

Now note that:

$$m = \tan\left(90° - \frac{\beta}{2}\right) = \cot\frac{\beta}{2} = \frac{1+\cos\beta}{\sin\beta} = \frac{1+\frac{3}{5}}{\frac{4}{5}} = 2$$

Using the point $(8,6)$, we use the point-slope formula:

$$y - 6 = 2(x-8)$$
$$y - 6 = 2x - 16$$
$$y = 2x - 10$$

So the bisectors at each vertex are:

A: $y = \frac{1}{3}x$

B: $y = -x + 8$

C: $y = 2x - 10$

(b) Setting A and B equal:
$$\tfrac{1}{3}x = -x + 8$$
$$\tfrac{4}{3}x = 8$$
$$x = 6$$
$$y = 2$$
Setting B and C equal:
$$-x + 8 = 2x - 10$$
$$18 = 3x$$
$$6 = x$$
$$2 = y$$
Setting A and C equal:
$$\tfrac{1}{3}x = 2x - 10$$
$$-\tfrac{5}{3}x = -10$$
$$x = 6$$
$$y = 2$$
So the angle bisectors are concurrent at the point $(6, 2)$.

41. The distance from the point (x, y) to $\left(0, \tfrac{1}{4}\right)$ is:
$$d = \sqrt{(x-0)^2 + \left(y - \tfrac{1}{4}\right)^2} = \sqrt{x^2 + y^2 - \tfrac{1}{2}y + \tfrac{1}{16}}$$

The distance from the point (x, y) to $y = -\tfrac{1}{4}$ is:
$$d = \frac{\left|0 - \tfrac{1}{4} - y\right|}{\sqrt{1+0}} = \left|y + \tfrac{1}{4}\right|$$
Setting these distances equal:
$$\left|y + \tfrac{1}{4}\right| = \sqrt{x^2 + y^2 - \tfrac{1}{2}y + \tfrac{1}{16}}$$
Squaring, we have:
$$y^2 + \tfrac{1}{2}y + \tfrac{1}{16} = x^2 + y^2 - \tfrac{1}{2}y + \tfrac{1}{16}$$
$$y = x^2$$
Thus, the points satisfy the equation $y = x^2$.

43. (a) We write the line in slope-intercept form:
$$Ax + By + C = 0$$
$$By = -Ax - C$$
$$y = -\tfrac{A}{B}x - \tfrac{C}{B}$$

So the slope is $-\tfrac{A}{B}$ and the y-intercept is $-\tfrac{C}{B}$.

(b)   Using the formula, we have:

$$d = \frac{|mx_0 + b - y_0|}{\sqrt{1 + m^2}} = \frac{\left|-\frac{Ax_0 + By_0 + C}{B}\right|}{\sqrt{1 + \frac{A^2}{B^2}}} = \frac{\frac{|Ax_0 + By_0 + C|}{|B|}}{\frac{\sqrt{A^2 + B^2}}{|B|}} = \frac{|Ax_0 + By_0 + C|}{\sqrt{A^2 + B^2}}$$

45.   Note that P and Q both lie on the circle $x^2 + y^2 = a^2$. We draw the figure:

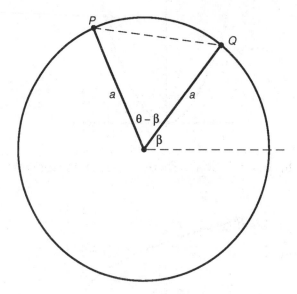

We have the right triangle:

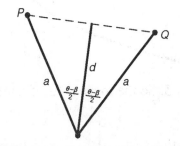

Using this triangle, we have:

$$\left|\cos\frac{\theta - \beta}{2}\right| = \frac{d}{a}$$

$$d = a\left|\cos\frac{\theta - \beta}{2}\right|$$

7.2 The Parabola

1. We note that $4p = 4$, so $p = 1$. So the focus is $(0, 1)$, the directrix is $y = -1$, and the focal width is 4.

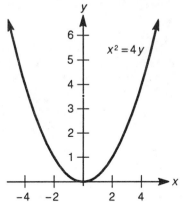

3. We note that $4p = 8$, so $p = 2$. So the focus is $(-2, 0)$, the directrix is $x = 2$, and the focal width is 8.

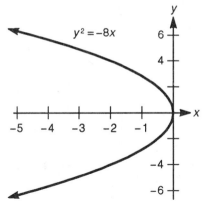

5. We note that $4p = 20$, so $p = 5$. So the focus is $(0, -5)$, the directrix is $y = 5$, and the focal width is 20.

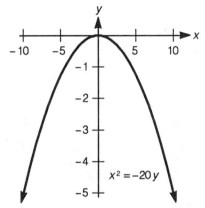

7.   Since $y^2 = -28x$, then $4p = 28$ so $p = 7$. So the focus is $(-7, 0)$, the directrix is $x = 7$, and the focal width is 28.

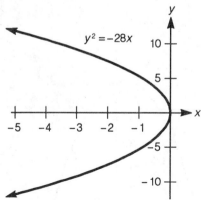

9.   We note that $4p = 6$, so $p = \frac{3}{2}$. So the focus is $\left(0, \frac{3}{2}\right)$, the directrix is $y = -\frac{3}{2}$, and the focal width is 6.

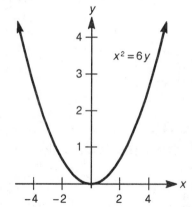

11.   Since $x^2 = \frac{7}{4}y$, then $4p = \frac{7}{4}$ so $p = \frac{7}{16}$. So the focus is $\left(0, \frac{7}{16}\right)$, the directrix is $y = -\frac{7}{16}$, and the focal width is $\frac{7}{4}$.

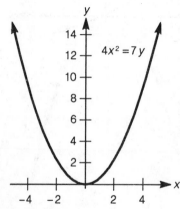

13. We note that $4p = 2$, so $p = \frac{1}{2}$. The vertex is $\left(-\frac{1}{2}, -1\right)$, the focus is $\left(-\frac{1}{2}, -\frac{3}{2}\right)$, the directrix is $y = -\frac{1}{2}$, and the focal width is 2.

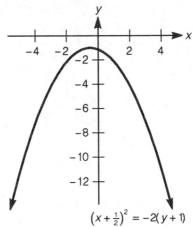

$$\left(x + \tfrac{1}{2}\right)^2 = -2(y + 1)$$

15. We first complete the square:

$$y^2 - 6y - 4x + 17 = 0$$
$$y^2 - 6y = 4x - 17$$
$$y^2 - 6y + 9 = 4x - 17 + 9$$
$$(y - 3)^2 = 4x - 8$$
$$(y - 3)^2 = 4(x - 2)$$

We note that $4p = 4$, so $p = 1$. The vertex is $(2, 3)$, the focus is $(3, 3)$, the directrix is $x = 1$, and the focal width is 4.

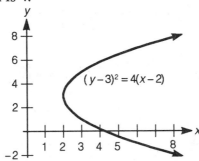

$(y - 3)^2 = 4(x - 2)$

17. We first complete the square:

$$x^2 - 8x - y + 18 = 0$$
$$x^2 - 8x = y - 18$$
$$x^2 - 8x + 16 = y - 18 + 16$$
$$(x - 4)^2 = y - 2$$

We note that $4p = 1$, so $p = \frac{1}{4}$. The vertex is $(4, 2)$, the focus is $\left(4, \frac{9}{4}\right)$, the directrix is $y = \frac{7}{4}$, and the focal width is 1.

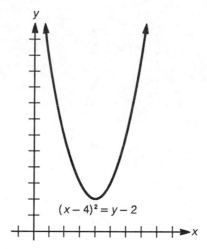

$(x - 4)^2 = y - 2$

19. We first complete the square:
$$y^2 + 2y - x + 1 = 0$$
$$y^2 + 2y = x - 1$$
$$y^2 + 2y + 1 = x - 1 + 1$$
$$(y + 1)^2 = x$$

We note that $4p = 1$, so $p = \frac{1}{4}$. The vertex is $(0, -1)$, the focus is $\left(\frac{1}{4}, -1\right)$, the directrix is $x = -\frac{1}{4}$, and the focal width is 1.

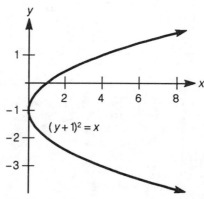

$(y + 1)^2 = x$

21. We first complete the square:
$$2x^2 - 12x - y + 18 = 0$$
$$2x^2 - 12x = y - 18$$
$$x^2 - 6x = \tfrac{1}{2}y - 9$$
$$x^2 - 6x + 9 = \tfrac{1}{2}y - 9 + 9$$
$$(x - 3)^2 = \tfrac{1}{2}y$$

We note that $4p = \tfrac{1}{2}$, so $p = \tfrac{1}{8}$. The vertex is $(3, 0)$, the focus is $\left(3, \tfrac{1}{8}\right)$, the directrix is $y = -\tfrac{1}{8}$, and the focal width is $\tfrac{1}{2}$.

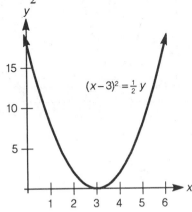

23. We first complete the square:
$$2x^2 - 16x - y + 33 = 0$$
$$2x^2 - 16x = y - 33$$
$$2(x - 4)^2 = y - 1$$
$$(x - 4)^2 = \tfrac{1}{2}(y - 1)$$

We note that $4p = \tfrac{1}{2}$, so $p = \tfrac{1}{8}$. The vertex is $(4, 1)$, the focus is $\left(4, \tfrac{9}{8}\right)$, the directrix is $y = \tfrac{7}{8}$, and the focal width is $\tfrac{1}{2}$.

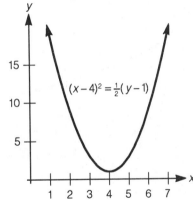

25. The line of symmetry has equation $y = 1$. The figure is graphed as:

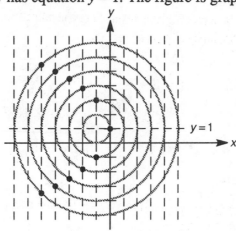

27. Given that the focus is at $(0, 3)$, we see that $p = 3$ and thus $4p = 12$. Since the parabola opens up, it has the $x^2 = 4py$ form, and its equation must be $x^2 = 12y$.

29. The directrix is $x = -32$, so $(32, 0)$ is the focus and the parabola opens to the right. So the parabola is of the form $y^2 = 4px$, where $p = 32$. Thus the equation is $y^2 = 128x$.

31. Since the parabola passes through the points $(5, 6)$ and $(5, -6)$, we know it is of the form $y^2 = 4px$. Using the points $(5, 6)$ and $(5, -6)$, we have:
$$36 = 4p(5)$$
$$36 = 20p$$
$$\tfrac{9}{5} = p$$

So the equation is $y^2 = \frac{36}{5}x$.

33. Since the parabola is symmetric about the x-axis, then $y^2 = \pm 4px$ is the form of the parabola. Since the x-coordinate of the focus is negative, then $y^2 = -4px$. Finally we are given $4p = 9$, so the equation is $y^2 = -9x$.

35. Since $4p = 4$, then $p = 1$. By symmetry, if $P(2, 1)$ is one endpoint of focal chord, then $Q(-2, 1)$ is the other endpoint.

37. We have $4p = 1$, so $p = \frac{1}{4}$. Thus the focus is at $\left(0, \frac{1}{4}\right)$. We find the slope of the focal chord PQ:
$$m = \frac{9 - \frac{1}{4}}{-3 - 0} = \frac{\frac{35}{4}}{-3} = -\frac{35}{12}$$

So $y = -\frac{35}{12}x + \frac{1}{4}$. We find where this line intersects the parabola (we already know P is one intersection point, but Q will be the other):

$$x^2 = -\frac{35}{12}x + \frac{1}{4}$$
$$12x^2 = -35x + 3$$
$$12x^2 + 35x - 3 = 0$$
$$(12x - 1)(x + 3) = 0$$
$$x = -3, \tfrac{1}{12}$$
$$y = 9, \tfrac{1}{144}$$

So the coordinates of Q are $Q\left(\frac{1}{12}, \frac{1}{144}\right)$. Since $P(-3, 9)$, we can use the distance formula:

$$PQ = \sqrt{\left(-3 - \tfrac{1}{12}\right)^2 + \left(9 - \tfrac{1}{144}\right)^2} = \tfrac{1369}{144}$$

39. We have $4p = 8$, so $p = 2$, and thus the coordinates of Q are $(2, -4)$. Thus the diameter of the circle is 8, its radius is thus 4, and its equation is $(x - 2)^2 + y^2 = 16$. Since the directrix is $x = -2$, we can see that the circle is tangent to the directrix.

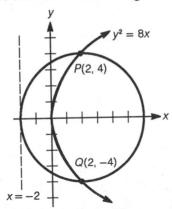

41. Call $(0, 0)$ the vertex, thus the ends of the arch are at $(20, -15)$ and $(-20, -15)$. The parabola is of the form $x^2 = -4py$, so we substitute:

$$400 = -4p(-15)$$
$$-\tfrac{80}{3} = -4p$$

So the equation is $x^2 = -\frac{80}{3}y$. We wish to find the y-coordinate when the base is 20 ft, so we substitute $x = \pm 10$ into the equation:

$$100 = -\tfrac{80}{3}y$$
$$-\tfrac{15}{4} = y$$

So the height above the base is $15 - \frac{15}{4} = \frac{45}{4} = 11.25$ ft.

43. (a)   We have $4p = 2$, so $p = \frac{1}{2}$, and thus the focus is $\left(0, \frac{1}{2}\right)$. To find A', we find the slope of the line containing A and the focus:

$$m = \frac{8 - \frac{1}{2}}{4 - 0} = \frac{\frac{15}{2}}{4} = \frac{15}{8}$$

So $y = \frac{15}{8}x + \frac{1}{2}$ is its equation. Now substitute into the parabola:

$$x^2 = 2\left(\frac{15}{8}x + \frac{1}{2}\right)$$
$$x^2 = \frac{15}{4}x + 1$$
$$4x^2 = 15x + 4$$
$$4x^2 - 15x - 4 = 0$$
$$x = -\frac{1}{4}, 4$$
$$y = \frac{1}{32}, 8$$

So $A' = \left(-\frac{1}{4}, \frac{1}{32}\right)$. To find B', we find the slope of the line containing B and the focus:

$$m = \frac{2 - \frac{1}{2}}{-2 - 0} = \frac{\frac{3}{2}}{-2} = -\frac{3}{4}$$

So $y = -\frac{3}{4}x + \frac{1}{2}$ is its equation. Now substitute into the parabola:

$$x^2 = 2\left(-\frac{3}{4}x + \frac{1}{2}\right)$$
$$x^2 = -\frac{3}{2}x + 1$$
$$2x^2 + 3x - 2 = 0$$
$$x = \frac{1}{2}, -2$$
$$y = \frac{1}{8}, 2$$

So $B' = \left(\frac{1}{2}, \frac{1}{8}\right)$. We now find the slope through A and B':

$$m = \frac{8 - \frac{1}{8}}{4 - \frac{1}{2}} = \frac{\frac{63}{8}}{\frac{7}{2}} = \frac{9}{4}$$

Using $(4, 8)$ in the point-slope formula:

$$y - 8 = \frac{9}{4}(x - 4)$$
$$y - 8 = \frac{9}{4}x - 9$$
$$y = \frac{9}{4}x - 1$$

(b)   We find the slope:

$$m = \frac{2 - \frac{1}{32}}{-2 + \frac{1}{4}} = \frac{\frac{63}{32}}{-\frac{7}{4}} = -\frac{9}{8}$$

Using $(-2, 2)$ in the point-slope formula:
$$y - 2 = -\tfrac{9}{8}(x + 2)$$
$$y - 2 = -\tfrac{9}{8}x - \tfrac{9}{4}$$
$$y = -\tfrac{9}{8}x - \tfrac{1}{4}$$

(c)   Setting the y-coordinates equal:
$$\tfrac{9}{4}x - 1 = -\tfrac{9}{8}x - \tfrac{1}{4}$$
$$18x - 8 = -9x - 2$$
$$27x = 6$$
$$x = \tfrac{2}{9}$$
$$y = \tfrac{1}{2} - 1 = -\tfrac{1}{2}$$

So the lines intersect at the point $\left(\tfrac{2}{9}, -\tfrac{1}{2}\right)$. Since the directrix is $y = -\tfrac{1}{2}$, then this point lies on the directrix.

45. (a)   Since $4p = 1$, $p = \tfrac{1}{4}$, thus the focus is $\left(0, \tfrac{1}{4}\right)$. We find the slope of the focal chord from P:
$$m = \frac{4 - \tfrac{1}{4}}{2 - 0} = \frac{\tfrac{15}{4}}{2} = \frac{15}{8}$$

So its equation is $y = \tfrac{15}{8}x + \tfrac{1}{4}$. We now find the intersection of this line with the parabola $y = x^2$:
$$x^2 = \tfrac{15}{8}x + \tfrac{1}{4}$$
$$8x^2 = 15x + 2$$
$$8x^2 - 15x - 2 = 0$$
$$(8x + 1)(x - 2) = 0$$
$$x = -\tfrac{1}{8}, 2$$
$$y = \tfrac{1}{64}, 4$$

So the coordinates of Q are $Q\left(-\tfrac{1}{8}, \tfrac{1}{64}\right)$.

(b)   We find the midpoint of PQ:
$$M = \left(\frac{2 - \tfrac{1}{8}}{2}, \frac{4 + \tfrac{1}{64}}{2}\right) = \left(\frac{15}{16}, \frac{257}{128}\right)$$

(c) The coordinates of S are $\left(0, \frac{257}{128}\right)$. We can find T, since the slope of this line (which

is perpendicular to PQ) is $-\frac{8}{15}$, and thus by the point-slope formula:

$$y - \tfrac{257}{128} = -\tfrac{8}{15}\left(x - \tfrac{15}{16}\right)$$
$$y - \tfrac{257}{128} = -\tfrac{8}{15}x + \tfrac{1}{2}$$
$$y = -\tfrac{8}{15}x + \tfrac{321}{128}$$

When $x = 0$, $y = \frac{321}{128}$, so the coordinates of T are $\left(0, \frac{321}{128}\right)$. Thus the length of ST is

$\frac{321}{128} - \frac{257}{128} = \frac{64}{128} = \frac{1}{2}$. Since the focal width is $4p = 1$, this verifies that ST is one-half the focal width.

47. The sketch in the book will have the following coordinates:

$$O(0,0),\ B\left(x, \frac{x^2}{4p}\right),\ A\left(-x, \frac{x^2}{4p}\right)$$

We require that $AB = OB$. So:

$$2x = \sqrt{(x-0)^2 + \left(\frac{x^2}{4p} - 0\right)^2}$$

$$2x = \sqrt{x^2 + \frac{x^4}{16p^2}}$$

Squaring, we have:

$$4x^2 = x^2 + \frac{x^4}{16p^2}$$

$$0 = -3x^2 + \frac{x^4}{16p^2}$$

$$0 = -48x^2p^2 + x^4$$

$$0 = x^2\left(-48p^2 + x^2\right)$$

So $x^2 = 0$ or $x = 0$ is one root and $x^2 = 48p^2$ or $x = \pm4\sqrt{3}p$ is the other. The distance from A to B is therefore:

$$2\left(4\sqrt{3}p\right) = 8\sqrt{3}p \text{ units}$$

To find the area we note that this is an equilateral triangle. Half of it is a $30°$-$60°$-$90°$ triangle whose height will be $\sqrt{3}$ times its base. The height is therefore $\sqrt{3} \bullet 4\sqrt{3}p = 12p$. The area is:

$$\tfrac{1}{2}\left(8\sqrt{3}p\right)(12p) = 48\sqrt{3}p^2 \text{ sq. units}$$

49. Denote the coordinates of P by $\left(a, \frac{a^2}{4p}\right)$. Then the coordinates of Q can be determined

from the fact that QV is perpendicular to PV. Since VQ has slope $-\frac{4p}{a}$, then Q is of the

form $\left(b, -\frac{4p}{a}b\right)$. Since $b \neq 0$ this yields $b = -\frac{16p^2}{a}$, so Q is $\left(-\frac{16p^2}{a}, \frac{64p^3}{a^2}\right)$. The

slope of QP is then found (after simplifying) to be $\frac{a^2-16p^2}{4ap}$. Now let y denote the
y-coordinate of G. Then since the slopes of QP and GP are equal, we have:

$$\frac{y - \frac{a^2}{4p}}{0 - a} = \frac{a^2 - 16p^2}{4ap}$$

By solving this equation for y, we find $y = 4p$, as required.

51. Since the focus is $(p, 0)$, then the slope of the focal chord is:

$$m = \frac{y_0 - 0}{\frac{y_0^2}{4p} - p} = \frac{4py_0}{y_0^2 - 4p^2}$$

Using the point-slope formula, the equation of this focal chord is:

$$y - y_0 = m\left(x - \frac{y_0^2}{4p}\right)$$

This line intersects the parabola when $x = \frac{y^2}{4p}$, so:

$$y - y_0 = m\left(\frac{y^2}{4p} - \frac{y_0^2}{4p}\right)$$
$$4p(y - y_0) = m(y + y_0)(y - y_0)$$
$$4p = m(y + y_0)$$

Substitute $m = \frac{4py_0}{y_0^2 - 4p^2}$:

$$4p = \frac{4py_0}{y_0^2 - yp^2}(y + y_0)$$
$$\frac{y_0^2 - 4p^2}{y_0} = y + y_0$$
$$-\frac{4p^2}{y_0} = y$$

Therefore:

$$x = \frac{1}{4p}\left(-\frac{4p^2}{y_0}\right)^2 = \frac{1}{4p}\left(\frac{16p^4}{y_0^2}\right) = \frac{4p^3}{y_0^2}$$

But $y_0^2 = 4px_0$, so:

$$x = \frac{4p^3}{4px_0} = \frac{p^2}{x_0}$$

Thus the coordinates of Q are $\left(\dfrac{p^2}{x_0}, -\dfrac{4p^2}{y_0}\right)$.

53. Let P have coordinates (x_0, y_0). Then, from Exercise 51, Q has coordinates $\left(\frac{p^2}{x_0}, -\frac{4p^2}{y_0}\right)$. The following results summarize the calculations required for this problem. Note: At each step we have replaced the quantity y_0^2 by $4px_0$. This simplifies matters a great deal.

Coordinates of M: $\left(\dfrac{x_0^2 + p^2}{2x_0}, \dfrac{2p(x_0 - p)}{y_0}\right)$

Coordinates of S: $\left(\dfrac{p^2 + x_0^2}{2x_0}, 0\right)$

Slope of PQ: $\dfrac{4px_0}{y_0(x_0 - p)}$

Slope of perpendicular: $\dfrac{y_0(p - x_0)}{4px_0}$

Since $-\frac{ST}{MS}$ is the slope, m, of the line through M and T, we have:

$$ST = -(MS)m = -\frac{2p(x_0 - p)}{y_0} \cdot \frac{4px_0}{y_0(p - x_0)} = \frac{8p^2x_0}{y_0^2} = 2p$$

Thus ST is one-half of the focal width.

55. Denote the coordinates of A and B by $\left(a, \frac{a^2}{4p}\right)$ and $\left(b, \frac{b^2}{4p}\right)$, respectively. Then by using the

result in Exercise 50, we find that the coordinates of A' and B' are $\left(-\frac{4p^2}{a}, \frac{4p^3}{a^2}\right)$ and

$\left(-\frac{4p^2}{b}, \frac{4p^3}{b^2}\right)$, respectively. Using these coordinates, the equations of the lines AB and

$A'B'$ can be determined. The results are as follows:

Line AB: $\qquad y = \dfrac{a+b}{4p}x - \dfrac{ab}{4p}$

Line $A'B'$: $\qquad y = \dfrac{-p(a+b)}{ab}x - \dfrac{4p^3}{ab}$

By solving the first of these two equations for x, and then using that result to substitute for x in the second equation, we obtain $y = -p$. That implies that the two lines intersect on the directrix, as required.

57. (a) Since the focus is $(0, p)$, then the slope of the focal chord is:

$$m = \frac{\frac{x_0^2}{4p} - p}{x_0 - 0} = \frac{x_0^2 - 4p^2}{4px_0}$$

The equation of the focal chord must be $y - p = mx$. This line intersects the parabola

when $y = \frac{x^2}{4p}$, so we have:

$$\frac{x^2}{4p} - p = mx$$

$$\frac{x^2}{4p} - p = \frac{x_0^2 - 4p^2}{4px_0} \bullet x$$

$$x^2 x_0 - 4p^2 x_0 = x x_0^2 - 4p^2 x$$

$$x x_0(x - x_0) = -4p^2(x - x_0)$$

$$x x_0 = -4p^2$$

$$x = -\frac{4p^2}{x_0}$$

Therefore:

$$y = \frac{1}{4p}\left(-\frac{4p^2}{x_0}\right)^2 = \frac{1}{4p}\left(\frac{16p^4}{x_0^2}\right) = \frac{4p^3}{x_0^2}$$

But $x_0^2 = 4py_0$, so:

$$y = \frac{4p^3}{4py_0} = \frac{p^2}{y_0}$$

Thus the coordinates of Q are $\left(-\dfrac{4p^2}{x_0}, \dfrac{p^2}{y_0}\right)$.

(b)   By the midpoint formula:

$$PQ = \left(\frac{x_0 - \frac{4p^2}{x_0}}{2}, \frac{y_0 + \frac{p^2}{y_0}}{2} \right) = \left(\frac{x_0^2 - 4p^2}{2x_0}, \frac{y_0^2 + p^2}{2y_0} \right)$$

(c)   The distance from P to the directrix is $y_0 + p$. Therefore $PF = y_0 + p$. The distance

from Q to the directrix is $\frac{p^2}{y_0} + p$. So we have $QF = \frac{p^2}{y_0} + p$ and consequently:

$$PQ = QF + PF = \frac{p^2}{y_0} + p + y_0 + p = \frac{p^2 + 2py_0 + y_0^2}{y_0} = \frac{(p + y_0)^2}{y_0}$$

(d)   The distance from the center of the circle to the directrix is found by adding p to the
      y-coordinate of the midpoint of PQ. This yields:

$$\frac{y_0^2 + p^2}{2y_0} + p = \frac{y_0^2 + p^2 + 2py_0}{2y_0} = \frac{(y_0 + p)^2}{2y_0}$$

Comparing this result with the expression for PQ determined in part (c), we
conclude that the distance from the center of the circle to the directrix equals the
radius of the circle. This implies that the circle is tangent to the directrix, as we
wished to show.

7.3 The Ellipse

1.   Dividing by 36, the standard form is $\frac{x^2}{9} + \frac{y^2}{4} = 1$. The length of the major axis is 6 and
     the length of the minor axis is 4. Using $c^2 = a^2 - b^2$, we find:
     $$c = \sqrt{a^2 - b^2} = \sqrt{9 - 4} = \sqrt{5}$$

So the foci are $\left(\pm\sqrt{5}, 0 \right)$ and the eccentricity is $\frac{\sqrt{5}}{3}$.

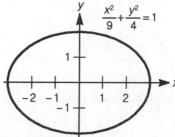

3.   Dividing by 16, the standard form is $\frac{x^2}{16} + \frac{y^2}{1} = 1$. The length of the major axis is 8 and
     the length of the minor axis is 2. Using $c^2 = a^2 - b^2$, we find:
     $$c = \sqrt{a^2 - b^2} = \sqrt{16 - 1} = \sqrt{15}$$

So the foci are $(\pm\sqrt{15},0)$ and the eccentricity is $\frac{\sqrt{15}}{4}$.

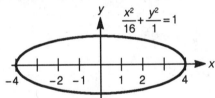

5.   Dividing by 2, the standard form is $\frac{x^2}{2} + \frac{y^2}{1} = 1$. The length of the major axis is $2\sqrt{2}$ and the length of the minor axis is 2. Using $c^2 = a^2 - b^2$, we find:

$$c = \sqrt{a^2 - b^2} = \sqrt{2-1} = 1$$

So the foci are $(\pm 1, 0)$ and the eccentricity is $\frac{\sqrt{2}}{2}$.

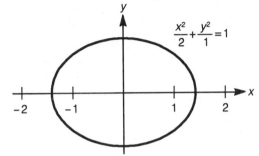

7.   Dividing by 144, the standard form is $\frac{x^2}{9} + \frac{y^2}{16} = 1$. The length of the major axis is 8 and the length of the minor axis is 6. Using $c^2 = a^2 - b^2$, we find:

$$c = \sqrt{a^2 - b^2} = \sqrt{16-9} = \sqrt{7}$$

So the foci are $(0, \pm\sqrt{7})$ and the eccentricity is $\frac{\sqrt{7}}{4}$.

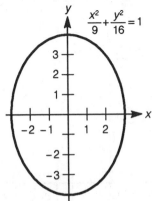

9.   Dividing by 5, the standard form is $\frac{x^2}{1/3} + \frac{y^2}{5/3} = 1$. The lengths of the major and minor axes are given by:

major axis: $2\sqrt{\frac{5}{3}} = \frac{2\sqrt{5}}{\sqrt{3}} = \frac{2\sqrt{15}}{3}$

minor axis: $2\sqrt{\frac{1}{3}} = \frac{2}{\sqrt{3}} = \frac{2\sqrt{3}}{3}$

Using $c^2 = a^2 - b^2$, we find:

$c = \sqrt{\frac{5}{3} - \frac{1}{3}} = \sqrt{\frac{4}{3}} = \frac{2}{\sqrt{3}} = \frac{2\sqrt{3}}{3}$

So the foci are $\left(0, \pm\frac{2\sqrt{3}}{3}\right)$. The eccentricity is given by:

$e = \frac{\frac{2\sqrt{3}}{3}}{\frac{\sqrt{15}}{3}} = \frac{2\sqrt{3}}{\sqrt{15}} = \frac{2}{\sqrt{5}} = \frac{2\sqrt{5}}{5}$

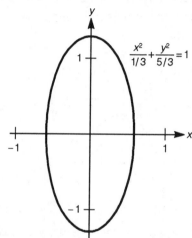

11.   Dividing by 4, the standard form is $\frac{x^2}{2} + \frac{y^2}{4} = 1$. The length of the major axis is 4 and the length of the minor axis is $2\sqrt{2}$. Using $c^2 = a^2 - b^2$, we find:

$c = \sqrt{a^2 - b^2} = \sqrt{4 - 2} = \sqrt{2}$

So the foci are $\left(0, \pm\sqrt{2}\right)$ and the eccentricity is $\frac{\sqrt{2}}{2}$.

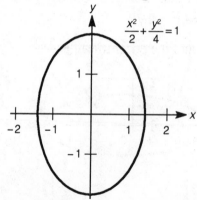

13. The equation is already in standard form with a center of $(5,-1)$. The length of the major axis is 10 and the length of the minor axis is 6. Using $c^2 = a^2 - b^2$, we find:

$$c = \sqrt{a^2 - b^2} = \sqrt{25-9} = \sqrt{16} = 4$$

So the foci are $(5+4,-1) = (9,-1)$ and $(5-4,-1) = (1,-1)$, and the eccentricity is $\frac{4}{5}$.

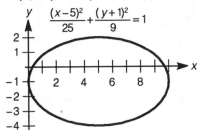

15. The equation is already in standard form with a center of $(1,2)$. The length of the major axis is 4 and the length of the minor axis is 2. Using $c^2 = a^2 - b^2$, we find:

$$c = \sqrt{a^2 - b^2} = \sqrt{4-1} = \sqrt{3}$$

So the foci are $\left(1, 2 \pm \sqrt{3}\right)$ and the eccentricity is $\frac{\sqrt{3}}{2}$.

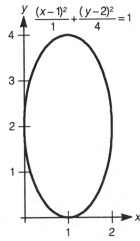

17. The equation is already in standard form with a center of $(-3,0)$. The length of the major axis is 6 and the length of the minor axis is 2. Using $c^2 = a^2 - b^2$, we find:

$$c = \sqrt{a^2 - b^2} = \sqrt{9-1} = \sqrt{8} = 2\sqrt{2}$$

So the foci are $\left(-3 \pm 2\sqrt{2}, 0\right)$ and the eccentricity is $\frac{2\sqrt{2}}{3}$.

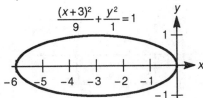

19. We complete the square to convert the equation to standard form:
$$3x^2 + 4y^2 - 6x + 16y + 7 = 0$$
$$3(x^2 - 2x) + 4(y^2 + 4y) = -7$$
$$3(x^2 - 2x + 1) + 4(y^2 + 4y + 4) = -7 + 3 + 16$$
$$3(x-1)^2 + 4(y+2)^2 = 12$$
$$\frac{(x-1)^2}{4} + \frac{(y+2)^2}{3} = 1$$

The center is $(1, -2)$, the length of the major axis is 4, and the length of the minor axis is $2\sqrt{3}$. Using $c^2 = a^2 - b^2$, we find:
$$c = \sqrt{a^2 - b^2} = \sqrt{4 - 3} = 1$$

So the foci are $(1 + 1, -2) = (2, -2)$ and $(1 - 1, -2) = (0, -2)$, and the eccentricity is $\frac{1}{2}$.

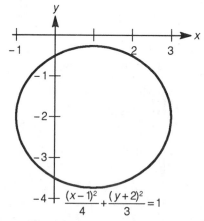

21. We complete the square to convert the equation to standard form:
$$5x^2 + 3y^2 - 40x - 36y + 188 = 0$$
$$5(x^2 - 8x) + 3(y^2 - 12y) = -188$$
$$5(x^2 - 8x + 16) + 3(y^2 - 12y + 36) = -188 + 80 + 108$$
$$5(x - 4)^2 + 3(y - 6)^2 = 0$$

Notice that the only solution to this is the center $(4, 6)$. This is called a degenerate ellipse, or, more commonly, a point!

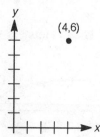

23. We complete the square to convert the equation to standard form:
$$16x^2 + 25y^2 - 64x - 100y + 564 = 0$$
$$16\left(x^2 - 4x\right) + 25\left(y^2 - 4y\right) = -564$$
$$16\left(x^2 - 4x + 4\right) + 25\left(y^2 - 4y + 4\right) = -564 + 64 + 100$$
$$16(x-2)^2 + 25(y-2)^2 = -400$$
Notice that there is no solution to this equation, since the left-hand side is non-negative. So there is no graph.

25. We are given $c = 3$ and $a = 5$, so we have the equation in the form:
$$\frac{x^2}{5^2} + \frac{y^2}{b^2} = 1$$
Since $c^2 = a^2 - b^2$, we find b:
$$9 = 25 - b^2$$
$$16 = b^2$$
$$4 = b$$
So the equation is $\frac{x^2}{25} + \frac{y^2}{16} = 1$, or $16x^2 + 25y^2 = 400$.

27. We are given $a = 4$, so the equation has a form of:
$$\frac{x^2}{16} + \frac{y^2}{b^2} = 1$$
Now $\frac{c}{a} = \frac{1}{4}$, so $\frac{c}{4} = \frac{1}{4}$ and thus $c = 1$.
We find b:
$$c^2 = a^2 - b^2$$
$$1 = 16 - b^2$$
$$b^2 = 15$$
So the equation is $\frac{x^2}{16} + \frac{y^2}{15} = 1$, or $15x^2 + 16y^2 = 240$.

29. We have $c = 2$ and $a = 5$, so the equation has a form of:
$$\frac{x^2}{b^2} + \frac{y^2}{25} = 1$$
We find b:
$$c^2 = a^2 - b^2$$
$$4 = 25 - b^2$$
$$21 = b^2$$
So the equation is $\frac{x^2}{21} + \frac{y^2}{25} = 1$, or $25x^2 + 21y^2 = 525$.

31. We know $a = 2b$ and that the equation has a form of:
$$\frac{x^2}{a^2} + \frac{y^2}{b^2} = 1$$

Using the point $\left(1, \sqrt{2}\right)$ and $a = 2b$, we have:
$$\frac{(1)^2}{(2b)^2} + \frac{\left(\sqrt{2}\right)^2}{b^2} = 1$$
$$\frac{1}{4b^2} + \frac{2}{b^2} = 1$$

Multiply by $4b^2$:
$$1 + 8 = 4b^2$$
$$\tfrac{9}{4} = b^2$$
$$\tfrac{3}{2} = b$$

Since $a = 2b$, then $a = 3$. So the equation is $\frac{x^2}{9} + \frac{y^2}{9/4} = 1$, or $x^2 + 4y^2 = 9$.

33. We begin by solving for y:
$$\frac{x^2}{9} + \frac{y^2}{4} = 1$$
$$\frac{y^2}{4} = 1 - \frac{x^2}{9}$$
$$\frac{y^2}{4} = \frac{9 - x^2}{9}$$
$$y^2 = \frac{36 - 4x^2}{9}$$
$$y = \pm\sqrt{\frac{36 - 4x^2}{9}}$$
$$y = \pm\frac{\sqrt{36 - 4x^2}}{3}$$

We use a calculator to fill in the table:

x	0	0.5	1.0	1.5	2.0	2.5	3.0
y	±2	±1.97	±1.89	±1.73	±1.49	±1.11	0

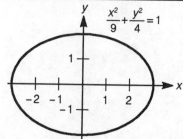

35. (a) Squaring both sides and simplifying, we obtain:

$$\sqrt{(x-c)^2 + y^2} = 2a - \sqrt{(x+c)^2 + y^2}$$
$$(x-c)^2 + y^2 = 4a^2 - 4a\sqrt{(x+c)^2 + y^2} + (x+c)^2 + y^2$$
$$x^2 - 2xc + c^2 + y^2 = 4a^2 - 4a\sqrt{(x+c)^2 + y^2} + x^2 + 2cx + c^2 + y^2$$
$$-4cx - 4a^2 = -4a\sqrt{(x+c)^2 + y^2}$$
$$a^2 + cx = a\sqrt{(x+c)^2 + y^2}$$

(b) Squaring each side again, we obtain:

$$\left(a^2 + cx\right)^2 = a^2\left((x+c)^2 + y^2\right)$$
$$a^4 + 2a^2cx + c^2x^2 = a^2\left(x^2 + 2cx + c^2 + y^2\right)$$
$$a^4 + 2a^2cx + c^2x^2 = a^2x^2 + 2a^2cx + a^2c^2 + a^2y^2$$
$$a^4 - a^2c^2 = a^2x^2 - c^2x^2 + a^2y^2$$

(c) We simply factor each side of the equation to obtain:

$$a^2\left(a^2 - c^2\right) = \left(a^2 - c^2\right)x^2 + a^2y^2$$

(d) Call $b^2 = a^2 - c^2$, so we can rewrite the equation from (c) as:

$$b^2x^2 + a^2y^2 = a^2b^2$$
$$\frac{x^2}{a^2} + \frac{y^2}{b^2} = 1$$

37. If we first multiply each equation by a^2b^2, we have:

$$b^2x^2 + a^2y^2 = a^2b^2$$
$$a^2x^2 + b^2y^2 = a^2b^2$$

Multiplying the first equation by $-a^2$ and the second equation by b^2 yields:

$$-a^2b^2x^2 - a^4y^2 = -a^4b^2$$
$$a^2b^2x^2 + b^4y^2 = a^2b^4$$

Adding, we have:

$$\left(b^4 - a^4\right)y^2 = a^2b^2\left(b^2 - a^2\right)$$
$$y^2 = \frac{a^2b^2}{b^2 + a^2}$$
$$y = \pm\frac{ab}{\sqrt{a^2 + b^2}}$$

Substituting into the first equation:

$$b^2x^2 + \frac{a^4b^2}{a^2+b^2} = a^2b^2$$

$$b^2x^2 = \frac{a^2b^4}{a^2+b^2}$$

$$x^2 = \frac{a^2b^2}{a^2+b^2}$$

$$x = \pm\frac{ab}{\sqrt{a^2+b^2}}$$

So there are four intersection points:

$$\left(\tfrac{ab}{A}, \tfrac{ab}{A}\right), \left(\tfrac{ab}{A}, -\tfrac{ab}{A}\right), \left(-\tfrac{ab}{A}, \tfrac{ab}{A}\right), \left(-\tfrac{ab}{A}, -\tfrac{ab}{A}\right), \text{ where } A = \sqrt{a^2+b^2}$$

We sketch the intersection:

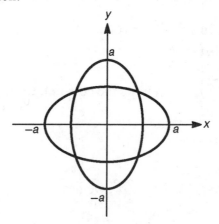

39. Substituting $x^2 = y$, we have:

$$b^2y + a^2y^2 = a^2b^2$$

$$a^2y^2 + b^2y - a^2b^2 = 0$$

$$y = \frac{-b^2 \pm \sqrt{b^4 - 4a^2\left(-a^2b^2\right)}}{2a^2}$$

$$y = \frac{-b^2 \pm b\sqrt{b^2 + 4a^4}}{2a^2}$$

Now clearly the negative sign is discarded (since $y = x^2$, then $y \geq 0$), so the intersection points are:

$$\left(\sqrt{A}, A\right) \text{ and } \left(-\sqrt{A}, A\right), \text{ where } A = \frac{-b^2 + b\sqrt{b^2 + 4a^4}}{2a^2}$$

41. Since $a = 5$, $b = 3$, we find $c = 4$, and so we can label the foci as $F_1(-4, 0)$ and $F_2(4, 0)$. If F_2PF_1 is a right angle, then F_1P and F_2P must be perpendicular. We compute slopes with $P(x, y)$:

$$F_1P: \quad m = \frac{0 - y}{-4 - x} = \frac{y}{x + 4}$$

$$F_2P: \quad m = \frac{0 - y}{4 - x} = \frac{y}{x - 4}$$

Since these are perpendicular, we have:

$$\frac{y}{x + 4} \bullet \frac{y}{x - 4} = -1$$

$$\frac{y^2}{x^2 - 16} = -1$$

$$y^2 = 16 - x^2$$

Now substitute into the ellipse:

$$9x^2 + 25(16 - x^2) = 225$$

$$9x^2 + 400 - 25x^2 = 225$$

$$-16x^2 = -175$$

$$x^2 = \frac{175}{16}$$

$$x = \frac{5\sqrt{7}}{4}$$

Substitute to find y:

$$y^2 = 16 - \frac{175}{16} = \frac{81}{16}, \text{ so } y = \frac{9}{4}$$

So the coordinates are $P\left(\frac{5\sqrt{7}}{4}, \frac{9}{4}\right)$.

43. By Exercise 40(a), $F_1P = a + ex$ and $PQ = x + \frac{a}{e}$. Thus:

$$\frac{F_1P}{PQ} = \frac{a + ex}{x + \frac{a}{e}} \bullet \frac{e}{e} = \frac{e(a + ex)}{ex + a} = e$$

7.4 The Hyperbola

1. Dividing by 4, the standard form is $\frac{x^2}{4} - \frac{y^2}{1} = 1$. The vertices are $(\pm 2, 0)$, the length of the transverse axis is 4, the length of the conjugate axis is 2, and the asymptotes are $y = \pm\frac{1}{2}x$. Using $c^2 = a^2 + b^2$, we find:

$$c = \sqrt{a^2 + b^2} = \sqrt{4 + 1} = \sqrt{5}$$

So the foci are $(\pm\sqrt{5},0)$ and the eccentricity is $\frac{\sqrt{5}}{2}$.

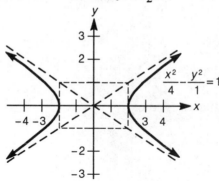

3.   Dividing by 4, the standard form is $\frac{y^2}{4} - \frac{x^2}{1} = 1$. The vertices are $(0,\pm 2)$, the length of the transverse axis is 4, the length of the conjugate axis is 2, and the asymptotes are $y = \pm 2x$. Using $c^2 = a^2 + b^2$, we find:
$$c = \sqrt{a^2 + b^2} = \sqrt{4+1} = \sqrt{5}$$

So the foci are $\left(0,\pm\sqrt{5}\right)$ and the eccentricity is $\frac{\sqrt{5}}{2}$.

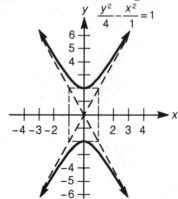

5.   Dividing by 400, the standard form is $\frac{x^2}{25} - \frac{y^2}{16} = 1$. The vertices are $(\pm 5,0)$, the length of the transverse axis is 10, the length of the conjugate axis is 8, and the asymptotes are

$y = \pm\frac{4}{5}x$. Using $c^2 = a^2 + b^2$, we find:
$$c = \sqrt{a^2 + b^2} = \sqrt{25+16} = \sqrt{41}$$

So the foci are $\left(\pm\sqrt{41},0\right)$ and the eccentricity is $\frac{\sqrt{41}}{5}$.

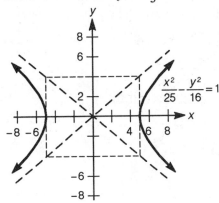

$\dfrac{x^2}{25}-\dfrac{y^2}{16}=1$

7. Rewriting the equation, the standard form is $\frac{y^2}{1/2}-\frac{x^2}{1/3}=1$. The vertices are

$\left(0,\pm\sqrt{\frac{1}{2}}\right)=\left(0,\pm\frac{\sqrt{2}}{2}\right)$, the length of the transverse axis is $\sqrt{2}$, the length of the conjugate

axis is $2\sqrt{\frac{1}{3}}=\frac{2\sqrt{3}}{3}$, and the asymptotes are $y=\pm\sqrt{\frac{3}{2}}x=\pm\frac{\sqrt{6}}{2}x$. Using $c^2=a^2+b^2$, we
find:

$$c=\sqrt{a^2+b^2}=\sqrt{\tfrac{1}{2}+\tfrac{1}{3}}=\sqrt{\tfrac{5}{6}}=\frac{\sqrt{5}}{\sqrt{6}}=\frac{\sqrt{30}}{6}$$

So the foci are $\left(0,\pm\frac{\sqrt{30}}{6}\right)$ and the eccentricity is:

$$e=\frac{\frac{\sqrt{30}}{6}}{\frac{\sqrt{2}}{2}}=\frac{\sqrt{30}}{3\sqrt{2}}=\frac{\sqrt{15}}{3}$$

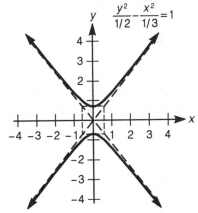

$\dfrac{y^2}{1/2}-\dfrac{x^2}{1/3}=1$

9. Dividing by 100, the standard form is $\frac{y^2}{25} - \frac{x^2}{4} = 1$. The vertices are $(0, \pm 5)$, the length of the transverse axis is 10, the length of the conjugate axis is 4, and the asymptotes are

$y = \pm \frac{5}{2}x$. Using $c^2 = a^2 + b^2$, we have:

$$c = \sqrt{a^2 + b^2} = \sqrt{25 + 4} = \sqrt{29}$$

So the foci are $\left(0, \pm \sqrt{29}\right)$ and the eccentricity is $\frac{\sqrt{29}}{5}$.

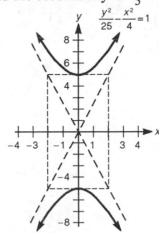

11. The equation is already in standard form with a center of $(5, -1)$. The vertices are $(5 + 5, -1) = (10, -1)$ and $(5 - 5, -1) = (0, -1)$, the length of the transverse axis is 10, and

the length of the conjugate axis is 6. The asymptotes have slopes of $\pm \frac{3}{5}$, so using the point-slope formula:

$$y + 1 = \frac{3}{5}(x - 5) \qquad\qquad y + 1 = -\frac{3}{5}(x - 5)$$
$$y + 1 = \frac{3}{5}x - 3 \qquad\qquad y + 1 = -\frac{3}{5}x + 3$$
$$y = \frac{3}{5}x - 4 \qquad\qquad y = -\frac{3}{5}x + 2$$

Using $c^2 = a^2 + b^2$, we have:

$$c = \sqrt{a^2 + b^2} = \sqrt{25 + 9} = \sqrt{34}$$

So the foci are $\left(5 \pm \sqrt{34}, -1\right)$ and the eccentricity is $\frac{\sqrt{34}}{5}$.

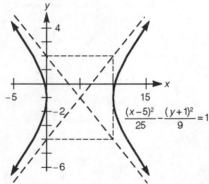

13. The equation is already in standard form with a center of $(1, 2)$. The vertices are $(1, 2 + 2) = (1, 4)$ and $(1, 2 - 2) = (1, 0)$, the length of the transverse axis is 4, and the length of the conjugate axis is 2. The asymptotes have slopes of ± 2, so using the point-slope formula:

$$y - 2 = 2(x - 1)$$
$$y - 2 = 2x - 2$$
$$y = 2x$$

$$y - 2 = -2(x - 1)$$
$$y - 2 = -2x + 2$$
$$y = -2x + 4$$

Using $c^2 = a^2 + b^2$, we have:

$$c = \sqrt{a^2 + b^2} = \sqrt{4 + 1} = \sqrt{5}$$

So the foci are $\left(1, 2 \pm \sqrt{5}\right)$ and the eccentricity is $\frac{\sqrt{5}}{2}$.

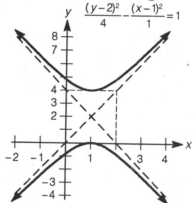

$$\frac{(y-2)^2}{4} - \frac{(x-1)^2}{1} = 1$$

15. The equation is already in standard form with a center of $(-3, 4)$. The vertices are $(-3 + 4, 4) = (1, 4)$ and $(-3 - 4, 4) = (-7, 4)$, the length of the transverse axis is 8, and the length of the conjugate axis is 8. The asymptotes have slopes of ± 1, so using the point-slope formula:

$$y - 4 = 1(x + 3)$$
$$y - 4 = x + 3$$
$$y = x + 7$$

$$y - 4 = -1(x + 3)$$
$$y - 4 = -x - 3$$
$$y = -x + 1$$

Using $c^2 = a^2 + b^2$, we have:

$$c = \sqrt{a^2 + b^2} = \sqrt{16 + 16} = \sqrt{32} = 4\sqrt{2}$$

So the foci are $\left(-3 \pm 4\sqrt{2}, 4\right)$ and the eccentricity is $\frac{4\sqrt{2}}{4} = \sqrt{2}$.

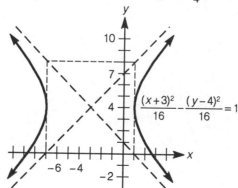

$$\frac{(x+3)^2}{16} - \frac{(y-4)^2}{16} = 1$$

17. We complete the square to convert the equation to standard form:
$$x^2 - y^2 + 2y - 5 = 0$$
$$x^2 - \left(y^2 - 2y\right) = 5$$
$$x^2 - \left(y^2 - 2y + 1\right) = 5 - 1$$
$$x^2 - (y-1)^2 = 4$$
$$\frac{x^2}{4} - \frac{(y-1)^2}{4} = 1$$

The center is $(0, 1)$, the vertices are $(0 + 2, 1) = (2, 1)$ and $(0 - 2, 1) = (-2, 1)$, and the lengths of both the transverse and conjugate axes are 4. The asymptotes have slopes of ± 1, so using the point-slope formula:

$$y - 1 = 1(x - 0) \qquad\qquad y - 1 = -1(x - 0)$$
$$y - 1 = x \qquad\qquad\qquad y - 1 = -x$$
$$y = x + 1 \qquad\qquad\qquad y = -x + 1$$

Using $c^2 = a^2 + b^2$, we have:
$$c = \sqrt{a^2 + b^2} = \sqrt{4 + 4} = \sqrt{8} = 2\sqrt{2}$$

So the foci are $\left(\pm 2\sqrt{2}, 1\right)$ and the eccentricity is $\frac{2\sqrt{2}}{2} = \sqrt{2}$.

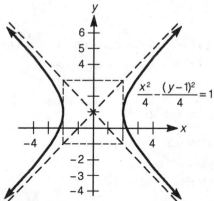

19. We complete the square to convert the equation to standard form:
$$x^2 - y^2 - 4x + 2y - 6 = 0$$
$$\left(x^2 - 4x\right) - \left(y^2 - 2y\right) = 6$$
$$\left(x^2 - 4x + 4\right) - \left(y^2 - 2y + 1\right) = 6 + 4 - 1$$
$$(x - 2)^2 - (y - 1)^2 = 9$$
$$\frac{(x - 2)^2}{9} - \frac{(y - 1)^2}{9} = 1$$

The center is (2, 1), the vertices are $(2 + 3, 1) = (5, 1)$ and $(2 - 3, 1) = (-1, 1)$, and the lengths of both the transverse and conjugate axes are 6. The asymptotes have slopes of ± 1, so using the point-slope formula:

$$y - 1 = 1(x - 2) \qquad\qquad y - 1 = -1(x - 2)$$
$$y - 1 = x - 2 \qquad\qquad\quad y - 1 = -x + 2$$
$$y = x - 1 \qquad\qquad\qquad y = -x + 3$$

Using $c^2 = a^2 + b^2$, we have:
$$c = \sqrt{a^2 + b^2} = \sqrt{9 + 9} = \sqrt{18} = 3\sqrt{2}$$

So the foci are $\left(2 \pm 3\sqrt{2}, 1\right)$ and the eccentricity is $\frac{3\sqrt{2}}{3} = \sqrt{2}$.

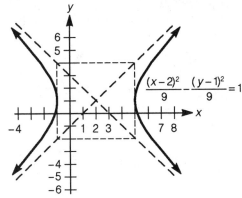

21. We complete the square to convert the equation to standard form:
$$y^2 - 25x^2 + 8y - 9 = 0$$
$$\left(y^2 + 8y\right) - 25x^2 = 9$$
$$\left(y^2 + 8y + 16\right) - 25x^2 = 9 + 16$$
$$(y + 4)^2 - 25x^2 = 25$$
$$\frac{(y + 4)^2}{25} - \frac{x^2}{1} = 1$$

The center is $(0, -4)$, the vertices are $(0, -4 + 5) = (0, 1)$ and $(0, -4 - 5) = (0, -9)$, the length of the transverse axis is 10, and the length of the conjugate axis is 2. The asymptotes have slopes of ± 5, so using the point-slope formula:

$$y + 4 = 5(x - 0) \qquad\qquad y + 4 = -5(x - 0)$$
$$y + 4 = 5x \qquad\qquad\qquad y + 4 = -5x$$
$$y = 5x - 4 \qquad\qquad\qquad y = -5x - 4$$

Using $c^2 = a^2 + b^2$, we have:
$$c = \sqrt{a^2 + b^2} = \sqrt{25 + 1} = \sqrt{26}$$

So the foci are $\left(0,-4\pm\sqrt{26}\right)$ and the eccentricity is $\frac{\sqrt{26}}{5}$.

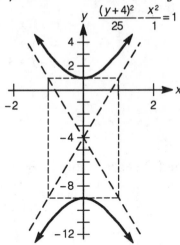

23. We complete the square to convert the equation to standard form:

$$x^2 + 7x - y^2 - y + 12 = 0$$
$$\left(x^2 + 7x\right) - \left(y^2 + y\right) = -12$$
$$\left(x^2 + 7x + \tfrac{49}{4}\right) - \left(y^2 + y + \tfrac{1}{4}\right) = -12 + \tfrac{49}{4} - \tfrac{1}{4}$$
$$\left(x + \tfrac{7}{2}\right)^2 - \left(y + \tfrac{1}{2}\right)^2 = 0$$

Notice that this is a degenerate hyperbola, and the graph consists of the "would-be" asymptotes with slopes ± 1:

$$y + \tfrac{1}{2} = 1\left(x + \tfrac{7}{2}\right) \qquad\qquad y + \tfrac{1}{2} = -1\left(x + \tfrac{7}{2}\right)$$
$$y + \tfrac{1}{2} = x + \tfrac{7}{2} \qquad\qquad\quad y + \tfrac{1}{2} = -x - \tfrac{7}{2}$$
$$y = x + 3 \qquad\qquad\qquad\quad y = -x - 4$$

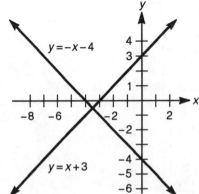

25. Since $P(x,y)$ lies on $\frac{x^2}{4} - \frac{y^2}{1} = 1$, we can find y in terms of x:

$$y^2 = \frac{x^2}{4} - 1$$

$$y^2 = \frac{x^2 - 4}{4}$$

Taking roots, we have $y = \frac{\sqrt{x^2-4}}{2}$, since $P(x,y)$ lies in the first quadrant. So the

coordinates of P are $\left(x, \frac{\sqrt{x^2-4}}{2}\right)$. Since $Q(x,y)$ lies in the first quadrant on the asymptote, we find the equation of the asymptote:

$$y - 0 = \tfrac{1}{2}(x - 0)$$

$$y = \tfrac{1}{2}x$$

So the coordinates of Q are $\left(x, \tfrac{1}{2}x\right)$. Since P and Q have the same x-coordinate PQ is the difference between their y-coordinates:

$$PQ = \frac{x}{2} - \frac{\sqrt{x^2-4}}{2} = \frac{x - \sqrt{x^2-4}}{2}$$

The order of subtraction is because the asymptote lies above the hyperbola in the first

quadrant, and thus $\frac{x}{2}$ is larger than $\frac{\sqrt{x^2-4}}{2}$. This proves the desired result.

27. Since the foci are $(\pm 4, 0)$ and the vertices are $(\pm 1, 0)$, then $c = 4$, $a = 1$, and the hyperbola has the form:

$$\frac{x^2}{1} - \frac{y^2}{b^2} = 1$$

We find b:

$$c^2 = a^2 + b^2$$

$$16 = 1 + b^2$$

$$15 = b^2$$

So the equation is $\frac{x^2}{1} - \frac{y^2}{15} = 1$, or $15x^2 - y^2 = 15$.

29. The slope of the asymptotes is $\pm\tfrac{1}{2}$, which tells us that the ratio $\frac{b}{a} = \tfrac{1}{2}$ in this hyperbola. Also, since the vertices are $(\pm 2, 0)$ then $a = 2$. The required ratio is therefore:

$$\frac{b}{2} = \frac{1}{2} \text{ and } b = 1$$

The equation is $\frac{x^2}{4} - \frac{y^2}{1} = 1$, or $x^2 - 4y^2 = 4$.

31. Since the asymptotes are $y = \pm\frac{\sqrt{10}}{5}x$, then $\frac{b}{a} = \frac{\sqrt{10}}{5}$, so $b = \frac{\sqrt{10}}{5}a$. Now, since the foci are $(\pm\sqrt{7}, 0)$, then $c = \sqrt{7}$ and the hyperbola has the form:

$$\frac{x^2}{a^2} - \frac{y^2}{b^2} = 1$$

Since $b = \frac{\sqrt{10}}{5}a$ and $c = \sqrt{7}$, we have:

$$c^2 = a^2 + b^2$$
$$7 = a^2 + \left(\frac{\sqrt{10}}{5}a\right)^2$$
$$7 = a^2 + \frac{2}{5}a^2$$
$$35 = 7a^2$$
$$5 = a^2$$
$$b^2 = \frac{2}{5}a^2 = \frac{2}{5}(5) = 2$$

So the equation is $\frac{x^2}{5} - \frac{y^2}{2} = 1$, or $2x^2 - 5y^2 = 10$.

33. The vertices are at $(0, \pm 7)$ so we know it is a "vertical" hyperbola. Its equation will be $\frac{y^2}{49} - \frac{x^2}{b^2} = 1$, but we also know that $(1, 9)$ is a point satisfying the equation. We use it to find b:

$$\frac{81}{49} - \frac{1}{b^2} = 1$$
$$81b^2 - 49 = 49b^2$$
$$32b^2 = 49$$
$$b^2 = \frac{49}{32}$$

So the equation is $\frac{y^2}{49} - \frac{x^2}{49/32} = 1$, or $y^2 - 32x^2 = 49$.

35. We have $2a = 6$, so $a = 3$. Also $2b = 2$, so $b = 1$. Since the foci are on the y-axis, the hyperbola will have the form:

$$\frac{y^2}{a^2} - \frac{x^2}{b^2} = 1$$

So the equation is $\frac{y^2}{9} - \frac{x^2}{1} = 1$, or $y^2 - 9x^2 = 9$.

37. Writing the equation as $\frac{x^2}{16} - \frac{y^2}{16} = 1$, we have $a = b = 4$. So the slopes of the asymptotes are $\pm\frac{b}{a} = \pm\frac{4}{4} = \pm 1$. But these are negative reciprocals of each other, so the asymptotes are perpendicular to each other.

39. (a) Substituting $P(5,6)$ into $5y^2 - 4x^2 = 80$, we have:
$$5(6)^2 - 4(5)^2 = 5(36) - 4(25) = 180 - 100 = 80$$
So $P(5,6)$ lies on the hyperbola.

(b) Dividing by 80 yields $\frac{y^2}{16} - \frac{x^2}{20} = 1$, so $a = 4$ and $b = 2\sqrt{5}$.
Using $c^2 = a^2 + b^2$, we have:
$$c = \sqrt{a^2 + b^2} = \sqrt{16 + 20} = \sqrt{36} = 6$$
So $c = 6$ and the foci are $(0, \pm 6)$.

(c) We compute the distance:
$$F_1 P = \sqrt{(5-0)^2 + (6-6)^2} = 5$$
$$F_2 P = \sqrt{(5-0)^2 + (6-(-6))^2} = \sqrt{25 + 144} = 13$$

(d) We verify the result:
$$\left| F_1 P - F_2 P \right| = |5 - 13| = |-8| = 8 = 2(4) = 2a$$

41. (a) Since the asymptotes must have slopes of $\pm \frac{b}{a}$, then:
$$-\frac{b}{a} = \frac{-1}{\frac{b}{a}}$$
$$\frac{b^2}{a^2} = 1$$
$$b^2 = a^2$$
$$b = a$$
Now, since $c^2 = a^2 + b^2 = 2a^2$, then the eccentricity is:
$$\frac{c}{a} = \frac{\sqrt{2a^2}}{a} = \frac{\sqrt{2}a}{a} = \sqrt{2}$$

(b) The slopes of the asymptotes are $\pm \frac{a}{a} = \pm 1$, so the hyperbola will have perpendicular

asymptotes. The eccentricity is $\frac{\sqrt{2}a}{a} = \sqrt{2}$.

43. (a) This equation is just $F_1 P - F_2 P = 2a$, the defining relation of a hyperbola. Since P is on the right-hand branch, $F_1 P > F_2 P$.

(b) Squaring each side of the equation, we obtain:
$$\sqrt{(x+c)^2 + y^2} = 2a + \sqrt{(x-c)^2 + y^2}$$
$$(x+c)^2 + y^2 = 4a^2 + 4a\sqrt{(x-c)^2 + y^2} + (x-c)^2 + y^2$$
$$4xc = 4a^2 + 4a\sqrt{(x-c)^2 + y^2}$$
$$xc - a^2 = a\sqrt{(x-c)^2 + y^2}$$
$$xc - a^2 = a\left(F_2 P\right)$$

(c)   Dividing by a, we have:
$$\frac{xc}{a} - a = F_2P$$
$$xe - a = F_2P$$

45.   For this hyperbola we find $a^2 = b^2 = k^2$, and $e = \sqrt{2}$.
Also $d^2 = x^2 + y^2 = x^2 + (x^2 - k^2) = 2x^2 - k^2$. Thus we want to show that

$F_1P \bullet F_2P = 2x^2 - k^2$. Using the formulas for F_1P and F_2P developed in Exercises 43 and 44, we have:
$$\begin{aligned}
F_1P \bullet F_2P &= (xe + a)(xe - a) \\
&= x^2e^2 - a^2 \\
&= x^2(2) - k^2 \\
&= 2x^2 - k^2 \\
&= d^2
\end{aligned}$$

47.   The coordinates of D are $\left(\frac{a}{e}, \frac{b}{e}\right)$ and those of F are $(c, 0)$. Let O denote the center $(0, 0)$.
We show that $\angle ODF$ is a right angle by showing $OD^2 + DF^2 = OF^2$:
$$OF^2 = c^2$$
$$OD^2 = \left(\frac{a}{e}\right)^2 + \left(\frac{b}{e}\right)^2 = \frac{a^2 + b^2}{e^2} = \frac{c^2}{e^2} = a^2$$
$$\begin{aligned}
DF^2 &= \left(\frac{a}{e} - c\right)^2 + \left(\frac{b}{e}\right)^2 \\
&= \frac{(a - ec)^2 + b^2}{e^2} \\
&= \frac{a^2 + b^2 - 2aec + e^2c^2}{e^2} \\
&= \frac{c^2 - 2c^2}{e^2} + c^2 \\
&= -\frac{c^2}{e^2} + c^2 \\
&= -a^2 + c^2
\end{aligned}$$
Thus $OD^2 + DF^2 = OF^2$, as required.

49.   (a)   Using the definition of the eccentricity e, we have:
$$e = \frac{c}{a} = \frac{\sqrt{a^2 + b^2}}{a} = \sqrt{\frac{a^2 + b^2}{a^2}} = \sqrt{1 + \frac{b^2}{a^2}}$$

(b) (i) Converting to standard form, we have $\frac{x^2}{1} - \frac{y^2}{0.0201} = 1$.

So $c^2 = 1 + 0.0201 = 1.0201$, thus $c = 1.01$. Then $e = \frac{c}{a} = 1.01$.

(ii) Converting to standard form, we have $\frac{x^2}{1} - \frac{y^2}{3} = 1$. So $c^2 = 1 + 3 = 4$, thus $c = 2$. Then $e = \frac{c}{a} = 2$.

(iii) Converting to standard form, we have $\frac{x^2}{1} - \frac{y^2}{8} = 1$. So $c^2 = 1 + 8 = 9$, thus $c = 3$. Then $e = \frac{c}{a} = 3$.

(iv) Converting to standard form, we have $\frac{x^2}{1} - \frac{y^2}{15} = 1$. So $c^2 = 1 + 15 = 16$, thus $c = 4$. Then $e = \frac{c}{a} = 4$.

(v) Converting to standard form, we have $\frac{x^2}{1} - \frac{y^2}{99} = 1$. So $c^2 = 1 + 99 = 100$, thus $c = 10$. Then $e = \frac{c}{a} = 10$.

(c) We draw the graphs:

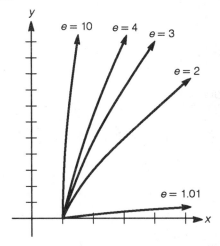

(d) The eccentricity affects the width of the hyperbola. That is, larger eccentricities result in a much wider curve than eccentricities closer to 1.

Section 7.2-7.4 TI-81 Graphing Calculator Exercises

1.   Entering $x^2 = 8y$ as $y = \frac{1}{8}x^2$ and using the ZOOM-6 settings, we have the parabola:

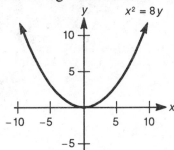

3.   Entering $y^2 = 8x$ as the two functions $y = \sqrt{8x}$ and $y = -\sqrt{8x}$, and using the ZOOM-6 settings, we have the parabola:

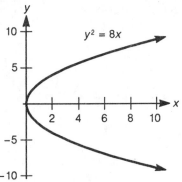

5.   (a)   Completing the square, we have:
$$4x + y^2 + 2y - 7 = 0$$
$$y^2 + 2y + 1 = -4x + 8$$
$$(y+1)^2 = -4(x-2)$$

(b)   Entering $(y+1)^2 = -4(x-2)$ as the two functions $y = -1 - \sqrt{-4(x-2)}$ and $y = -1 + \sqrt{-4(x-2)}$, and using the ZOOM-6 settings, we have the parabola:

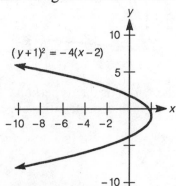

The graph is consistent with Figure 9.

7. Since the tangent line passes through the points $(4, 2)$ and $(0, -2)$, we can find the slope:
$$m = \frac{-2-2}{0-4} = \frac{-4}{-4} = 1$$
Since the y-intercept is -2, the equation of the tangent line is $y = x - 2$. Using the ZOOM-6 settings, we graph the parabola and tangent line:

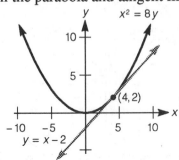

9. Since the tangent line passes through the points $(-3, 9)$ and $(0, -9)$, we can find the slope:
$$m = \frac{-9-9}{0-(-3)} = \frac{-18}{3} = -6$$
Since the y-intercept is -9, the equation of the tangent line is $y = -6x - 9$. Using the ZOOM-6 settings, we graph the parabola and tangent line:

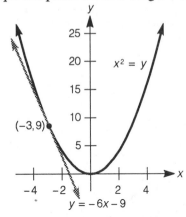

11. (a) Solving for y, we obtain:
$$\frac{x^2}{9} + \frac{y^2}{4} = 1$$
$$4x^2 + 9y^2 = 36$$
$$9y^2 = 36 - 4x^2$$
$$y^2 = \frac{36 - 4x^2}{9}$$
$$y = \pm\tfrac{1}{3}\sqrt{36 - 4x^2}$$

Graphing the two functions with the indicated settings:

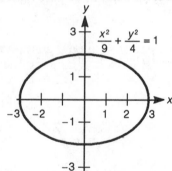

The graph appears to have symmetry with respect to the x-axis, the y-axis, and the origin.

(b)   Solving for y, we obtain:

$$\frac{x^2}{1} + \frac{y^2}{16} = 1$$
$$16x^2 + y^2 = 16$$
$$y^2 = 16 - 16x^2$$
$$y = \pm 4\sqrt{1 - x^2}$$

Graphing the two functions with the indicated settings:

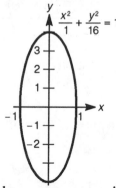

Again, the graph appears to have symmetry with respect to the x-axis, the y-axis, and the origin.

13. (a)   Dividing by $16b^2$, we have:

$$\frac{x^2}{16} + \frac{y^2}{b^2} = 1$$

So the horizontal axis has length $2(4) = 8$.

(b)   When $b^2 = 16$, then $c^2 = 16 - 16 = 0$, so $c = 0$ and thus:

$$e = \frac{c}{a} = \frac{0}{4} = 0$$

When $b^2 = 13.44$, then $c^2 = 16 - 13.44 = 2.56$, so $c = 1.6$ and thus:

$$e = \frac{c}{a} = \frac{1.6}{4} = 0.4$$

When $b^2 = 10.24$, then $c^2 = 16 - 10.24 = 5.76$, so $c = 2.4$ and thus:

$$e = \frac{c}{a} = \frac{2.4}{4} = 0.6$$

When $b^2 = 5.76$, then $c^2 = 16 - 5.76 = 10.24$, so $c = 3.2$ and thus:

$$e = \frac{c}{a} = \frac{3.2}{4} = 0.8$$

(c)   Solving for y, we have:

$$b^2 x^2 + 16y^2 = 16b^2$$
$$16y^2 = 16b^2 - b^2 x^2$$
$$y^2 = \frac{b^2 (16 - x^2)}{16}$$
$$y = \pm 0.25\sqrt{b^2 (16 - x^2)}$$

Using the indicated settings, the graphs of all four ellipses appear as:

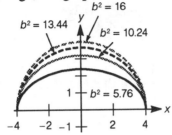

15.   Dividing the ellipse equation by 12, we have:

$$\frac{x^2}{4} + \frac{y^2}{12} = 1$$

So $a^2 = 4$ and $b^2 = 12$. Using $(x_0, y_0) = (1, -3)$ in the tangent line formula yields:

$$\frac{1x}{4} + \frac{-3y}{12} = 1$$
$$3x - 3y = 12$$
$$x - y = 4$$
$$y = x - 4$$

Now graph the ellipse $\frac{x^2}{4} + \frac{y^2}{12} = 1$ and tangent line $y = x - 4$:

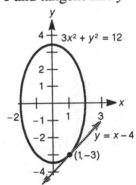

17. (a) Dividing by 12, we graph the ellipse $\frac{x^2}{12} + \frac{y^2}{4} = 1$:

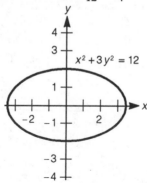

(b) Since $a^2 = 12$ and $b^2 = 4$, then $a = 2\sqrt{3}$ and $b = 2$. Therefore:
$c^2 = a^2 - b^2 = 12 - 4 = 8$, so $c = \sqrt{8} = 2\sqrt{2}$

(c) The auxiliary circle has an equation $x^2 + y^2 = 12$. We graph the ellipse and auxiliary circle:

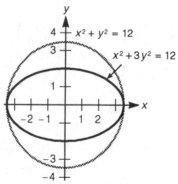

(d) We verify that the point $P(3, 1)$ lies on the ellipse:
$(3)^2 + 3(1)^2 = 9 + 3 = 12$
Using $(x_0, y_0) = (3, 1)$, we find the equation of the tangent line to the ellipse at P:
$$\frac{3x}{12} + \frac{1y}{4} = 1$$
$$\frac{x}{4} + \frac{y}{4} = 1$$
$$x + y = 4$$
Now we graph the upper halves of the ellipse and circle, and the line $y = -x + 4$:

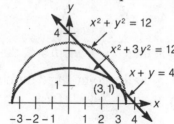

(e)   The perpendicular to the tangent at $P(3,1)$ has slope $m = 1$. Using the point $(-2\sqrt{2},0)$ and the point-slope formula, we have:
$$y - 0 = 1(x + 2\sqrt{2})$$
$$y = x + 2\sqrt{2}$$
Now we graph the upper halves of the ellipse and circle, as well as the lines $y = -x + 4$ and $y = x + 2\sqrt{2}$:

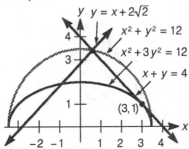

19. (a)   Solving for y, we have:
$$\frac{(y-3)^2}{25} - \frac{(x-4)^2}{9} = 1$$
$$9(y-3)^2 - 25(x-4)^2 = 225$$
$$9(y-3)^2 = 25(x-4)^2 + 225$$
$$(y-3)^2 = \tfrac{25}{9}(x-4)^2 + 25$$
$$y - 3 = \pm 5\sqrt{1 + \tfrac{1}{9}(x-4)^2}$$
$$y = 3 \pm 5\sqrt{1 + \tfrac{1}{9}(x-4)^2}$$
Entering these two functions and making the suggested range settings, we graph the hyperbola:

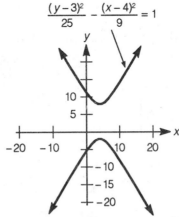

(b)   Since $a^2 = 25$ and $b^2 = 9$, then $a = 5$ and $b = 3$, thus the slopes of the asymptotes are $\pm\tfrac{5}{3}$. Using the point $(4,3)$ in the point-slope formula, we have:
$$y - 3 = \pm\tfrac{5}{3}(x-4), \text{ so } y = \pm\tfrac{5}{3}(x-4) + 3$$

Now we graph the hyperbola and the two asymptotes:

$$\frac{(y-3)^2}{25} - \frac{(x-4)^2}{9} = 1$$

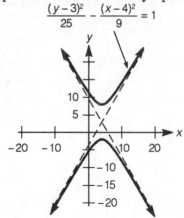

21. (a) Using the quadratic formula to solve $y^2 + xy + (x^2 - 3x - 1) = 0$, we have:

$$y = \frac{-x \pm \sqrt{x^2 - 4(1)(x^2 - 3x - 1)}}{2(1)}$$

$$= \frac{-x \pm \sqrt{x^2 - 4x^2 + 12x + 4}}{2}$$

$$= \frac{-x \pm \sqrt{-3x^2 + 12x + 4}}{2}$$

(b) Entering these two functions and making the suggested range settings, we have:

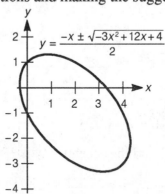

$$y = \frac{-x \pm \sqrt{-3x^2 + 12x + 4}}{2}$$

(c) The center appears to be approximately $(2, -1)$.

(d) Given $A = 1, B = 1, C = 1, D = -3, E = 0$ and $F = -1$, we compute the center using the given formulas:

$$h = \frac{2CD - BE}{B^2 - 4AC} = \frac{2(1)(-3) - (1)(0)}{(1)^2 - 4(1)(1)} = \frac{-6}{-3} = 2$$

$$k = \frac{2AE - BD}{B^2 - 4AC} = \frac{2(1)(0) - (1)(-3)}{(1)^2 - 4(1)(1)} = \frac{3}{-3} = -1$$

The center is $(2, -1)$, agreeing with our estimate from part (c).

7.5 The Focus-Directrix Property of Conics

1. In order to use the formulas for the focal radii, we must find a and e. Dividing by 76, the standard form is $\frac{x^2}{76} + \frac{y^2}{76/3} = 1$. So $a = \sqrt{76} = 2\sqrt{19}$. To find the eccentricity, we first find c:

$$c = \sqrt{a^2 - b^2} = \sqrt{76 - \frac{76}{3}} = \sqrt{\frac{152}{3}} = \frac{2\sqrt{38}}{\sqrt{3}} = \frac{2\sqrt{114}}{3}$$

So the eccentricity is:

$$e = \frac{c}{a} = \frac{\frac{2\sqrt{114}}{3}}{2\sqrt{19}} = \frac{\sqrt{6}}{3}$$

So the focal radii are given by:

$$F_1 P = a + ex = 2\sqrt{19} + \frac{\sqrt{6}}{3}(-8) = 2\sqrt{19} - \frac{8\sqrt{6}}{3} = \frac{6\sqrt{19} - 8\sqrt{6}}{3}$$

$$F_2 P = a - ex = 2\sqrt{19} - \frac{\sqrt{6}}{3}(-8) = 2\sqrt{19} + \frac{8\sqrt{6}}{3} = \frac{6\sqrt{19} + 8\sqrt{6}}{3}$$

3. The equation is already in standard form with $a = 15$ and $b = 5$. We find c:

$$c = \sqrt{a^2 - b^2} = \sqrt{225 - 25} = \sqrt{200} = 10\sqrt{2}$$

So the eccentricity is:

$$e = \frac{c}{a} = \frac{10\sqrt{2}}{15} = \frac{2\sqrt{2}}{3}$$

So the focal radii are given by:

$$F_1 P = a + ex = 15 + \frac{2\sqrt{2}}{3}(9) = 15 + 6\sqrt{2}$$

$$F_2 P = a - ex = 15 - \frac{2\sqrt{2}}{3}(9) = 15 - 6\sqrt{2}$$

5. (a) The ellipse is already in standard form with $a = 4$ and $b = 3$. We find c:

$$c = \sqrt{a^2 - b^2} = \sqrt{16 - 9} = \sqrt{7}$$

So the foci are $(\pm\sqrt{7}, 0)$ and the eccentricity is $\frac{\sqrt{7}}{4}$. The directricies are given by:

$$x = \pm\frac{a}{e} = \pm\frac{4}{\frac{\sqrt{7}}{4}} = \pm\frac{16}{\sqrt{7}} = \pm\frac{16\sqrt{7}}{7}$$

(b) The hyperbola is already in standard form with $a = 4$ and $b = 3$. We find c:

$$c = \sqrt{a^2 + b^2} = \sqrt{16 + 9} = \sqrt{25} = 5$$

So the foci are $(\pm 5, 0)$ and the eccentricity is $\frac{5}{4}$. The directricies are given by:

$$x = \pm\frac{a}{e} = \pm\frac{4}{\frac{5}{4}} = \pm\frac{16}{5}$$

7. (a) Dividing by 156, the standard form for the ellipse is $\frac{x^2}{13} + \frac{y^2}{12} = 1$, and so $a = \sqrt{13}$ and $b = \sqrt{12} = 2\sqrt{3}$. We find c:

$$c = \sqrt{a^2 - b^2} = \sqrt{13 - 12} = \sqrt{1} = 1$$

So the foci are $(\pm 1, 0)$ and the eccentricity is $\frac{1}{\sqrt{13}} = \frac{\sqrt{13}}{13}$. The directricies are given by:

$$x = \pm \frac{a}{e} = \pm \frac{\sqrt{13}}{\frac{\sqrt{13}}{13}} = \pm 13$$

(b) Dividing by 156, the standard form for the hyperbola is $\frac{x^2}{13} - \frac{y^2}{12} = 1$, and so $a = \sqrt{13}$ and $b = \sqrt{12} = 2\sqrt{3}$. We find c:

$$c = \sqrt{a^2 + b^2} = \sqrt{13 + 12} = \sqrt{25} = 5$$

So the foci are $(\pm 5, 0)$ and the eccentricity is $\frac{5}{\sqrt{13}} = \frac{5\sqrt{13}}{13}$. The directricies are given by:

$$x = \pm \frac{a}{e} = \pm \frac{\sqrt{13}}{\frac{5\sqrt{13}}{13}} = \pm \frac{13}{5}$$

9. (a) Dividing by 900, the standard form for the ellipse is $\frac{x^2}{36} + \frac{y^2}{25} = 1$, and so $a = 6$ and $b = 5$. We find c:

$$c = \sqrt{a^2 - b^2} = \sqrt{36 - 25} = \sqrt{11}$$

So the foci are $(\pm\sqrt{11}, 0)$ and the eccentricity is $\frac{\sqrt{11}}{6}$. The directricies are given by:

$$x = \pm \frac{a}{e} = \pm \frac{6}{\frac{\sqrt{11}}{6}} = \pm \frac{36}{\sqrt{11}} = \pm \frac{36\sqrt{11}}{11}$$

(b) Dividing by 900, the standard form for the hyperbola is $\frac{x^2}{36} - \frac{y^2}{25} = 1$, and so $a = 6$ and $b = 5$. We find c:

$$c = \sqrt{a^2 + b^2} = \sqrt{36 + 25} = \sqrt{61}$$

So the foci are $(\pm\sqrt{61}, 0)$ and the eccentricity is $\frac{\sqrt{61}}{6}$. The directricies are given by:

$$x = \pm \frac{a}{e} = \pm \frac{6}{\frac{\sqrt{61}}{6}} = \pm \frac{36}{\sqrt{61}} = \pm \frac{36\sqrt{61}}{61}$$

11. Since the foci are $(\pm 1, 0)$ then $c = 1$. Since the directricies are $x = \pm 4$, then $\frac{a}{e} = 4$, so

$a = 4e$. But since $e = \frac{c}{a} = \frac{1}{a}$, we have:

$a = 4 \bullet \frac{1}{a}$

$a^2 = 4$

$a = 2$ (since $a > 0$)

Since $a^2 - b^2 = c^2$, we can find b^2:

$2^2 - b^2 = 1^2$

$-b^2 = -3$

$b^2 = 3$

So the equation of the ellipse is $\frac{x^2}{4} + \frac{y^2}{3} = 1$, or $3x^2 + 4y^2 = 12$.

13. Since the foci are $(\pm 2, 0)$ then $c = 2$. Since the directricies are $x = \pm 1$, then $\frac{a}{e} = 1$, so

$a = e$. But since $e = \frac{c}{a} = \frac{2}{a}$, we have:

$a = \frac{2}{a}$

$a^2 = 2$

Since $a^2 + b^2 = c^2$, we can find b^2:

$2 + b^2 = 2^2$

$b^2 = 2$

So the equation of the hyperbola is $\frac{x^2}{2} - \frac{y^2}{2} = 1$, or $x^2 - y^2 = 2$.

15. (a) By the distance formula, we have:

$d_1 = \sqrt{(x+c)^2 + (y-0)^2} = \sqrt{(x+c)^2 + y^2}$

$d_2 = \sqrt{(x-c)^2 + (y-0)^2} = \sqrt{(x-c)^2 + y^2}$

Squaring, we have:

$d_1^2 = (x+c)^2 + y^2$

$d_2^2 = (x-c)^2 + y^2$

(b) Working from the left-hand side, we have:

$d_1^2 - d_2^2 = (x+c)^2 - (x-c)^2$

$= x^2 + 2cx + c^2 - x^2 + 2cx - c^2$

$= 4cx$

(c) Since d_1 and d_2 represent the distances from the foci to a point on the ellipse, then $d_1 + d_2 = 2a$ by the definition of an ellipse.

(d) Factoring, we have:

$$d_1{}^2 - d_2{}^2 = 4cx$$
$$(d_1 + d_2)(d_1 - d_2) = 4cx$$
$$2a(d_1 - d_2) = 4cx$$
$$d_1 - d_2 = \frac{2cx}{a}$$

(e) Adding the two equations, we have:

$$2d_1 = 2a + \frac{2cx}{a}$$
$$d_1 = a + \frac{c}{a}x$$
$$d_1 = a + ex$$

(f) Substituting the result from (e), we have:

$$a + ex + d_2 = 2a$$
$$d_2 = a - ex$$

7.6 Rotation of Axes

1. For x, we have:

$$x = x'\cos\theta - y'\sin\theta$$
$$= \sqrt{3}\cos 30° - 2\sin 30°$$
$$= \sqrt{3} \bullet \tfrac{\sqrt{3}}{2} - 2 \bullet \tfrac{1}{2}$$
$$= \tfrac{3}{2} - 1$$
$$= \tfrac{1}{2}$$

For y, we have:

$$y = x'\sin\theta + y'\cos\theta$$
$$= \sqrt{3}\sin 30° + 2\cos 30°$$
$$= \sqrt{3} \bullet \tfrac{1}{2} + 2 \bullet \tfrac{\sqrt{3}}{2}$$
$$= \tfrac{\sqrt{3}}{2} + \sqrt{3}$$
$$= \tfrac{3\sqrt{3}}{2}$$

So the coordinates in the x-y system are $\left(\tfrac{1}{2}, \tfrac{3\sqrt{3}}{2}\right)$.

3. For x, we have:
$$x = x'\cos\theta - y'\sin\theta$$
$$= \sqrt{2}\cos 45° + \sqrt{2}\sin 45°$$
$$= \sqrt{2}\cdot\tfrac{1}{\sqrt{2}} + \sqrt{2}\cdot\tfrac{1}{\sqrt{2}}$$
$$= 1+1$$
$$= 2$$
For y, we have:
$$y = x'\sin\theta + y'\cos\theta$$
$$= \sqrt{2}\sin 45° - \sqrt{2}\cos 45°$$
$$= \sqrt{2}\cdot\tfrac{1}{\sqrt{2}} - \sqrt{2}\cdot\tfrac{1}{\sqrt{2}}$$
$$= 1-1$$
$$= 0$$
So the coordinates in the x-y system are $(2,0)$.

5. For x', we have:
$$x' = x\cos\theta + y\sin\theta$$
$$= -3\cos\left[\sin^{-1}\left(\tfrac{5}{13}\right)\right] + 1\sin\left[\sin^{-1}\left(\tfrac{5}{13}\right)\right]$$
$$= -3\cdot\tfrac{12}{13} + 1\cdot\tfrac{5}{13}$$
$$= -\tfrac{31}{13}$$
For y', we have:
$$y' = -x\sin\theta + y\cos\theta$$
$$= 3\sin\left[\sin^{-1}\left(\tfrac{5}{13}\right)\right] + 1\cos\left[\sin^{-1}\left(\tfrac{5}{13}\right)\right]$$
$$= 3\cdot\tfrac{5}{13} + 1\cdot\tfrac{12}{13}$$
$$= \tfrac{27}{13}$$

So the coordinates in the x'-y' system are $\left(-\tfrac{31}{13}, \tfrac{27}{13}\right)$.

7. We have:
$$\cot 2\theta = \frac{A-C}{B} = \frac{25-18}{-24} = -\frac{7}{24}, \text{ so } \tan 2\theta = -\frac{24}{7}$$
$$\sec^2 2\theta = 1 + \tan^2 2\theta = 1 + \left(-\frac{24}{7}\right)^2 = \frac{625}{49}$$

So $\sec 2\theta = -\tfrac{25}{7}$ (second quadrant, since $\cot 2\theta < 0$), and thus $\cos 2\theta = -\tfrac{7}{25}$. Since θ is in the first quadrant, we have:
$$\sin\theta = \sqrt{\frac{1-\cos 2\theta}{2}} = \sqrt{\frac{1+\frac{7}{25}}{2}} = \sqrt{\frac{16}{25}} = \frac{4}{5}$$
$$\cos\theta = \sqrt{\frac{1+\cos 2\theta}{2}} = \sqrt{\frac{1-\frac{7}{25}}{2}} = \sqrt{\frac{9}{25}} = \frac{3}{5}$$

9.   We have:

$$\cot 2\theta = \frac{1-8}{-24} = \frac{-7}{-24} = \frac{7}{24}, \text{ so } \tan 2\theta = \frac{24}{7}$$

$$\sec^2 2\theta = 1 + \tan^2 2\theta = 1 + \left(\frac{24}{7}\right)^2 = \frac{625}{49}$$

So $\sec 2\theta = \frac{25}{7}$ (first quadrant, since $\cot 2\theta > 0$), and thus $\cos 2\theta = \frac{7}{25}$. Since θ is in the first quadrant, we have:

$$\sin\theta = \sqrt{\frac{1-\cos 2\theta}{2}} = \sqrt{\frac{1-\frac{7}{25}}{2}} = \sqrt{\frac{9}{25}} = \frac{3}{5}$$

$$\cos\theta = \sqrt{\frac{1+\cos 2\theta}{2}} = \sqrt{\frac{1+\frac{7}{25}}{2}} = \sqrt{\frac{16}{25}} = \frac{4}{5}$$

11.   We have:

$$\cot 2\theta = \frac{A-C}{B} = \frac{1-(-1)}{-2\sqrt{3}} = -\frac{1}{\sqrt{3}}, \text{ so } \tan 2\theta = -\sqrt{3}$$

Therefore $2\theta = 120°$, and thus $\theta = 60°$. So:

$$\sin\theta = \sin 60° = \frac{\sqrt{3}}{2}$$

$$\cos\theta = \cos 60° = \frac{1}{2}$$

13.   We have:

$$\cot 2\theta = \frac{A-C}{B} = \frac{0-(-240)}{161} = \frac{240}{161}, \text{ so } \tan 2\theta = \frac{161}{240}$$

$$\sec^2 2\theta = 1 + \tan^2 2\theta = 1 + \left(\frac{161}{240}\right)^2 = \frac{83521}{57600}$$

So $\sec 2\theta = \frac{289}{240}$ (first quadrant, since $\cot 2\theta > 0$), and thus $\cos 2\theta = \frac{240}{289}$. Since θ is in the first quadrant, we have:

$$\sin\theta = \sqrt{\frac{1-\cos 2\theta}{2}} = \sqrt{\frac{1-\frac{240}{289}}{2}} = \sqrt{\frac{49}{578}} = \frac{7\sqrt{2}}{34}$$

$$\cos\theta = \sqrt{\frac{1+\cos 2\theta}{2}} = \sqrt{\frac{1+\frac{240}{289}}{2}} = \sqrt{\frac{529}{578}} = \frac{23\sqrt{2}}{34}$$

15.   Using the rotation equations:

$$x = x'\cos\theta - y'\sin\theta = x'\cos 45° - y'\sin 45° = \frac{\sqrt{2}}{2}x' - \frac{\sqrt{2}}{2}y'$$

$$y = x'\sin\theta + y'\cos\theta = x'\sin 45° + y'\cos 45° = \frac{\sqrt{2}}{2}x' + \frac{\sqrt{2}}{2}y'$$

So the equation $2xy = 9$ becomes:

$$2\left(\frac{\sqrt{2}}{2}x' - \frac{\sqrt{2}}{2}y'\right)\left(\frac{\sqrt{2}}{2}x' + \frac{\sqrt{2}}{2}y'\right) = 9$$

$$2\left(\frac{1}{2}x'^2 - \frac{1}{2}y'^2\right) = 9$$

$$x'^2 - y'^2 = 9$$

We graph the equation:

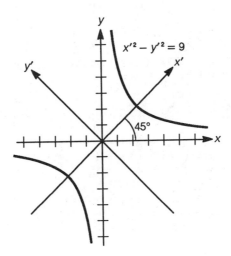

$x'^2 - y'^2 = 9$

45°

17. We find:

$$\cot 2\theta = \frac{7-1}{8} = \frac{3}{4}, \text{ so } \tan 2\theta = \frac{4}{3}$$

$$\sec^2 2\theta = 1 + \tan^2 2\theta = 1 + \left(\frac{4}{3}\right)^2 = \frac{25}{9}, \text{ so } \sec 2\theta = \frac{5}{3} \quad (\text{since } 2\theta < 90°)$$

Thus $\cos 2\theta = \frac{3}{5}$, and therefore:

$$\sin\theta = \sqrt{\frac{1-\cos 2\theta}{2}} = \sqrt{\frac{1-\frac{3}{5}}{2}} = \sqrt{\frac{1}{5}} = \frac{\sqrt{5}}{5}$$

$$\cos\theta = \sqrt{\frac{1+\cos 2\theta}{2}} = \sqrt{\frac{1+\frac{3}{5}}{2}} = \sqrt{\frac{4}{5}} = \frac{2\sqrt{5}}{5}$$

Thus $\theta = \sin^{-1}\left(\frac{\sqrt{5}}{5}\right) \approx 26.6°$. Now:

$$x = x'\cos\theta - y'\sin\theta = \frac{2\sqrt{5}}{5}x' - \frac{\sqrt{5}}{5}y'$$

$$y = x'\sin\theta + y'\cos\theta = \frac{\sqrt{5}}{5}x' + \frac{2\sqrt{5}}{5}y'$$

Making the substitutions into $7x^2 + 8xy + y^2 - 1 = 0$, we have:

$$7\left(\frac{4}{5}x'^2 - \frac{4}{5}x'y' + \frac{1}{5}y'^2\right) + 8\left(\frac{2}{5}x'^2 + \frac{3}{5}x'y' - \frac{2}{5}y'^2\right) + \left(\frac{1}{5}x'^2 + \frac{4}{5}x'y' + \frac{4}{5}y'^2\right) - 1 = 0$$

$$\frac{28}{5}x'^2 - \frac{28}{5}x'y' + \frac{7}{5}y'^2 + \frac{16}{5}x'^2 + \frac{24}{5}x'y' - \frac{16}{5}y'^2 + \frac{1}{5}x'^2 + \frac{4}{5}x'y' + \frac{4}{5}y'^2 - 1 = 0$$

$$9x'^2 - y'^2 = 1$$

$$\frac{x'^2}{\frac{1}{9}} - \frac{y'^2}{1} = 1$$

So, rotating $\theta = 26.6°$, we sketch the hyperbola:

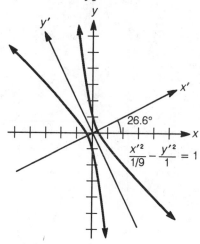

19. We can use an alternate approach here:

$$x^2 + 4xy + 4y^2 = 1$$
$$(x + 2y)^2 = 1$$

$$x + 2y = 1 \qquad\qquad \text{or} \qquad x + 2y = -1$$
$$y = -\tfrac{1}{2}x + \tfrac{1}{2} \qquad\qquad\qquad y = -\tfrac{1}{2}x - \tfrac{1}{2}$$

So the graph consists of two lines:

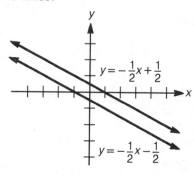

21. We find:

$$\cot 2\theta = \frac{9 - 16}{-24} = \frac{7}{24}, \text{ so } \tan 2\theta = \frac{24}{7}$$

$$\sec^2 2\theta = 1 + \tan^2 2\theta = 1 + \left(\frac{24}{7}\right)^2 = \frac{625}{49}, \text{ so } \sec 2\theta = \frac{25}{7} \quad \text{(since } 2\theta < 90°)$$

Thus $\cos 2\theta = \frac{7}{25}$, and therefore:

$$\sin \theta = \sqrt{\frac{1 - \cos 2\theta}{2}} = \sqrt{\frac{1 - \frac{7}{25}}{2}} = \sqrt{\frac{9}{25}} = \frac{3}{5}$$

$$\cos \theta = \sqrt{\frac{1 + \cos 2\theta}{2}} = \sqrt{\frac{1 + \frac{7}{25}}{2}} = \sqrt{\frac{16}{25}} = \frac{4}{5}$$

Thus $\theta = \sin^{-1}\left(\frac{3}{5}\right) \approx 36.9°$. Now:

$$x = x'\cos\theta - y'\sin\theta = \tfrac{4}{5}x' - \tfrac{3}{5}y'$$
$$y = x'\sin\theta + y'\cos\theta = \tfrac{3}{5}x' + \tfrac{4}{5}y'$$

Making the substitutions into $9x^2 - 24xy + 16y^2 - 400x - 300y = 0$ and collecting like terms, we have:

$$25y'^2 - 500x' = 0$$
$$y'^2 = 20x'$$

So, rotating 36.9°, we sketch the parabola:

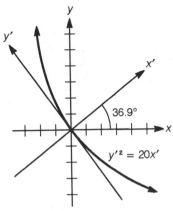

23. We find:

$$\cot 2\theta = \frac{0-3}{4} = -\frac{3}{4}, \text{ so } \tan 2\theta = -\frac{4}{3}$$

$$\sec^2 2\theta = 1 + \tan^2 2\theta = 1 + \left(-\frac{4}{3}\right)^2 = \frac{25}{9}$$

$$\sec 2\theta = -\frac{5}{3} \quad (\text{since } 2\theta > 90°)$$

Thus $\cos 2\theta = -\frac{3}{5}$, and therefore:

$$\sin\theta = \sqrt{\frac{1-\cos 2\theta}{2}} = \sqrt{\frac{1+\frac{3}{5}}{2}} = \sqrt{\frac{4}{5}} = \frac{2\sqrt{5}}{5}$$

$$\cos\theta = \sqrt{\frac{1+\cos 2\theta}{2}} = \sqrt{\frac{1-\frac{3}{5}}{2}} = \sqrt{\frac{1}{5}} = \frac{\sqrt{5}}{5}$$

Thus $\theta = \cos^{-1}\left(\frac{\sqrt{5}}{5}\right) \approx 63.4°$. Now:

$$x = x'\cos\theta - y'\sin\theta = \frac{\sqrt{5}}{5}x' - \frac{2\sqrt{5}}{5}y'$$
$$y = x'\sin\theta + y'\cos\theta = \frac{2\sqrt{5}}{5}x' + \frac{\sqrt{5}}{5}y'$$

Making the substitutions into $4xy + 3y^2 + 4x + 6y = 1$ and completing the square on x' and y' terms, we have:

$$\frac{\left(x' + \frac{2\sqrt{5}}{5}\right)^2}{1} - \frac{\left(y' + \frac{\sqrt{5}}{5}\right)^2}{4} = 1$$

So, rotating $63.4°$, we sketch the hyperbola:

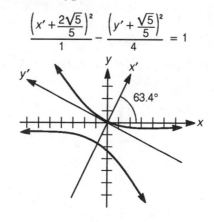

$$\frac{\left(x' + \frac{2\sqrt{5}}{5}\right)^2}{1} - \frac{\left(y' + \frac{\sqrt{5}}{5}\right)^2}{4} = 1$$

25. We find:

$$\cot 2\theta = \frac{3-3}{-2} = 0, \text{ so } 2\theta = 90° \text{ and thus } \theta = 45°$$

Therefore:

$$x = x'\cos\theta - y'\sin\theta = x'\cos 45° - y'\sin 45° = \frac{\sqrt{2}}{2}x' - \frac{\sqrt{2}}{2}y'$$

$$y = x'\sin\theta + y'\cos\theta = x'\sin 45° + y'\cos 45° = \frac{\sqrt{2}}{2}x' + \frac{\sqrt{2}}{2}y'$$

Making the substitutions into $3x^2 - 2xy + 3y^2 - 6\sqrt{2}x + 2\sqrt{2}y + 4 = 0$ and completing the square on x' and y' terms, we have:

$$\frac{(x'-1)^2}{1} + \frac{(y'+1)^2}{\frac{1}{2}} = 1$$

So, rotating $45°$, we sketch the ellipse:

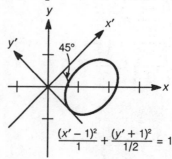

$$\frac{(x'-1)^2}{1} + \frac{(y'+1)^2}{1/2} = 1$$

27. We first multiply out, to get:
$$x^2 - 2xy + y^2 = 8y - 48$$
$$x^2 - 2xy + y^2 - 8y + 48 = 0$$
Now:
$$\cot 2\theta = \frac{1-1}{-2} = 0, \text{ so } 2\theta = 90° \text{ and thus } \theta = 45°$$
Therefore:
$$x = x'\cos\theta - y'\sin\theta = \tfrac{\sqrt{2}}{2}x' - \tfrac{\sqrt{2}}{2}y'$$
$$y = x'\sin\theta + y'\cos\theta = \tfrac{\sqrt{2}}{2}x' + \tfrac{\sqrt{2}}{2}y'$$
Making the substitutions and completing the square on x' and y' terms yields:
$$\left(y' - \sqrt{2}\right)^2 = 2\sqrt{2}\left(x' - \tfrac{11\sqrt{2}}{2}\right)$$
So, rotating $45°$, we sketch the parabola:

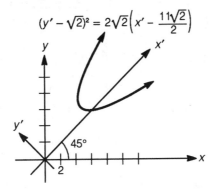

29. We find:
$$\cot 2\theta = \frac{3-6}{4} = -\frac{3}{4}, \text{ so } \tan 2\theta = -\frac{4}{3}$$
$$\sec^2 2\theta = 1 + \tan^2 2\theta = 1 + \left(-\frac{4}{3}\right)^2 = \frac{25}{9}$$

So $\sec 2\theta = -\frac{5}{3}$ (since $2\theta > 90°$), and thus $\cos 2\theta = -\frac{3}{5}$.
Therefore:
$$\sin\theta = \sqrt{\frac{1-\cos 2\theta}{2}} = \sqrt{\frac{1+\frac{3}{5}}{2}} = \frac{2\sqrt{5}}{5}$$
$$\cos\theta = \sqrt{\frac{1+\cos 2\theta}{2}} = \sqrt{\frac{1-\frac{3}{5}}{2}} = \frac{\sqrt{5}}{5}$$

Thus $\theta = \cos^{-1}\left(\frac{\sqrt{5}}{5}\right) \approx 63.4°$. Now:
$$x = x'\cos\theta - y'\sin\theta = \tfrac{\sqrt{5}}{5}x' - \tfrac{2\sqrt{5}}{5}y'$$
$$y = x'\sin\theta + y'\cos\theta = \tfrac{2\sqrt{5}}{5}x' + \tfrac{\sqrt{5}}{5}y'$$

Making the substitutions into $3x^2 + 4xy + 6y^2 = 7$, we have:

$$\frac{x'^2}{1} + \frac{y'^2}{\frac{7}{2}} = 1$$

So, rotating $63.4°$, we sketch the ellipse:

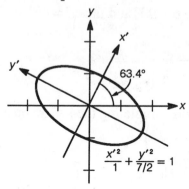

31. We find:

$$\cot 2\theta = \frac{17-8}{-12} = -\frac{3}{4}, \text{ so } \tan 2\theta = -\frac{4}{3}$$

As with Exercise 29 we find $\sin \theta = \frac{2\sqrt{5}}{5}$ and $\cos \theta = \frac{\sqrt{5}}{5}$, so:

$$x = \frac{\sqrt{5}}{5}x' - \frac{2\sqrt{5}}{5}y'$$

$$y = \frac{2\sqrt{5}}{5}x' + \frac{\sqrt{5}}{5}y'$$

Substituting into $17x^2 - 12xy + 8y^2 - 80 = 0$ yields:

$$\frac{x'^2}{16} + \frac{y'^2}{4} = 1$$

So, rotating $63.4°$, we sketch the ellipse:

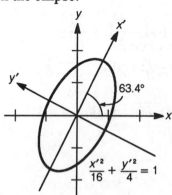

33. We find:

$$\cot 2\theta = \frac{0+4}{3} = \frac{4}{3}, \text{ so } \tan 2\theta = \frac{3}{4}$$

$$\sec^2 2\theta = 1 + \tan^2 2\theta = 1 + \frac{9}{16} = \frac{25}{16}, \text{ so } \sec 2\theta = \frac{5}{4}$$

Then $\cos 2\theta = \frac{4}{5}$, and thus:

$$\sin\theta = \sqrt{\frac{1-\cos 2\theta}{2}} = \sqrt{\frac{1-\frac{4}{5}}{2}} = \frac{\sqrt{10}}{10}$$

$$\cos\theta = \sqrt{\frac{1+\cos 2\theta}{2}} = \sqrt{\frac{1+\frac{4}{5}}{2}} = \frac{3\sqrt{10}}{10}$$

Then $\theta = \sin^{-1}\left(\frac{\sqrt{10}}{10}\right) \approx 18.4°$, and:

$$x = \frac{3\sqrt{10}}{10}x' - \frac{\sqrt{10}}{10}y'$$
$$y = \frac{\sqrt{10}}{10}x' + \frac{3\sqrt{10}}{10}y'$$

Substituting into $3xy - 4y^2 + 18 = 0$ results in:

$$\frac{y'^2}{4} - \frac{x'^2}{36} = 1$$

So, rotating $18.4°$, we sketch the hyperbola:

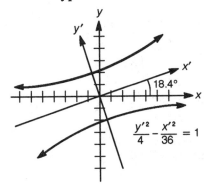

35. We first multiply out terms to obtain:

$$x^2 + 2xy + y^2 + 4\sqrt{2}x - 4\sqrt{2}y = 0$$

We find:

$$\cot 2\theta = \frac{1-1}{2} = 0, \text{ so } 2\theta = 90° \text{ and thus } \theta = 45°$$

Now:

$$x = \frac{\sqrt{2}}{2}x' - \frac{\sqrt{2}}{2}y'$$
$$y = \frac{\sqrt{2}}{2}x' + \frac{\sqrt{2}}{2}y'$$

Substituting into $x^2 + 2xy + y^2 + 4\sqrt{2}x - 4\sqrt{2}y = 0$ results in:

$$x'^2 = 4y'$$

So, rotating 45°, we sketch the parabola:

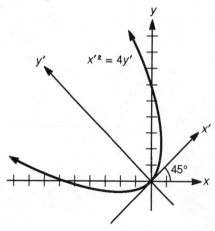

37. We find:

$$\cot 2\theta = \frac{3-2}{-\sqrt{15}} = -\frac{1}{\sqrt{15}}, \text{ so } \tan 2\theta = -\sqrt{15}$$

$$\sec^2 2\theta = 1 + \tan^2 2\theta = 1 + 15 = 16, \text{ so } \sec 2\theta = -4 \quad (\text{since } 2\theta > 90°)$$

Thus $\cos 2\theta = -\frac{1}{4}$ and we find:

$$\sin\theta = \sqrt{\frac{1-\cos 2\theta}{2}} = \sqrt{\frac{1+\frac{1}{4}}{2}} = \sqrt{\frac{5}{8}} = \frac{\sqrt{10}}{4}$$

$$\cos\theta = \sqrt{\frac{1+\cos 2\theta}{2}} = \sqrt{\frac{1-\frac{1}{4}}{2}} = \sqrt{\frac{3}{8}} = \frac{\sqrt{6}}{4}$$

Then $\theta = \cos^{-1}\left(\frac{\sqrt{6}}{4}\right) \approx 52.2°$, and:

$$x = \frac{\sqrt{6}}{4}x' - \frac{\sqrt{10}}{4}y'$$
$$y = \frac{\sqrt{10}}{4}x' + \frac{\sqrt{6}}{4}y'$$

Substituting into $3x^2 - \sqrt{15}xy + 2y^2 = 3$ results in:

$$\frac{x'^2}{6} + \frac{y'^2}{\frac{2}{3}} = 1$$

So, rotating 52.2°, we sketch the ellipse:

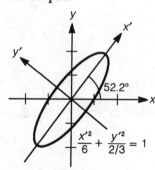

39. We find:
$$\cot 2\theta = \frac{3-3}{-2} = 0, \text{ so } 2\theta = 90° \text{ and thus } \theta = 45°$$
Now:
$$x = \tfrac{\sqrt{2}}{2}x' - \tfrac{\sqrt{2}}{2}y'$$
$$y = \tfrac{\sqrt{2}}{2}x' + \tfrac{\sqrt{2}}{2}y'$$
Substituting into $3x^2 - 2xy + 3y^2 + 2 = 0$ results in:
$$x'^2 + 2y'^2 = -1$$
But clearly this is impossible, so there is no graph.

41. Multiplying the first equation by $\sin\theta$ and the second equation by $\cos\theta$ yields:
$$(\sin\theta\cos\theta)x + (\sin^2\theta)y = x'\sin\theta$$
$$(-\sin\theta\cos\theta)x + (\cos^2\theta)y = y'\cos\theta$$
Adding, we get:
$$(\sin^2\theta + \cos^2\theta)y = x'\sin\theta + y'\cos\theta$$
$$y = x'\sin\theta + y'\cos\theta$$
Multiplying the first equation by $\cos\theta$ and the second equation by $-\sin\theta$ yields:
$$(\cos^2\theta)x + (\sin\theta\cos\theta)y = x'\cos\theta$$
$$(\sin^2\theta)x - (\sin\theta\cos\theta)y = -y'\sin\theta$$
Adding, we get:
$$(\sin^2\theta + \cos^2\theta)x = x'\cos\theta - y'\sin\theta$$
$$x = x'\cos\theta - y'\sin\theta$$

43. (a)   Since $\cot 2\theta = -\frac{7}{24}$, $\tan 2\theta = -\frac{24}{7}$ so:
$$\sec^2 2\theta = 1 + \left(\frac{24}{7}\right)^2 = \frac{49}{49} + \frac{576}{49} = \frac{625}{49}$$

Thus $\sec 2\theta = \pm\frac{25}{7}$. Since 2θ is in either the first or second quadrant, and $\cot 2\theta < 0$, then 2θ is in the second quadrant, so $\sec 2\theta = -\frac{25}{7}$ and thus $\cos 2\theta = -\frac{7}{25}$.

(b)   We compute $\sin\theta$ and $\cos\theta$:
$$\sin\theta = \sqrt{\frac{1-\cos 2\theta}{2}} = \sqrt{\frac{1+\frac{7}{25}}{2}} = \sqrt{\frac{32}{50}} = \sqrt{\frac{16}{25}} = \frac{4}{5}$$
$$\cos\theta = \sqrt{\frac{1+\cos 2\theta}{2}} = \sqrt{\frac{1-\frac{7}{25}}{2}} = \sqrt{\frac{18}{50}} = \sqrt{\frac{9}{25}} = \frac{3}{5}$$

(c)   Making the substitutions and simplifying, we have:

$$16\left(\tfrac{1}{5}(3x'-4y')\right)^2 - 24\left(\tfrac{1}{5}(3x'-4y')\right)\left(\tfrac{1}{5}(4x'+3y')\right)$$

$$+9\left(\tfrac{1}{5}(4x'+3y')\right)^2 +110\left(\tfrac{1}{5}(3x'-4y')\right)-20\bullet\tfrac{1}{5}(4x'+3y')+100 = 0$$

$$\tfrac{16}{25}\left(9x'^2 -24x'y'+16y'^2\right)-\tfrac{24}{25}\left(12x'^2 -7x'y'-12y'^2\right)$$

$$+\tfrac{9}{25}\left(16x'^2 +24x'y'+9y'^2\right)+22(3x'-4y')-4(4x'+3y')+100 = 0$$

$$\frac{144-288+144}{25}x'^2 +\frac{-384+168+216}{25}x'y'+\frac{256+288+81}{25}y'^2$$

$$+(66-16)x'+(-88-12)y'+100 = 0$$

$$25y'^2 +50x'-100y'+100 = 0$$

$$y'^2 +2x'-4y'+4 = 0$$

45.   Working from the right-hand side, we have:
$$A'+C' = A\cos^2\theta + B\sin\theta\cos\theta + C\sin^2\theta + A\sin^2\theta - B\sin\theta\cos\theta + C\cos^2\theta$$
$$= (A+C)\left(\cos^2\theta + \sin^2\theta\right)$$
$$= A+C$$

7.7 Introduction to Polar Coordinates

1.   (a)   Using $x = r\cos\theta$ and $y = r\sin\theta$, we have:
$$x = 3\cos\tfrac{2\pi}{3} = 3\left(-\tfrac{1}{2}\right)=-\tfrac{3}{2}$$
$$y = 3\sin\tfrac{2\pi}{3} = 3\left(\tfrac{\sqrt{3}}{2}\right)=\tfrac{3\sqrt{3}}{2}$$

So the rectangular coordinates are $\left(-\tfrac{3}{2},\tfrac{3\sqrt{3}}{2}\right)$.

(b)   Using $x = r\cos\theta$ and $y = r\sin\theta$, we have:
$$x = 4\cos\tfrac{11\pi}{6} = 4\left(\tfrac{\sqrt{3}}{2}\right)=2\sqrt{3}$$
$$y = 4\sin\tfrac{11\pi}{6} = 4\left(-\tfrac{1}{2}\right)=-2$$

So the rectangular coordinates are $\left(2\sqrt{3},-2\right)$.

(c)   Using $x = r\cos\theta$ and $y = r\sin\theta$, we have:
$$x = 4\cos\left(-\tfrac{\pi}{6}\right)= 4\left(\tfrac{\sqrt{3}}{2}\right)=2\sqrt{3}$$
$$y = 4\sin\left(-\tfrac{\pi}{6}\right)= 4\left(-\tfrac{1}{2}\right)=-2$$

So the rectangular coordinates are $\left(2\sqrt{3},-2\right)$.

3. (a) Using $x = r\cos\theta$ and $y = r\sin\theta$, we have:

$$x = 1\cos\tfrac{\pi}{2} = 1(0) = 0$$
$$y = 1\sin\tfrac{\pi}{2} = 1(1) = 1$$

So the rectangular coordinates are $(0, 1)$.

(b) Using $x = r\cos\theta$ and $y = r\sin\theta$, we have:

$$x = 1\cos\tfrac{5\pi}{2} = 1(0) = 0$$
$$y = 1\sin\tfrac{5\pi}{2} = 1(1) = 1$$

So the rectangular coordinates are $(0, 1)$.

(c) Using $x = r\cos\theta$ and $y = r\sin\theta$, we have (using the half-angle formulas for $\sin\theta$ and $\cos\theta$):

$$x = -1\cos\tfrac{\pi}{8} = -1\left(\sqrt{\frac{1+\frac{\sqrt{2}}{2}}{2}}\right) = -\sqrt{\frac{2+\sqrt{2}}{4}} = -\frac{\sqrt{2+\sqrt{2}}}{2}$$

$$y = -1\sin\tfrac{\pi}{8} = -1\left(\sqrt{\frac{1-\frac{\sqrt{2}}{2}}{2}}\right) = -\sqrt{\frac{2-\sqrt{2}}{4}} = -\frac{\sqrt{2-\sqrt{2}}}{2}$$

So the rectangular coordinates are $\left(-\frac{\sqrt{2+\sqrt{2}}}{2}, -\frac{\sqrt{2-\sqrt{2}}}{2}\right)$.

5. We have:

$$r^2 = 1+1 = 2, \text{ so } r = \sqrt{2}$$
$$\theta = \tan^{-1}\left(\tfrac{-1}{-1}\right) + \pi = \tfrac{\pi}{4} + \pi = \tfrac{5\pi}{4}$$

So the polar form is $\left(\sqrt{2}, \tfrac{5\pi}{4}\right)$.

7. If we first multiply by r, we have:

$$r^2 = 2r\cos\theta$$
$$x^2 + y^2 = 2x$$
$$x^2 - 2x + y^2 = 0$$
$$x^2 - 2x + 1 + y^2 = 1$$
$$(x-1)^2 + y^2 = 1$$

9. Substituting for r and $\tan\theta$, we have:

$$\sqrt{x^2 + y^2} = \frac{y}{x}$$
$$x^2 + y^2 = \frac{y^2}{x^2}$$
$$x^4 + x^2 y^2 = y^2$$
$$x^4 + x^2 y^2 - y^2 = 0$$

11. Using the double-angle formula for $\cos 2\theta$, we have:
$$r = 3(\cos^2 \theta - \sin^2 \theta)$$
Multiplying by r^2:
$$r^3 = 3(r^2 \cos^2 \theta - r^2 \sin^2 \theta)$$
$$(x^2 + y^2)^{3/2} = 3(x^2 - y^2)$$
$$(x^2 + y^2)^3 = 9(x^2 - y^2)^2$$

13. Multiplying each side by $2 - \sin^2 \theta$ yields:
$$2r^2 - r^2 \sin^2 \theta = 8$$
$$2(x^2 + y^2) - y^2 = 8$$
$$2x^2 + 2y^2 - y^2 = 8$$
$$2x^2 + y^2 = 8$$
$$\frac{x^2}{4} + \frac{y^2}{8} = 1$$

15. Using the addition formula for $\cos(s - t)$, we have:
$$r\cos\left(\theta - \tfrac{\pi}{6}\right) = 2$$
$$r\left(\cos\theta \cos\tfrac{\pi}{6} + \sin\theta \sin\tfrac{\pi}{6}\right) = 2$$
$$\tfrac{\sqrt{3}}{2} r\cos\theta + \tfrac{1}{2} r\sin\theta = 2$$
$$\tfrac{\sqrt{3}}{2} x + \tfrac{1}{2} y = 2$$
$$\sqrt{3}x + y = 4$$
$$y = -\sqrt{3}x + 4$$

17. Substituting $x = r\cos\theta$ and $y = r\sin\theta$, we have:
$$3r\cos\theta - 4r\sin\theta = 2$$
$$r(3\cos\theta - 4\sin\theta) = 2$$
$$r = \frac{2}{3\cos\theta - 4\sin\theta}$$

19. Substituting $x = r\cos\theta$ and $y = r\sin\theta$, we have:
$$r^2 \sin^2 \theta = r^3 \cos^3 \theta$$
$$\sin^2 \theta = r\cos^3 \theta$$
$$r = \frac{\sin^2 \theta}{\cos^3 \theta}$$
$$r = \tan^2 \theta \sec\theta$$

21. Substituting $x = r\cos\theta$ and $y = r\sin\theta$, we have:
$$2(r\cos\theta)(r\sin\theta) = 1$$
$$r^2(2\sin\theta\cos\theta) = 1$$
$$r^2\sin 2\theta = 1$$
$$r^2 = \frac{1}{\sin 2\theta}$$
$$r^2 = \csc 2\theta$$

23. Substituting $x = r\cos\theta$ and $y = r\sin\theta$, we have:
$$9r^2\cos^2\theta + r^2\sin^2\theta = 9$$
$$r^2(9\cos^2\theta + \sin^2\theta) = 9$$
$$r^2 = \frac{9}{9\cos^2\theta + \sin^2\theta}$$

25. (a) Since $r\cos\theta = x$, this is the graph of $x = 3$:

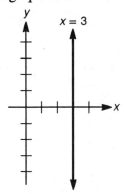

(b) Since $r\cos\theta = x$, this is the graph of $x = -2$:

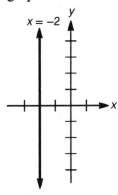

27. Since $r \cos \theta = x$, this is the graph of $x = 5$:

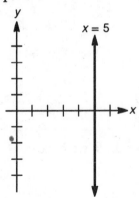

29. We plot points in polar form, noting that $\pm \theta$ values result in the same r value.

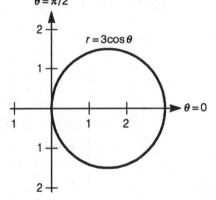

31. We plot points in polar form, noting that $\pm \theta$ values result in the same r value.

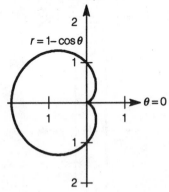

33. We plot points in polar form, noting that $\pm r$ values result in the same θ value. Also note that $r^2 \geq 0$, and thus $0 \leq \theta \leq \frac{\pi}{2}$ or $\pi \leq \theta \leq \frac{3\pi}{2}$.

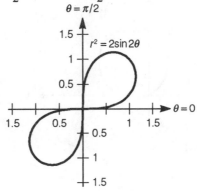

35. This is a circle of radius 1 centered at the origin:

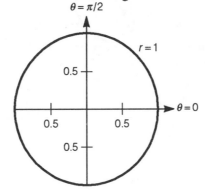

37. We plot points in polar form, noting that values of r do not affect the graph.

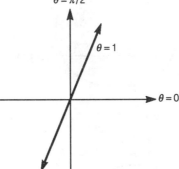

39. We first multiply by r:

$$r^2 = 4r\sin\theta + 2r\cos\theta$$
$$x^2 + y^2 = 4y + 2x$$
$$x^2 - 2x + y^2 - 4y = 0$$
$$x^2 - 2x + 1 + y^2 - 4y + 4 = 1 + 4$$
$$(x-1)^2 + (y-2)^2 = 5$$

The graph is a circle with center $(1, 2)$ and radius $= \sqrt{5}$:

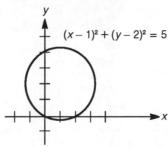

41. We plot points in polar form:

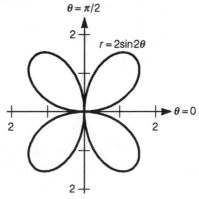

43. We plot points in polar form:

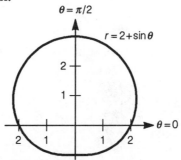

45. We plot points in polar form, noting that $\pm\theta$ values result in the same r value:

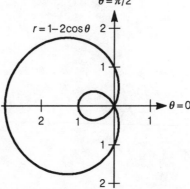

47. We plot points in polar form, noting that $\pm r$ values result in the same θ value:

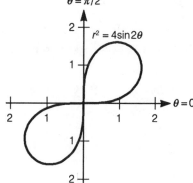

49. We plot points in polar form:

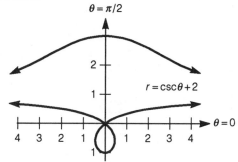

51. (a) The graph is C, since $\theta = \frac{\pi}{2}$ corresponds to $r = 6$.

(b) The graph is B, since $\theta = \frac{\pi}{2}$ corresponds to $r = 0$.

(c) The graph is D, since $\theta = \frac{\pi}{2}$ corresponds to $r = 3$.

(d) The graph is A, since $\theta = 0$ corresponds to $r = 6$.

53. We first draw the figure:

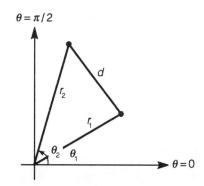

By the law of cosines:
$$d^2 = r_1^2 + r_2^2 - 2r_1r_2 \cos(\theta_2 - \theta_1)$$
$$d = \sqrt{r_1^2 + r_2^2 - 2r_1r_2 \cos(\theta_2 - \theta_1)}$$

55. We first establish an identity for $\sin 3\theta$:

$$\sin 3\theta = \sin(2\theta + \theta)$$
$$= \sin 2\theta \cos \theta + \cos 2\theta \sin \theta$$
$$= (2\sin \theta \cos \theta)\cos \theta + (\cos^2 \theta - \sin^2 \theta)\sin \theta$$
$$= 3\sin \theta \cos^2 \theta - \sin^3 \theta$$

Now we multiply the equation $r = a\sin 3\theta$ by r^3 and substitute:

$$r^4 = a\big((3r\sin \theta)(r^2 \cos^2 \theta) - r^3 \sin^3 \theta\big)$$
$$\left(x^2 + y^2\right)^2 = a\left(3yx^2 - y^3\right)$$
$$\left(x^2 + y^2\right)^2 = ay\left(3x^2 - y^2\right)$$

57. Substituting $x = r\cos \theta$ and $y = r\sin \theta$ results in the equation:

$$\frac{r^2 \cos^2 \theta}{a^2} - \frac{r^2 \sin^2 \theta}{b^2} = 1$$
$$\frac{r^2\left(b^2 \cos^2 \theta - a^2 \sin^2 \theta\right)}{a^2 b^2} = 1$$
$$r^2 = \frac{a^2 b^2}{b^2 \cos^2 \theta - a^2 \sin^2 \theta}$$

59. (a) Note the following figure:

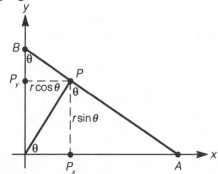

Using triangle AP_xP, we have:

$$\cos \theta = \frac{r\sin \theta}{AP}, \text{ so } AP = r\frac{\sin \theta}{\cos \theta}$$

Using triangle BP_yP, we have:

$$\sin \theta = \frac{r\cos \theta}{BP}, \text{ so } BP = r\frac{\cos \theta}{\sin \theta}$$

Since $AP + BP = 2k$, we have:

$$r\left(\frac{\sin \theta}{\cos \theta} + \frac{\cos \theta}{\sin \theta}\right) = 2k$$
$$r\left(\sin^2 \theta + \cos^2 \theta\right) = k \bullet 2\sin \theta \cos \theta$$
$$r = k\sin 2\theta$$

(b) We plot points in polar form:

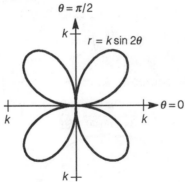

(c) Using the double-angle identity, we have:
$$r = 2k \sin \theta \cos \theta$$
Multiplying by r^2:
$$r^3 = 2k(r \sin \theta)(r \cos \theta)$$
$$\left(x^2 + y^2\right)^{3/2} = 2k(xy)$$
$$\left(x^2 + y^2\right)^3 = 4k^2 x^2 y^2$$

7.8 The Conics in Polar Coordinates

1. (a) Comparing the given equation with the four basic types, it appears this is the type associated with Figure 2. We divide both numerator and denominator by 3 to obtain:
$$r = \frac{2}{1 + \frac{2}{3}\cos\theta} = \frac{\frac{2}{3} \bullet 3}{1 + \frac{2}{3}\cos\theta}$$

Therefore $e = \frac{2}{3}$ and $d = 3$. Since $e < 1$, this confirms the given conic is an ellipse.

The eccentricity is $\frac{2}{3}$ and the directrix is $x = 3$. Computing the values of r when

$\theta = 0, \frac{\pi}{2}, \pi$ and $\frac{3\pi}{2}$, we have:

θ	0	$\frac{\pi}{2}$	π	$\frac{3\pi}{2}$
r	$\frac{6}{5}$	2	6	2

Since the major axis of this ellipse lies along the x-axis, the length of the major axis is:
$$2a = \frac{6}{5} + 6 = \frac{36}{5}, \text{ so } a = \frac{18}{5}$$

The endpoints of the major axis are at $\left(\frac{6}{5}, 0\right)$ and $(-6, 0)$, so the x-coordinate of the center is:
$$\frac{1}{2}\left(-6 + \frac{6}{5}\right) = \frac{1}{2}\left(-\frac{24}{5}\right) = -\frac{12}{5}$$

So the center is $\left(-\frac{12}{5},0\right)$. Finally, we calculate b:

$$b = a\sqrt{1-e^2} = \frac{18}{5}\sqrt{1-\frac{4}{9}} = \frac{18}{5}\sqrt{\frac{5}{9}} = \frac{18\sqrt{5}}{15} = \frac{6\sqrt{5}}{5}$$

So the endpoints of the minor axis are $\left(-\frac{12}{5},\pm\frac{6\sqrt{5}}{5}\right)$. We graph the ellipse:

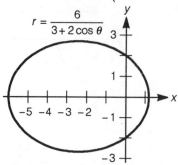

(b)   Comparing the given equation with the four basic types, it appears this is the type associated with Figure 3. We divide both numerator and denominator by 3 to obtain:

$$r = \frac{2}{1-\frac{2}{3}\cos\theta} = \frac{\frac{2}{3}\bullet 3}{1-\frac{2}{3}\cos\theta}$$

Therefore $e = \frac{2}{3}$ and $d = 3$. Since $e < 1$, this confirms the given conic is an ellipse.

The eccentricity is $\frac{2}{3}$ and the directrix is $x = -3$. Computing the values of r when

$\theta = 0, \frac{\pi}{2}, \pi$ and $\frac{3\pi}{2}$, we have:

θ	0	$\frac{\pi}{2}$	π	$\frac{3\pi}{2}$
r	6	2	$\frac{6}{5}$	2

Since the major axis of this ellipse lies along the x-axis, the length of the major axis is:

$$2a = \frac{6}{5}+6 = \frac{36}{5}, \text{ so } a = \frac{18}{5}$$

The endpoints of the major axis are at $(6,0)$ and $\left(-\frac{6}{5},0\right)$, so the x-coordinate of the center is:

$$\tfrac{1}{2}\left(6-\tfrac{6}{5}\right) = \tfrac{1}{2}\left(\tfrac{24}{5}\right) = \tfrac{12}{5}$$

So the center is $\left(\frac{12}{5},0\right)$. Finally, we calculate b:

$$b = a\sqrt{1-e^2} = \frac{18}{5}\sqrt{1-\frac{4}{9}} = \frac{18}{5}\sqrt{\frac{5}{9}} = \frac{18\sqrt{5}}{15} = \frac{6\sqrt{5}}{5}$$

So the endpoints of the minor axis are $\left(\frac{12}{5}, \pm \frac{6\sqrt{5}}{5}\right)$. We graph the ellipse:

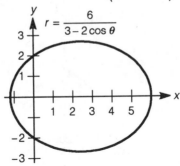

$$r = \frac{6}{3 - 2\cos\theta}$$

3. (a) Comparing the given equation with the four basic types, it appears this is the type associated with Figure 2. We divide both numerator and denominator by 2 to obtain:

$$r = \frac{\frac{5}{2}}{1 + \cos\theta} = \frac{1 \cdot \frac{5}{2}}{1 + \cos\theta}$$

Therefore $e = 1$ and $d = \frac{5}{2}$. Since $e = 1$, this confirms the given conic is a parabola.

The directrix is $x = \frac{5}{2}$. Computing the value of r when $\theta = 0$ yields $r = \frac{5}{4}$, so the

vertex is $\left(\frac{5}{4}, 0\right)$. We graph the parabola:

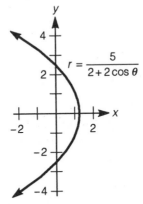

$$r = \frac{5}{2 + 2\cos\theta}$$

(b) Comparing the given equation with the four basic types, it appears this is the type associated with Figure 3. We divide both numerator and denominator by 2 to obtain:

$$r = \frac{\frac{5}{2}}{1 - \cos\theta} = \frac{1 \cdot \frac{5}{2}}{1 - \cos\theta}$$

Therefore $e = 1$ and $d = \frac{5}{2}$. Since $e = 1$, this confirms the given conic is a parabola.

The directrix is $x = -\frac{5}{2}$. Computing the value of r when $\theta = \pi$ yields $r = \frac{5}{4}$, so the

vertex is $\left(-\frac{5}{4}, 0\right)$. We graph the parabola:

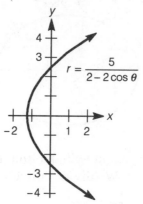

$$r = \frac{5}{2 - 2\cos\theta}$$

5.   (a)   Comparing the given equation with the four basic types, it appears this is the type
associated with Figure 2. We divide both numerator and denominator by 2 to
obtain:

$$r = \frac{3}{2 + 4\cos\theta} = \frac{\frac{3}{2}}{1 + 2\cos\theta} = \frac{2 \cdot \frac{3}{4}}{1 + 2\cos\theta}$$

Therefore $e = 2$ and $d = \frac{3}{4}$. Since $e > 1$, this confirms the given conic is a hyperbola.

The eccentricity is 2 and the directrix is $x = \frac{3}{4}$. Computing the values of r when

$\theta = 0, \frac{\pi}{2}, \pi$ and $\frac{3\pi}{2}$, we have:

θ	0	$\frac{\pi}{2}$	π	$\frac{3\pi}{2}$
r	$\frac{1}{2}$	$\frac{3}{2}$	$-\frac{3}{2}$	$\frac{3}{2}$

Since the two vertices $\left(\frac{1}{2}, 0\right)$ and $\left(\frac{3}{2}, 0\right)$ lie on the transverse axis, then:
$$2a = \frac{3}{2} - \frac{1}{2} = 1, \text{ so } a = \frac{1}{2}$$
The center of the hyperbola is the midpoint of these two vertices, which is $(1, 0)$.
Since a focus is $(0, 0)$, then $c = 1$. Finally, we find b:
$$b = \sqrt{c^2 - a^2} = \sqrt{1 - \frac{1}{4}} = \sqrt{\frac{3}{4}} = \frac{1}{2}\sqrt{3}$$
Notice that we could also find b from the eccentricity:
$$b = a\sqrt{e^2 - 1} = \frac{1}{2}\sqrt{4 - 1} = \frac{1}{2}\sqrt{3}$$

We graph the hyperbola:

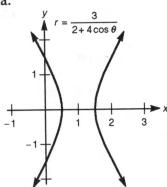

(b)   Comparing the given equation with the four basic types, it appears this is the type associated with Figure 3. We divide both numerator and denominator by 2 to obtain:

$$r = \frac{3}{2 - 4\cos\theta} = \frac{\frac{3}{2}}{1 - 2\cos\theta} = \frac{2 \bullet \frac{3}{4}}{1 - 2\cos\theta}$$

Therefore $e = 2$ and $d = \frac{3}{4}$. Since $e > 1$, this confirms the given conic is a hyperbola.

The eccentricity is 2 and the directrix is $x = -\frac{3}{4}$. Computing the values of r when

$\theta = 0, \frac{\pi}{2}, \pi$ and $\frac{3\pi}{2}$, we have:

θ	0	$\frac{\pi}{2}$	π	$\frac{3\pi}{2}$
r	$-\frac{3}{2}$	$\frac{3}{2}$	$\frac{1}{2}$	$\frac{3}{2}$

Since the two vertices $\left(-\frac{3}{2}, 0\right)$ and $\left(-\frac{1}{2}, 0\right)$ lie on the transverse axis, then:

$$2a = -\frac{1}{2} + \frac{3}{2} = 1, \text{ so } a = \frac{1}{2}$$

The center of the hyperbola is the midpoint of these two vertices, which is $(-1, 0)$. Since a focus is $(0, 0)$, then $c = 1$. Finally, we find b:

$$b = \sqrt{c^2 - a^2} = \sqrt{1 - \frac{1}{4}} = \sqrt{\frac{3}{4}} = \frac{1}{2}\sqrt{3}$$

Notice that we could also find b from the eccentricity:

$$b = a\sqrt{e^2 - 1} = \frac{1}{2}\sqrt{4 - 1} = \frac{1}{2}\sqrt{3}$$

We graph the hyperbola:

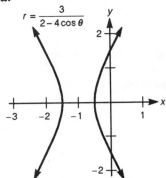

7. Comparing the given equation with the four basic types, it appears this is the type associated with Figure 3. We divide both numerator and denominator by 2 to obtain:

$$r = \frac{24}{2-3\cos\theta} = \frac{12}{1-\frac{3}{2}\cos\theta} = \frac{\frac{3}{2}\bullet 8}{1-\frac{3}{2}\cos\theta}$$

Since the eccentricity is $e = \frac{3}{2} > 1$, this conic is a hyperbola. Computing the values of r when $\theta = 0, \frac{\pi}{2}, \pi$ and $\frac{3\pi}{2}$, we have:

θ	0	$\frac{\pi}{2}$	π	$\frac{3\pi}{2}$
r	-24	12	$\frac{24}{5}$	12

Since the two vertices $(-24,0)$ and $\left(-\frac{24}{5},0\right)$ lie on the transverse axis, its length must be:

$$2a = -\tfrac{24}{5} + 24 = \tfrac{96}{5}, \text{ so } a = \tfrac{48}{5}$$

The center of the hyperbola is the midpoint of these two vertices, which is $\left(-\frac{72}{5},0\right)$.

Since a focus is $(0,0)$, then $c = \frac{72}{5}$ and thus:

$$b = \sqrt{c^2 - a^2} = \sqrt{\tfrac{5184}{25} - \tfrac{2304}{25}} = \sqrt{\tfrac{2880}{25}} = \tfrac{24\sqrt{5}}{5}$$

So the length of the conjugate axis is $2b = \frac{48\sqrt{5}}{5}$. We graph the hyperbola:

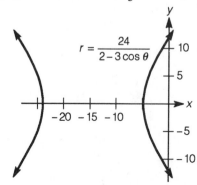

9. Comparing the given equation with the four basic types, it appears this is the type associated with Figure 4. We divide both numerator and denominator by 5 to obtain:

$$r = \frac{8}{5+3\sin\theta} = \frac{\frac{8}{5}}{1+\frac{3}{5}\sin\theta} = \frac{\frac{3}{5}\bullet\frac{8}{3}}{1+\frac{3}{5}\sin\theta}$$

Since the eccentricity is $e = \frac{3}{5} < 1$, this conic is an ellipse. Computing the values of r when $\theta = 0, \frac{\pi}{2}, \pi$ and $\frac{3\pi}{2}$, we have:

θ	0	$\frac{\pi}{2}$	π	$\frac{3\pi}{2}$
r	$\frac{8}{5}$	1	$\frac{8}{5}$	4

Since the two vertices $(0,1)$ and $(0,-4)$ lie on the major axis, its length must be:

$2a = 1 + 4 = 5$, so $a = \frac{5}{2}$

The center of the ellipse is the midpoint of these two vertices, which is $\left(0,-\frac{3}{2}\right)$. Since a

focus is $(0,0)$, then $c = \frac{3}{2}$ and thus:

$b = \sqrt{a^2 - c^2} = \sqrt{\frac{25}{4} - \frac{9}{4}} = \sqrt{\frac{16}{4}} = 2$

So the length of the minor axis is $2b = 4$. We graph the ellipse:

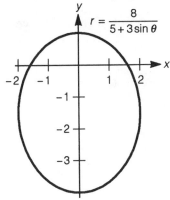

11. Comparing the given equation with the four basic types, it appears this is the type
associated with Figure 5. We divide both numerator and denominator by 5 to obtain:

$$r = \frac{12}{5 - 5\sin\theta} = \frac{\frac{12}{5}}{1 - \sin\theta} = \frac{1 \bullet \frac{12}{5}}{1 - \sin\theta}$$

Since the eccentricity is $e = 1$, this conic is a parabola with directrix $y = -\frac{12}{5}$. Since the

focus is $(0,0)$, the vertex must be the midpoint of $(0,0)$ and $\left(0,-\frac{12}{5}\right)$, which is $\left(0,-\frac{6}{5}\right)$.
We graph the parabola:

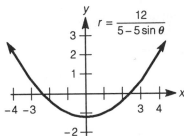

13. Comparing the given equation with the four basic types, it appears this is the type
associated with Figure 2. We divide both numerator and denominator by 7 to obtain:

$$r = \frac{12}{7 + 5\cos\theta} = \frac{\frac{12}{7}}{1 + \frac{5}{7}\cos\theta} = \frac{\frac{5}{7} \bullet \frac{12}{5}}{1 + \frac{5}{7}\cos\theta}$$

Since the eccentricity is $e = \frac{5}{7} < 1$, this conic is an ellipse. Computing the values of r when $\theta = 0, \frac{\pi}{2}, \pi$ and $\frac{3\pi}{2}$, we have:

θ	0	$\frac{\pi}{2}$	π	$\frac{3\pi}{2}$
r	1	$\frac{12}{7}$	6	$\frac{12}{7}$

Since the two vertices $(1,0)$ and $(-6,0)$ lie on the major axis, its length must be:
$$2a = 1 + 6 = 7, \text{ so } a = \frac{7}{2}$$

The center of the ellipse is the midpoint of these two vertices, which is $\left(-\frac{5}{2}, 0\right)$. Since a

focus is $(0,0)$, then $c = \frac{5}{2}$ and thus:
$$b = \sqrt{a^2 - c^2} = \sqrt{\frac{49}{4} - \frac{25}{4}} = \sqrt{6}$$
So the length of the minor axis is $2b = 2\sqrt{6}$. We graph the ellipse:

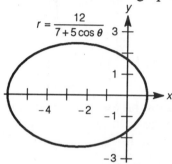

15. Comparing the given equation with the four basic types, it appears this is the type associated with Figure 4. We divide both numerator and denominator by 5 to obtain:
$$r = \frac{4}{5 + 5\sin\theta} = \frac{\frac{4}{5}}{1 + \sin\theta} = \frac{1 \bullet \frac{4}{5}}{1 + \sin\theta}$$

Since the eccentricity is $e = 1$, this conic is a parabola with directrix $y = \frac{4}{5}$. Since the

focus is $(0,0)$, the vertex must be the midpoint of $(0,0)$ and $\left(0, \frac{4}{5}\right)$, which is $\left(0, \frac{2}{5}\right)$. We graph the parabola:

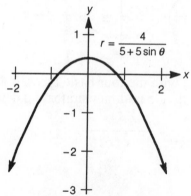

17. Comparing the given equation with the four basic types, it appears this is the type associated with Figure 3. The equation is already in standard form with eccentricity $e = 2 > 1$, so this conic is a hyperbola. Computing the values of r when

$\theta = 0, \frac{\pi}{2}, \pi$ and $\frac{3\pi}{2}$, we have:

θ	0	$\frac{\pi}{2}$	π	$\frac{3\pi}{2}$
r	-9	9	3	9

Since the two vertices $(-9,0)$ and $(-3,0)$ lie on the transverse axis, its length must be:
$$2a = -3 + 9 = 6, \text{ so } a = 3$$
The center of the hyperbola is the midpoint of these two vertices, which is $(-6,0)$. Since a focus is $(0,0)$, then $c = 6$ and thus:
$$b = \sqrt{c^2 - a^2} = \sqrt{36 - 9} = \sqrt{27} = 3\sqrt{3}$$
So the length of the conjugate axis is $2b = 6\sqrt{3}$. We graph the hyperbola:

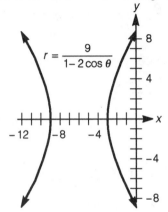

19. Since the coordinates of P are (r, θ), then the coordinates of Q are $(r, \theta + \pi)$. We now find FP and FQ, noting that $\cos(\theta + \pi) = -\cos\theta$

$$FP = r = \frac{ed}{1 - e\cos\theta}$$

$$FQ = r = \frac{ed}{1 - e\cos(\theta + \pi)} = \frac{ed}{1 + e\cos\theta}$$

Therefore:
$$\frac{1}{FP} + \frac{1}{FQ} = \frac{1 - e\cos\theta}{ed} + \frac{1 + e\cos\theta}{ed} = \frac{2}{ed}$$

This is remarkable in that $\frac{2}{ed}$ is a constant, even though P is a variable point.

21. Draw a focal chord $\overline{AB}$, with A representing the endpoint on the "left". Since $\overline{AB}$ is a $90°$ rotation from $\overline{PQ}$, then A corresponds to the polar coordinates $A\left(r, \theta + \frac{\pi}{2}\right)$ and B corresponds to the coordinates $B\left(r, \theta + \frac{3\pi}{2}\right)$.

Using the identities $\cos\left(\theta+\frac{\pi}{2}\right)=-\sin\theta$ and $\cos\left(\theta+\frac{3\pi}{2}\right)=\sin\theta$, we have:

$$AF = r = \frac{ed}{1-e\cos\left(\theta+\frac{\pi}{2}\right)} = \frac{ed}{1+e\sin\theta}$$

$$FB = r = \frac{ed}{1-e\cos\left(\theta+\frac{3\pi}{2}\right)} = \frac{ed}{1-e\sin\theta}$$

Since $AB = AF + FB$, we have:

$$\begin{aligned}
AB &= \frac{ed}{1+e\sin\theta} + \frac{ed}{1-e\sin\theta} \\
&= \frac{ed(1-e\sin\theta)+ed(1+e\sin\theta)}{(1+e\sin\theta)(1-e\sin\theta)} \\
&= \frac{ed-e^2d\sin\theta+ed+e^2d\sin\theta}{1-e^2\sin^2\theta} \\
&= \frac{2ed}{1-e^2\sin^2\theta}
\end{aligned}$$

Using the result from Exercise 10, we show the required sum is constant:

$$\frac{1}{PQ}+\frac{1}{AB} = \frac{1-e^2\cos^2\theta}{2ed}+\frac{1-e^2\sin^2\theta}{2ed} = \frac{2-e^2\left(\sin^2\theta+\cos^2\theta\right)}{2ed} = \frac{2-e^2}{2ed}$$

But since e and d are constants, we have proven the desired result.

7.9 Parametric Equations

1.   We find the x- and y-coordinates corresponding to $t = 0$:
$$x = 2-4(0) = 2$$
$$y = 3-5(0) = 3$$
The point corresponding to $t = 0$ is $(2,3)$.

3.   We find the x- and y-coordinates corresponding to $t = \frac{\pi}{6}$:
$$x = 5\cos\frac{\pi}{6} = 5\bullet\frac{\sqrt{3}}{2} = \frac{5\sqrt{3}}{2}$$
$$y = 2\sin\frac{\pi}{6} = 2\bullet\frac{1}{2} = 1$$

The point corresponding to $t = \frac{\pi}{6}$ is $\left(\frac{5\sqrt{3}}{2},1\right)$.

5.   We find the x- and y-coordinates corresponding to $t = \frac{\pi}{4}$:
$$x = 3\sin^3\frac{\pi}{4} = 3\left(\frac{\sqrt{2}}{2}\right)^3 = 3\bullet\frac{\sqrt{2}}{4} = \frac{3\sqrt{2}}{4}$$
$$y = 3\cos^3\frac{\pi}{4} = 3\left(\frac{\sqrt{2}}{2}\right)^3 = 3\bullet\frac{\sqrt{2}}{4} = \frac{3\sqrt{2}}{4}$$

The point corresponding to $t = \frac{\pi}{4}$ is $\left(\frac{3\sqrt{2}}{4},\frac{3\sqrt{2}}{4}\right)$.

7.   Solving $x = t + 1$ for t yields $t = x - 1$, now substituting:
$$y = t^2 = (x-1)^2$$
We graph the parabola which has a vertex at $(1, 0)$:

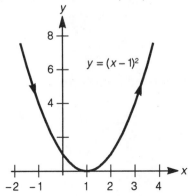

9.   Solving $y = t + 1$ for t yields $t = y - 1$, now substituting:
$$x = (y-1)^2 + 1$$
$$x + 1 = (y-1)^2$$
We graph the parabola which has a vertex at $(-1, 1)$:

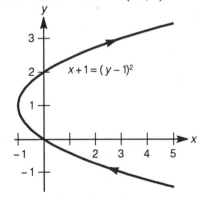

11. Multiplying the first equation by 2 and the second equation by 5 yields $2x = 10\cos t$ and $5y = 10\sin t$, so:
$$(2x)^2 + (5y)^2 = 100\cos^2 t + 100\sin^2 t$$
$$4x^2 + 25y^2 = 100$$
$$\frac{x^2}{25} + \frac{y^2}{4} = 1$$
We graph the ellipse which is centered at the origin:

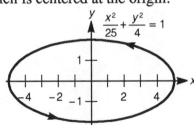

13. Multiplying the first equation by 3 and the second equation by 2 yields $3x = 12\cos 2t$ and $2y = 12\sin 2t$, so:

$$(3x)^2 + (2y)^2 = 144\cos^2 2t + 144\sin^2 2t$$
$$9x^2 + 4y^2 = 144$$
$$\frac{x^2}{16} + \frac{y^2}{36} = 1$$

We graph the ellipse which is centered at the origin:

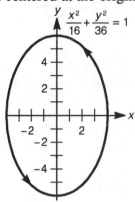

15.   (a)   We compute:

$$x^2 + y^2 = 4\cos^2 t + 4\sin^2 t$$
$$x^2 + y^2 = 4$$

We graph the circle which is centered at the origin:

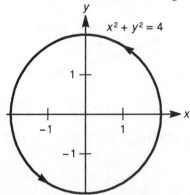

(b)   Multiplying the second equation by 2 yields $x = 4\cos t$ and $2y = 4\sin t$, so:

$$x^2 + (2y)^2 = 16\cos^2 t + 16\sin^2 t$$
$$x^2 + 4y^2 = 16$$
$$\frac{x^2}{16} + \frac{y^2}{4} = 1$$

We graph the ellipse which is centered at the origin:

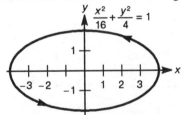

(c) Multiplying the second equation by 2 yields $x - 2 = 4\cos t$ and $2(y - 2) = 4\sin t$, so:

$$(x-2)^2 + [2(y-2)]^2 = 16\cos^2 t + 16\sin^2 t$$
$$(x-2)^2 + 4(y-2)^2 = 16$$
$$\frac{(x-2)^2}{16} + \frac{(y-2)^2}{4} = 1$$

We graph the ellipse which is centered at $(2, 2)$:

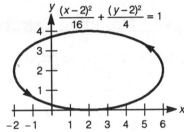

17. (a) When $t = 1$, we compute the x- and y-coordinates:

$$x = (100\cos 70°) \bullet 1 = 100\cos 70° \approx 34.2$$
$$y = 5 + (100\sin 70°) \bullet 1 - 16(1)^2 = 100\sin 70° - 11 \approx 83.0$$

When $t = 2$, we compute the x- and y-coordinates:

$$x = (100\cos 70°) \bullet 2 = 200\cos 70° \approx 68.4$$
$$y = 5 + (100\sin 70°) \bullet 2 - 16(2)^2 = 200\sin 70° - 59 \approx 128.9$$

When $t = 3$, we compute the x- and y-coordinates:

$$x = (100\cos 70°) \bullet 3 = 300\cos 70° \approx 102.6$$
$$y = 5 + (100\sin 70°) \bullet 3 - 16(3)^2 = 300\sin 70° - 139 \approx 142.9$$

(b) We will find the value of t when $y = 0$:

$$5 + (100\sin 70°)t - 16t^2 = 0$$
$$16t^2 - (100\sin 70°)t - 5 = 0$$

Using the quadratic formula, we have:

$$t = \frac{100\sin 70° \pm \sqrt{10000\sin^2 70° + 320}}{32} \approx 5.93, -0.05$$

We discard the negative value and conclude that the ball is in flight for approximately 5.93 seconds. When $t \approx 5.93$, we have:

$$x \approx (100\cos 70°)(5.93) = 593\cos 70° \approx 203$$

The total horizontal distance traveled is approximately 203 feet. Note that this is consistent with the figure.

19. We solve for t:
$$6 + (88\sin 35°)t - 16t^2 = 0$$
$$16t^2 - (88\sin 35°) - 6 = 0$$
Using the quadratic formula, we have:
$$t = \frac{88\sin 35° \pm \sqrt{7744\sin^2 35° + 384}}{32} \approx 3.27, -0.11$$
The solutions are verified.

21. Using the hint, we raise each side to the 2/3 power to obtain $x^{2/3} = \cos^2 t$ and $y^{2/3} = \sin^2 t$, so:
$$x^{2/3} + y^{2/3} = \cos^2 t + \sin^2 t = 1$$
The x-y equation for the curve is $x^{2/3} + y^{2/3} = 1$.

Section 7.7-7.9 TI-81 Graphing Calculator Exercises

1.   (a)   Using the indicated settings, we obtain the graph:

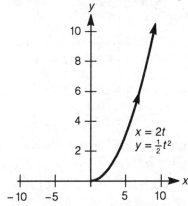

   (b)   Changing $t_{\min} = -6.28$, we obtain the graph:

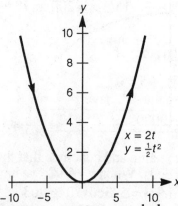

Notice the graph now appears more as a parabola.

(c)   Setting $t_{min} = 2$ and $t_{max} = 3$, we obtain the graph:

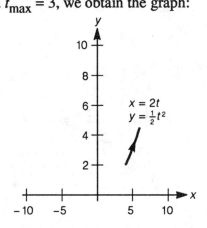

(d)   Using TRACE verifies the endpoint has coordinates $(4, 2)$ when $t = 2$.

3.   Using settings $x_{min} = -4$, $x_{max} = 4$, $y_{min} = -4$, $y_{max} = 4$, we obtain the graph:

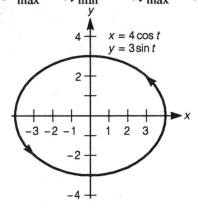

5.   Using the settings $x_{min} = 0$, $x_{max} = 4$, $y_{min} = 0$, $y_{max} = 3$, we obtain the graph:

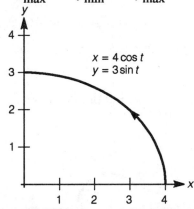

7.   Using the settings $x_{min} = -3$, $x_{max} = 3$, $y_{min} = -3$, $y_{max} = 3$, we obtain the graph:

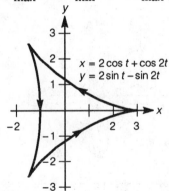

$$x = 2\cos t + \cos 2t$$
$$y = 2\sin t - \sin 2t$$

9.   Using the settings $x_{min} = -4$, $x_{max} = 4$, $y_{min} = 0$, $y_{max} = 3$, we obtain the graph:

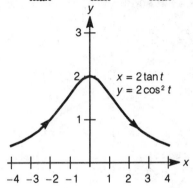

$$x = 2\tan t$$
$$y = 2\cos^2 t$$

11.   (a)   Graphing the polar curve, we obtain:

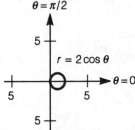

$$r = 2\cos \theta$$

(b)   Using the indicated settings, we obtain the curve:

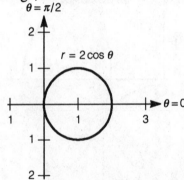

$$r = 2\cos \theta$$

13. (a)   We graph the polar curve $x = \cos 4\theta \cos \theta$, $y = \cos 4\theta \sin \theta$:

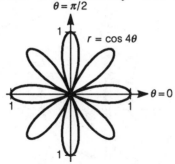

(b)   We graph the polar curve $x = \sin 4\theta \cos \theta$, $y = \sin 4\theta \sin \theta$:

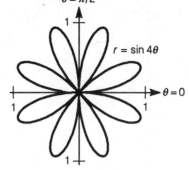

15. (a)   We graph the polar curve $x = \cos 5\theta \cos \theta$, $y = \cos 5\theta \sin \theta$:

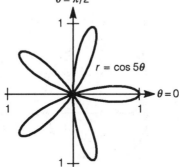

(b)   We graph the polar curve $x = \sin 5\theta \cos \theta$, $y = \sin 5\theta \sin \theta$:

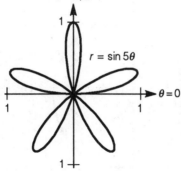

17.   We graph the polar curve $x = \theta \cos \theta$, $y = \theta \sin \theta$:

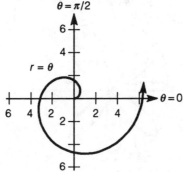

19.   (a)   We graph the polar curve $x = (2 \cos \theta + 1) \cos \theta$, $y = (2 \cos \theta + 1) \sin \theta$:

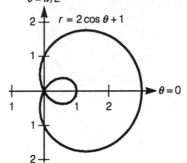

   (b)   We graph the polar curve $x = (2 \cos \theta - 1) \cos \theta$, $y = (2 \cos \theta - 1) \sin \theta$:

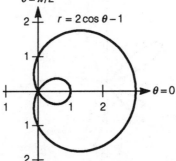

   (c)   We graph the polar curve $x = (\cos \theta + 2) \cos \theta$, $y = (\cos \theta + 2) \sin \theta$:

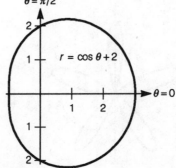

(d)   We graph the polar curve $x = (\cos\theta - 2)\cos\theta$, $y = (\cos\theta - 2)\sin\theta$:

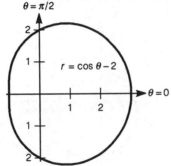

(e)   We graph the polar curve $x = (2\cos\theta + 2)\cos\theta$, $y = (2\cos\theta + 2)\sin\theta$:

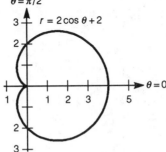

(f)   We graph the polar curve $x = \cos\theta\cos\theta$, $y = \cos\theta\sin\theta$:

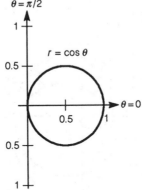

21.   (a)   Using $t_{max} = 10$, we graph the polar curve $x = e^{\theta/10}\cos\theta$, $y = e^{\theta/10}\sin\theta$:

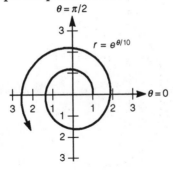

(b)   Using $t_{max} = 20$, we graph the polar curve $x = e^{\theta/10}\cos\theta$, $y = e^{\theta/10}\sin\theta$:

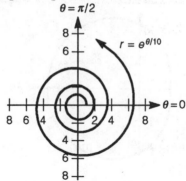

(c)   Using $t_{max} = 25$, we graph the polar curve $x = e^{\theta/10}\cos\theta$, $y = e^{\theta/10}\sin\theta$:

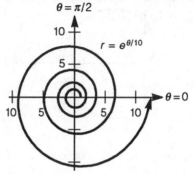

23.   Using the indicated settings, we obtain the graph:

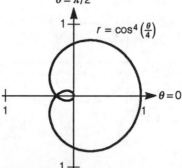

Using ZOOM-2 several times, we find that the inner loop near the origin is not simple, but rather a cardioid type shape which passes through both the first and fourth quadrants.

Chapter Seven Review Exercises

1. The x-axis contains AB and has equation $y = 0$. The slope of the line containing BC is $-\frac{1}{b}$ and the equation is:
$$y = -\tfrac{1}{b}(x - 6b)$$
$$by = 6b - x$$
$$x + by = 6b$$

   The slope of the line containing AC is $-\frac{1}{a}$ and the equation is:
$$y = -\tfrac{1}{a}(x - 6a)$$
$$x + ay = 6a$$

3. Using the equations from Exercise 2, we show $G(2a + 2b, 2)$ lies on all three medians:
$$2(2a + 2b) + (a + b)(2) = 4a + 4b + 2a + 2b = 6(a + b)$$
$$2a + 2b - (b - 2a)(2) = 2a + 2b - 2b + 4a = 6a$$
$$2a + 2b - (a - 2b)(2) = 2a + 2b - 2a + 4b = 6b$$

5. Using the equations from Exercise 4, we show $H(0, -6ab)$ lies on each altitude:
$$x = 0$$
$$-6ab = b \bullet 0 - 6ab = -6ab$$
$$-6ab = a \bullet 0 - 6ab = -6ab$$

7. Using the equations from Exercise 6, we show $O(3a + 3b, 3ab + 3)$ lies on each perpendicular bisector:
$$x = 3a + 3b$$
$$b(3a + 3b) - (3ab + 3) = 3ab + 3b^2 - 3ab - 3 = 3b^2 - 3$$
$$a(3a + 3b) - (3ab + 3) = 3a^2 + 3ab - 3ab - 3 = 3a^2 - 3$$

9. We find each distance and use the result from Exercise 8:
$$p = \sqrt{(6b)^2 + 6^2} = 6\sqrt{b^2 + 1}$$
$$q = \sqrt{(6a)^2 + 6^2} = 6\sqrt{a^2 + 1}$$
$$r = 6(a - b)$$
$$R = 3\sqrt{(a^2 + 1)(b^2 + 1)}$$

   Thus:
$$\frac{pqr}{4R} = \frac{\left(6\sqrt{b^2 + 1}\right)\left(6\sqrt{a^2 + 1}\right)(6(a - b))}{4 \bullet 3\sqrt{(a^2 + 1)(b^2 + 1)}} = 18(a - b)$$

   Area of $\triangle ABC = \tfrac{1}{2}(6(a - b))(6) = 18(a - b)$

   So the area is $\frac{pqr}{4R}$, as required.

11. We compute each side of the identity:
$$OH^2 = (3a+3b-0)^2 + (3ab+3+6ab)^2$$
$$= 9(a^2+2ab+b^2) + 9(3ab+1)^2$$
$$= 9(a^2+2ab+b^2+9a^2b^2+6ab+1)$$
$$= 81a^2b^2+9a^2+9b^2+72ab+9$$
$$9R^2 - (p^2+q^2+r^2)$$
$$= 9[9(a^2+1)(b^2+1)] - 36(b^2+1) - 36(a^2+1) - 36(a^2-2ab+b^2)$$
$$= 81a^2b^2+81a^2+81b^2+81-36b^2-36-36a^2-36-36a^2+72ab-36b^2$$
$$= 81a^2b^2+9a^2+9b^2+72ab+9$$
Thus $OH^2 = 9R^2 - (p^2+q^2+r^2)$.

13. We first compute the squares:
$$HA^2 = (6a)^2 + (-6ab)^2 = 36a^2 + 36a^2b^2$$
$$HB^2 = (6b)^2 + (-6ab)^2 = 36b^2 + 36a^2b^2$$
$$HC^2 = 0^2 + (6+6ab)^2 = 36+72ab+36a^2b^2$$
So $HA^2 + HB^2 + HC^2 = 108a^2b^2 + 36a^2 + 36b^2 + 72ab + 36$.
Now compute the right-hand side:
$$12R^2 - (p^2+q^2+r^2)$$
$$= 12[9(a^2+1)(b^2+1)] - [36(b^2+1)+36(a^2+1)+36(a^2-2ab+b^2)]$$
$$= 108a^2b^2+108a^2+108b^2+108-36b^2-36-36a^2-36-36a^2+72ab-36b^2$$
$$= 108a^2b^2+36a^2+36b^2+72ab+36$$
Thus $HA^2 + HB^2 + HC^2 = 12R^2 - (p^2+q^2+r^2)$.

15. In Exercise 12 we saw that:
$$GH^2 = 4(9a^2b^2+a^2+b^2+8ab+1)$$
Therefore:
$$GH = 2\sqrt{9a^2b^2+a^2+b^2+8ab+1}$$
Now:
$$2GO = 2\sqrt{(2a+2b-3a-3b)^2+(2-3ab-3)^2}$$
$$= 2\sqrt{a^2+2ab+b^2+1+6ab+9a^2b^2}$$
$$= 2\sqrt{9a^2b^2+a^2+b^2+8ab+1}$$
Thus $GH = 2GO$.

17. Since $\tan\theta = -\frac{2}{3}$, then $\theta \approx 146.3°$.

19. We have $(x_0, y_0) = (-1,-3)$, $A = 5$, $B = 6$ and $C = -30$, so using the distance formula from a point to a line yields:
$$d = \frac{|5(-1)+6(-3)-30|}{\sqrt{5^2+6^2}} = \frac{53}{\sqrt{61}} = \frac{53\sqrt{61}}{61}$$

21.   Label $A(-6,0)$, $B(6,0)$ and $C(0,6\sqrt{3})$. The height of the triangle is $6\sqrt{3}$. Since the line AB is the x-axis, the distance from $(1,2)$ to AB is 2. The line containing AC has slope $\sqrt{3}$ and equation:
$$y - 0 = \sqrt{3}(x+6)$$
$$y = \sqrt{3}x + 6\sqrt{3}$$
The distance from $(1,2)$ to AC is thus:
$$\frac{\left|\sqrt{3}(1)+6\sqrt{3}-2\right|}{\sqrt{1^2 + \left(-\sqrt{3}\right)^2}} = \frac{7\sqrt{3}-2}{2}$$
The line containing BC has slope $-\sqrt{3}$ and equation:
$$y - 0 = -\sqrt{3}(x-6)$$
$$y = -\sqrt{3}x + 6\sqrt{3}$$
The distance from $(1,2)$ to BC is thus:
$$\frac{\left|-\sqrt{3}(1)+6\sqrt{3}-2\right|}{\sqrt{1^2 + \left(\sqrt{3}\right)^2}} = \frac{5\sqrt{3}-2}{2}$$
The sum of these distances is $6\sqrt{3}$, which is also the height.

23.   (a)   We have the form $y^2 = 4px$, where $p = 4$. Thus the equation is $y^2 = 16x$.

(b)   We have the form $x^2 = 4py$, where $p = 4$. Thus the equation is $x^2 = 16y$.

25.   Since the parabola is symmetric about the positive y-axis, its equation must be of the form $x^2 = 4py$, where $p > 0$. Now the focal width is 12, so $4p = 12$. Thus the equation is $x^2 = 12y$.

27.   We have $c = 2$ and $a = 8$, and the ellipse must have the form:
$$\frac{x^2}{8^2} + \frac{y^2}{b^2} = 1$$
Since $c^2 = a^2 - b^2$, we can find b:
$$4 = 64 - b^2$$
$$b^2 = 60$$
So the equation is $\frac{x^2}{64} + \frac{y^2}{60} = 1$, or $15x^2 + 16y^2 = 960$.

29.   Since one end of the minor axis is $(-6,0)$, then $b = 6$ and the ellipse has a form of:
$$\frac{x^2}{36} + \frac{y^2}{a^2} = 1$$

Now $\frac{c}{a} = \frac{4}{5}$, so $c = \frac{4}{5}a$. We find a:

$$c^2 = a^2 - b^2$$
$$\left(\tfrac{4}{5}a\right)^2 = a^2 - 36$$
$$\tfrac{16}{25}a^2 = a^2 - 36$$
$$36 = \tfrac{9}{25}a^2$$
$$100 = a^2$$

So the equation is $\frac{x^2}{36} + \frac{y^2}{100} = 1$, or $25x^2 + 9y^2 = 900$.

31. Since the foci are $(\pm 6, 0)$ and the vertices are $(\pm 2, 0)$, then $c = 6$, $a = 2$, and the equation has the form:

$$\frac{x^2}{4} - \frac{y^2}{b^2} = 1$$

We find b:
$$c^2 = a^2 + b^2$$
$$36 = 4 + b^2$$
$$32 = b^2$$

So the equation is $\frac{x^2}{4} - \frac{y^2}{32} = 1$, or $8x^2 - y^2 = 32$.

33. Since the foci are $(\pm 3, 0)$, then $c = 3$ and the equation has the form:

$$\frac{x^2}{a^2} - \frac{y^2}{b^2} = 1$$

Now $\frac{c}{a} = 4$, so $\frac{3}{a} = 4$, thus $a = \frac{3}{4}$. We substitute to find b:

$$c^2 = a^2 + b^2$$
$$9 = \tfrac{9}{16} + b^2$$
$$144 = 9 + 16b^2$$
$$135 = 16b^2$$
$$\tfrac{135}{16} = b^2$$

So the equation is $\dfrac{x^2}{\frac{9}{16}} - \dfrac{y^2}{\frac{135}{16}} = 1$, or $240x^2 - 16y^2 = 135$.

35. We note that $4p = 10$, so $p = \frac{5}{2}$. So the vertex is $(0,0)$, the focus is $\left(0,\frac{5}{2}\right)$, the directrix is $y = -\frac{5}{2}$, and the focal width is 10.

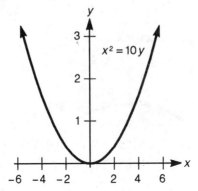

37. We note that $4p = 12$, so $p = 3$. So the vertex is $(0,3)$, the focus is $(0,0)$, the directrix is $y = 6$, and the focal width is 12.

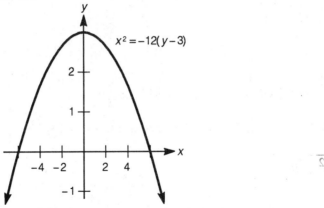

39. We note that $4p = 4$, so $p = 1$. So the vertex is $(1,1)$, the focus is $(0,1)$, the directrix is $x = 2$, and the focal width is 4.

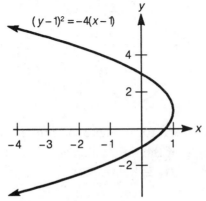

41. Dividing by 4, the standard form is $\frac{x^2}{4} + \frac{y^2}{2} = 1$. The center is $(0,0)$, the length of the major axis is 4, and the length of the minor axis is $2\sqrt{2}$. Using $c^2 = a^2 - b^2$, we find:
$$c = \sqrt{a^2 - b^2} = \sqrt{4-2} = \sqrt{2}$$
So the foci are $\left(\pm\sqrt{2},0\right)$ and the eccentricity is $\frac{\sqrt{2}}{2}$.

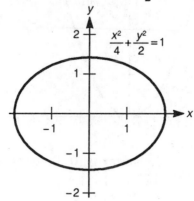

43. Dividing by 441, the standard form is $\frac{x^2}{9} + \frac{y^2}{49} = 1$. The center is $(0,0)$, the length of the major axis is 14, and the length of the minor axis is 6. Using $c^2 = a^2 - b^2$, we find:
$$c = \sqrt{a^2 - b^2} = \sqrt{49-9} = \sqrt{40} = 2\sqrt{10}$$
So the foci are $\left(0, \pm 2\sqrt{10}\right)$ and the eccentricity is $\frac{2\sqrt{10}}{7}$.

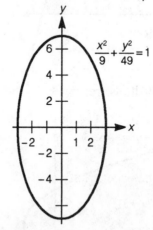

45. The equation is already in standard form where the center is $(1,-2)$, the length of the major axis is 10, and the length of the minor axis is 6. Using $c^2 = a^2 - b^2$, we find:
$$c = \sqrt{a^2 - b^2} = \sqrt{25-9} = \sqrt{16} = 4$$

So the foci are $(1 + 4, -2) = (5, -2)$ and $(1 - 4, -2) = (-3, -2)$, and the eccentricity is $\frac{4}{5}$.

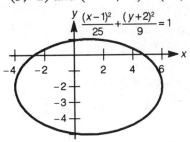

47. Dividing by 4, the standard form is $\frac{x^2}{4} - \frac{y^2}{2} = 1$. The center is $(0, 0)$, the vertices are

$(\pm 2, 0)$, and the asymptotes are $y = \pm \frac{\sqrt{2}}{2} x$. Using $c^2 = a^2 + b^2$, we have:
$$c = \sqrt{a^2 + b^2} = \sqrt{4 + 2} = \sqrt{6}$$

So the foci are $(\pm 6, 0)$ and the eccentricity is $\frac{\sqrt{6}}{2}$.

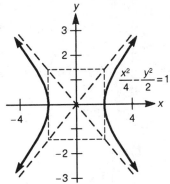

49. Dividing by 441, the standard form is $\frac{y^2}{9} - \frac{x^2}{49} = 1$. The center is $(0, 0)$, the vertices are

$(0, \pm 3)$, and the asymptotes are $y = \pm \frac{3}{7} x$. Using $c^2 = a^2 + b^2$, we have:
$$c = \sqrt{a^2 + b^2} = \sqrt{9 + 49} = \sqrt{58}$$

So the foci are $\left(0, \pm \sqrt{58}\right)$ and the eccentricity is $\frac{\sqrt{58}}{3}$.

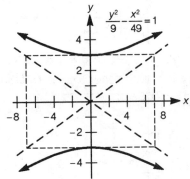

51. The equation is already in standard form where the center is $(1, -2)$ and the vertices are $(1 + 5, -2) = (6, -2)$ and $(1 - 5, -2) = (-4, -2)$. The asymptotes have slopes of $\pm\frac{3}{5}$, so using the point-slope formula:

$$y - (-2) = \tfrac{3}{5}(x - 1) \qquad\qquad y - (-2) = -\tfrac{3}{5}(x - 1)$$
$$y + 2 = \tfrac{3}{5}x - \tfrac{3}{5} \qquad\qquad y + 2 = -\tfrac{3}{5}x + \tfrac{3}{5}$$
$$y = \tfrac{3}{5}x - \tfrac{13}{5} \qquad\qquad y = -\tfrac{3}{5}x - \tfrac{7}{5}$$

Using $c^2 = a^2 + b^2$, we have:
$$c = \sqrt{a^2 + b^2} = \sqrt{25 + 9} = \sqrt{34}$$

So the foci are $\left(1 \pm \sqrt{34}, -2\right)$ and the eccentricity is $\frac{\sqrt{34}}{5}$.

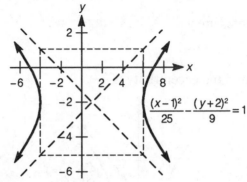

53. We complete the square to convert the equation to standard form:
$$3x^2 - 6x + 4y^2 + 16 = -7$$
$$3\left(x^2 - 2x + 1\right) + 4\left(y^2 + 4y + 4\right) = -7 + 3 + 16$$
$$3(x - 1)^2 + 4(y + 2)^2 = 12$$
$$\frac{(x - 1)^2}{4} + \frac{(y + 2)^2}{3} = 1$$

This is the equation of an ellipse. Its center is $(1, -2)$, the length of the major axis is 4, and the length of the minor axis is $2\sqrt{3}$. Using $c^2 = a^2 - b^2$, we find:
$$c = \sqrt{a^2 - b^2} = \sqrt{4 - 3} = 1$$

So the foci are $(1 + 1, -2) = (2, -2)$ and $(1 - 1, -2) = (0, -2)$.

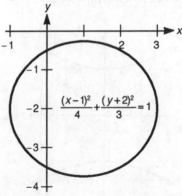

55. We complete the square to convert the equation to standard form:
$$y^2 + 2y = -4x + 15$$
$$y^2 + 2y + 1 = -4x + 15 + 1$$
$$(y+1)^2 = -4(x-4)$$

This is the equation of a parabola. Since $4p = 4$, then $p = 1$. So the vertex is $(4, -1)$, the axis of symmetry is $y = -1$, the focus is $(3, -1)$, and the directrix is $x = 5$.

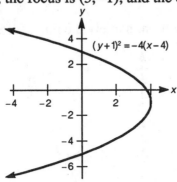

57. We complete the square to convert the equation to standard form:
$$16(x^2 - 2x) - 9(y^2 - 10y) = 353$$
$$16(x^2 - 2x + 1) - 9(y^2 - 10y + 25) = 353 + 16 - 225$$
$$16(x-1)^2 - 9(y-5)^2 = 144$$
$$\frac{(x-1)^2}{9} - \frac{(y-5)^2}{16} = 1$$

This is the equation of a hyperbola. Its center is $(1, 5)$ and its vertices are

$(1 + 3, 5) = (4, 5)$ and $(1 - 3, 5) = (-2, 5)$. The asymptotes have slopes of $\pm\frac{4}{3}$, so using the point-slope formula:

$$y - 5 = \tfrac{4}{3}(x-1) \qquad\qquad y - 5 = -\tfrac{4}{3}(x-1)$$
$$y - 5 = \tfrac{4}{3}x - \tfrac{4}{3} \qquad\qquad y - 5 = -\tfrac{4}{3}x + \tfrac{4}{3}$$
$$y = \tfrac{4}{3}x + \tfrac{11}{3} \qquad\qquad y = -\tfrac{4}{3}x + \tfrac{19}{3}$$

Using $c^2 = a^2 + b^2$, we have:
$$c = \sqrt{a^2 + b^2} = \sqrt{9 + 16} = \sqrt{25} = 5$$
So the foci are $(1 + 5, 5) = (6, 5)$ and $(1 - 5, 5) = (-4, 5)$.

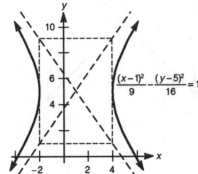

59. We complete the square to convert the equation to standard form:
$$5x^2 + 3y^2 - 40x - 36y + 188 = 0$$
$$5\left(x^2 - 8x\right) + 3\left(y^2 - 12y\right) = -188$$
$$5(x-4)^2 + 3(y-6)^2 = -188 + 80 + 108$$
$$5(x-4)^2 + 3(y-6)^2 = 0$$

This equation has only one solution, namely the point $(4, 6)$.

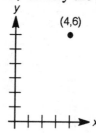

61. We complete the square to convert the equation to standard form:
$$9x^2 - 90x - 16y^2 + 32y = -209$$
$$9\left(x^2 - 10x\right) - 16\left(y^2 - 2y\right) = -209$$
$$9(x-5)^2 - 16(y-1)^2 = -209 + 225 - 16$$
$$9(x-5)^2 - 16(y-1)^2 = 0$$
$$9(x-5)^2 = 16(y-1)^2$$
$$\pm 3(x-5) = 4(y-1)$$

The graph of this equation consists of just two lines:

$$3(x-5) = 4(y-1) \qquad\qquad -3(x-5) = 4(y-1)$$
$$3x - 15 = 4y - 4 \qquad\qquad -3x + 15 = 4y - 4$$
$$3x - 4y = 11 \qquad\qquad 3x + 4y = 19$$

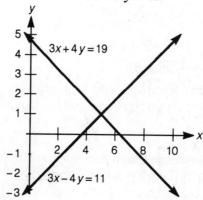

63. We complete the square to convert the equation to standard form:

$$y^2 + 8y - 25x^2 = 9$$
$$(y+4)^2 - 25x^2 = 25$$
$$\frac{(y+4)^2}{25} - \frac{x^2}{1} = 1$$

This is the equation of a hyperbola. Its center is $(0, -4)$ and its vertices are $(0, -4+5) = (0, 1)$ and $(0, -4-5) = (0, -9)$. The asymptotes have slopes of ± 5, so using the point-slope formula:

$$y - (-4) = 5(x - 0) \qquad\qquad y - (-4) = -5(x - 0)$$
$$y + 4 = 5x \qquad\qquad\qquad y + 4 = -5x$$
$$y = 5x - 4 \qquad\qquad\qquad y = -5x - 4$$

Using $c^2 = a^2 + b^2$, we have:

$$c = \sqrt{a^2 + b^2} = \sqrt{25 + 1} = \sqrt{26}$$

So the foci are $\left(0, -4 \pm \sqrt{26}\right)$.

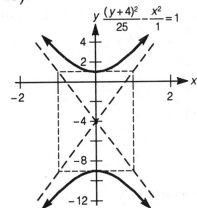

65. We complete the square to convert the equation to standard form:

$$16x^2 - 64x + 25y^2 - 100y = -564$$
$$16\left(x^2 - 4x\right) + 25\left(y^2 - 4y\right) = -564$$
$$16(x-2)^2 + 25(y-2)^2 = -564 + 64 + 100$$
$$16(x-2)^2 + 25(y-2)^2 = -400$$

Since the left-hand side of this equation is the sum of two squares, it must be non-negative and thus there are no points (x, y) which satisfy the equation. Thus there is no graph for this equation.

67. (a) Substituting $P(6, 5)$ into the equation $5x^2 - 4y^2 = 80$:

$$5(6)^2 - 4(5)^2 = 5(36) - 4(25) = 180 - 100 = 80$$

So $P(6, 5)$ lies on the hyperbola.

(b)   Since P lies on the hyperbola with foci F_1, F_2, then the quantity $F_1P - F_2P = \pm 2a$ (by the definition of the hyperbola). Thus $(F_1P - F_2P)^2 = 4a^2$. It remains to find a:

$$5x^2 - 4y^2 = 80$$

$$\frac{x^2}{16} - \frac{y^2}{20} = 1$$

So $a^2 = 16$, thus $(F_1P - F_2P)^2 = 4a^2 = 4(16) = 64$.

69. Completing the square, we obtain:

$$A\left(x^2 + \frac{D}{A}x + \frac{D^2}{4A^2}\right) + C\left(y^2 + \frac{E}{C}y + \frac{E^2}{4C^2}\right) = \frac{D^2}{4A} + \frac{E^2}{4C} - F$$

$$A\left(x + \frac{D}{2A}\right)^2 + C\left(y + \frac{E}{2C}\right)^2 = \frac{CD^2 + AE^2 - 4ACF}{4AC}$$

Now if the curve represents an ellipse or a hyperbola, this last equation shows that the center is $\left(-\frac{D}{2A}, -\frac{E}{2C}\right)$.

71. For the parabola, the vertex is $(5, 0)$, so we know $y^2 = -4p(x - 5)$. Since the focus is $(3, 0)$, then $p = 2$, so the equation is $y^2 = -8(x - 5)$. For the ellipse, $a = 5$ and $c = 3$, so $b^2 = a^2 - c^2 = 25 - 9 = 16$, so $b = 4$. Thus the equation of the ellipse is $\frac{x^2}{25} + \frac{y^2}{16} = 1$.

73. We have:

$$\cot 2\theta = \frac{4 - 1}{4} = \frac{3}{4}, \text{ so } \tan 2\theta = \frac{4}{3}$$

$$\sec^2 2\theta = 1 + \tan^2 2\theta = 1 + \frac{16}{9} = \frac{25}{9}$$

So $\sec 2\theta = \frac{5}{3}$ and $\cos 2\theta = \frac{3}{5}$. Thus:

$$\sin \theta = \sqrt{\frac{1 - \cos 2\theta}{2}} = \sqrt{\frac{1 - \frac{3}{5}}{2}} = \sqrt{\frac{1}{5}} = \frac{\sqrt{5}}{5}$$

$$\cos \theta = \sqrt{\frac{1 + \cos 2\theta}{2}} = \sqrt{\frac{1 + \frac{3}{5}}{2}} = \sqrt{\frac{4}{5}} = \frac{2\sqrt{5}}{5}$$

So $\theta = \sin^{-1}\left(\frac{\sqrt{5}}{5}\right) \approx 26.6°$, and:

$$x = \frac{2\sqrt{5}}{5}x' - \frac{\sqrt{5}}{5}y'$$

$$y = \frac{\sqrt{5}}{5}x' + \frac{2\sqrt{5}}{5}y'$$

Substituting, we obtain:

$$\left(x' + \frac{3\sqrt{5}}{5}\right)^2 = \frac{8\sqrt{5}}{5}\left(y' + \frac{9\sqrt{5}}{40}\right)$$

So, rotating 26.6°, we graph the parabola:

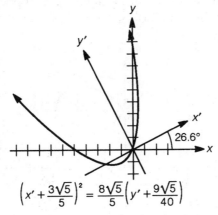

$$\left(x' + \frac{3\sqrt{5}}{5}\right)^2 = \frac{8\sqrt{5}}{5}\left(y' + \frac{9\sqrt{5}}{40}\right)$$

75. We have:

$$\cot 2\theta = \frac{13 - 13}{10} = 0, \text{ so } 2\theta = 90° \text{ and } \theta = 45°$$

Therefore:

$$\sin\theta = \frac{\sqrt{2}}{2} \text{ and } \cos\theta = \frac{\sqrt{2}}{2}$$
$$x = \frac{\sqrt{2}}{2}x' - \frac{\sqrt{2}}{2}y'$$
$$y = \frac{\sqrt{2}}{2}x' + \frac{\sqrt{2}}{2}y'$$

Substituting, we obtain:

$$\frac{x'^2}{16} + \frac{\left(y' - \sqrt{2}\right)^2}{36} = 1$$

So, rotating 45°, we graph the ellipse:

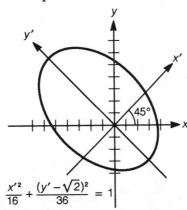

$$\frac{x'^2}{16} + \frac{(y' - \sqrt{2})^2}{36} = 1$$

77. **(a)** We plot points in polar form, noting that $r^2 \geq 0$ implies $0 \leq \theta \leq \frac{\pi}{4}, \frac{3\pi}{4} \leq \theta \leq \frac{5\pi}{4}$, and $\frac{7\pi}{4} \leq \theta \leq 2\pi$:

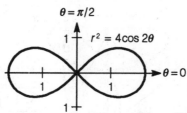

(b) We plot points in polar form, noting that $r^2 \geq 0$ implies $0 \leq \theta \leq \frac{\pi}{2}$ and $\pi \leq \theta \leq \frac{3\pi}{2}$:

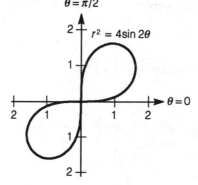

79. **(a)** We plot points in polar form:

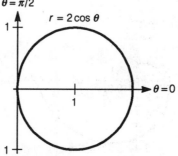

(b) We plot points in polar form:

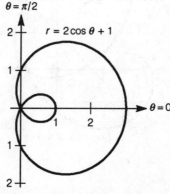

81. (a)   We plot points in polar form:

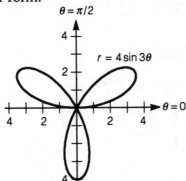

(b)   We plot points in polar form:

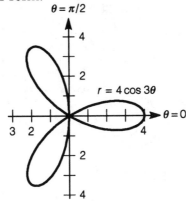

83. (a)   We plot points in polar form:

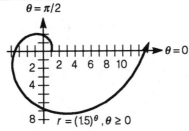

(b)   We plot points in polar form (the graph starts at $r = 1$ on the $\theta = 0$ axis, then spirals inward indefinitely):

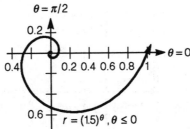

85. Comparing the given equation with the four basic types, it appears this is the type associated with Figure 3 (Section 7.8). We divide both numerator and denominator by 2 to obtain:

$$r = \frac{3}{2 - 4\cos\theta} = \frac{\frac{3}{2}}{1 - 2\cos\theta}$$

Since the eccentricity is $e = 2 > 1$, this conic is a hyperbola. Computing values of r when $\theta = 0, \frac{\pi}{2}, \pi$ and $\frac{3\pi}{2}$, we have:

θ	0	$\frac{\pi}{2}$	π	$\frac{3\pi}{2}$
r	$-\frac{3}{2}$	$\frac{3}{2}$	$\frac{1}{2}$	$\frac{3}{2}$

So the vertices are $\left(-\frac{3}{2}, 0\right)$ and $\left(-\frac{1}{2}, 0\right)$, and thus the center is $(-1, 0)$. Note that $a = \frac{1}{2}$ and since a focus is $(0, 0)$, then $c = 1$ and thus:

$$b = \sqrt{c^2 - a^2} = \sqrt{1 - \frac{1}{4}} = \sqrt{\frac{3}{4}} = \frac{\sqrt{3}}{2}$$

We graph the hyperbola:

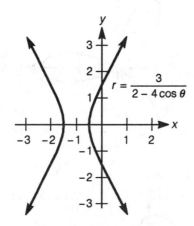

$$r = \frac{3}{2 - 4\cos\theta}$$

87. Comparing the given equation with the four basic types, it appears this is the type associated with Figure 3 (Section 7.8). We divide both numerator and denominator by 4 to obtain:

$$r = \frac{4}{4 - \cos\theta} = \frac{1}{1 - \frac{1}{4}\cos\theta}$$

Since the eccentricity is $e = \frac{1}{4} < 1$, this conic is an ellipse. Computing the values of r when $\theta = 0, \frac{\pi}{2}, \pi$ and $\frac{3\pi}{2}$, we have:

θ	0	$\frac{\pi}{2}$	π	$\frac{3\pi}{2}$
r	$\frac{4}{3}$	1	$\frac{4}{5}$	1

So the vertices on the major axis are $\left(\frac{4}{3}, 0\right)$ and $\left(-\frac{4}{5}, 0\right)$, and thus the center is $\left(\frac{4}{15}, 0\right)$. The major axis has length:

$$2a = \frac{4}{3} - \left(-\frac{4}{5}\right) = \frac{32}{15}, \text{ so } a = \frac{16}{15}$$

Since a focus is $(0,0)$, then $c = \frac{4}{15}$ and thus:
$$b = \sqrt{a^2 - c^2} = \sqrt{\left(\tfrac{16}{15}\right)^2 - \left(\tfrac{4}{15}\right)^2} = \sqrt{\tfrac{240}{225}} = \tfrac{4\sqrt{15}}{15}$$
We graph the ellipse:

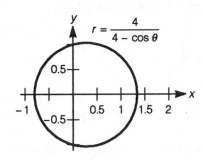

$$r = \frac{4}{4 - \cos \theta}$$

89. Comparing the given equation with the four basic types, it appears this is the type associated with Figure 5 (Section 7.8). We divide both numerator and denominator by 3 to obtain:
$$r = \frac{20}{3 - 5\sin \theta} = \frac{\frac{20}{3}}{1 - \frac{5}{3}\sin \theta}$$

Since the eccentricity is $e = \frac{5}{3} > 1$, this conic is a hyperbola. Computing the values of r when $\theta = 0, \frac{\pi}{2}, \pi$ and $\frac{3\pi}{2}$, we have:

θ	0	$\frac{\pi}{2}$	π	$\frac{3\pi}{2}$
r	$\frac{20}{3}$	-10	$\frac{20}{3}$	$\frac{5}{2}$

So the vertices are $(0,-10)$ and $\left(0,-\frac{5}{2}\right)$, and thus the center is $\left(0,-\frac{25}{4}\right)$. The transverse axis has length:
$$2a = -\tfrac{5}{2} + 10 = \tfrac{15}{2}, \text{ so } a = \tfrac{15}{4}$$

Since a focus is $(0,0)$, then $c = \frac{25}{4}$ and thus:
$$b = \sqrt{c^2 - a^2} = \sqrt{\left(\tfrac{25}{4}\right)^2 - \left(\tfrac{15}{4}\right)^2} = \sqrt{\tfrac{400}{16}} = 5$$
We graph the hyperbola:

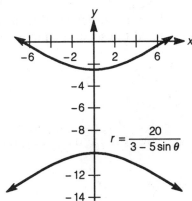

$$r = \frac{20}{3 - 5\sin \theta}$$

91. Solving $y = 1 + t$ for t yields $t = y - 1$, now substituting:
$$x = 3 - 5(y - 1)$$
$$x = 3 - 5y + 5$$
$$x + 5y = 8$$
So the given parametric equations determine a line.

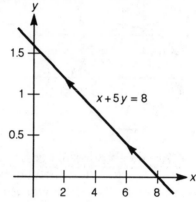

93. Multiplying the first equation by 2 yields $2x = 6 \sin t$ and $y = 6 \cos t$, so:
$$(2x)^2 + y^2 = 36 \sin^2 t + 36 \cos^2 t$$
$$4x^2 + y^2 = 36$$
$$\frac{x^2}{9} + \frac{y^2}{36} = 1$$
So the given parametric equations determine an ellipse.

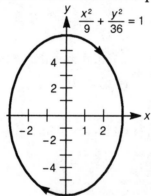

95. Multiplying the first equation by 3 and the second equation by 4 yields $3x = 12 \sec t$ and $4y = 12 \tan t$, so:
$$(3x)^2 - (4y)^2 = 144 \sec^2 t - 144 \tan^2 t$$
$$9x^2 - 16y^2 = 144$$
$$\frac{x^2}{16} - \frac{y^2}{9} = 1$$

So the given parametric equations determine a hyperbola.

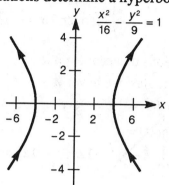

$$\frac{x^2}{16} - \frac{y^2}{9} = 1$$

Chapter Seven Test

1. Since $4p = 12$, then $p = 3$. So the focus is $(-3, 0)$ and the directrix is $x = 3$.

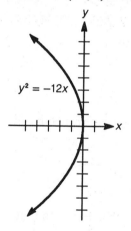

$$y^2 = -12x$$

2. Dividing by 4, the standard form is $\frac{x^2}{4} - \frac{y^2}{1} = 1$. So the asymptotes are $y = \pm\frac{1}{2}x$. Using $c^2 = a^2 + b^2$, we have:
$$c = \sqrt{a^2 + b^2} = \sqrt{4 + 1} = \sqrt{5}$$
So the foci are $\left(\pm\sqrt{5}, 0\right)$.

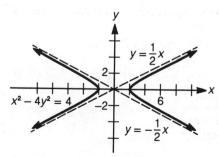

$$y = \frac{1}{2}x$$
$$x^2 - 4y^2 = 4$$
$$y = -\frac{1}{2}x$$

3. (a) We have:
$$\cot 2\theta = \frac{A-C}{B} = \frac{1-3}{2\sqrt{3}} = -\frac{1}{\sqrt{3}}, \text{ so } \tan 2\theta = -\sqrt{3}$$
Thus $2\theta = 120°$, or $\theta = 60°$.

(b) Applying the rotation formulas, we have:
$$x = x'\cos 60° - y'\sin 60° = \tfrac{1}{2}x' - \tfrac{\sqrt{3}}{2}y'$$
$$y = x'\sin 60° + y'\cos 60° = \tfrac{\sqrt{3}}{2}x' + \tfrac{1}{2}y'$$
Substituting into $x^2 + 2\sqrt{3}xy + 3y^2 - 12\sqrt{3}x + 12y = 0$ yields:
$$x'^2 = -6y'$$
Rotating $60°$, we graph the parabola:

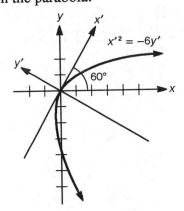

4. Since $\tan\theta = \frac{1}{\sqrt{3}}$, then $\theta = 30°$.

5. Since $e = \frac{c}{a}$, then $\frac{c}{a} = \frac{1}{2}$, so $a = 2c$. Since the foci are $(0, \pm 2)$, then $c = 2$ and thus $a = 4$.
    Now we find b^2:
$$c^2 = a^2 - b^2$$
$$4 = 16 - b^2$$
$$b^2 = 12$$
    Thus the equation of the ellipse is $\frac{x^2}{12} + \frac{y^2}{16} = 1$.

6. Call m the slope of the tangents, so:
$$y - 0 = m(x + 4)$$
$$y = mx + 4m$$
    We now find the distance from the center of the circle $(0,0)$ to this line:
$$r = \frac{|0 + 0 - 4m|}{\sqrt{1+m^2}} = \frac{|4m|}{\sqrt{1+m^2}}$$
    Since this radius is 1, we have:
$$1 = \frac{|4m|}{\sqrt{1+m^2}}$$

Squaring, we have:
$$1 = \frac{16m^2}{1+m^2}$$
$$1 + m^2 = 16m^2$$
$$1 = 15m^2$$
$$m^2 = \frac{1}{15}$$
$$m = \pm\frac{\sqrt{15}}{15}$$

7. The slope is given by $m = \tan 60° = \sqrt{3}$. Using the point $(2, 0)$ in the point-slope formula:
$$y - 0 = \sqrt{3}(x - 2)$$
$$y = \sqrt{3}x - 2\sqrt{3}$$
$$\sqrt{3}x - y - 2\sqrt{3} = 0$$

8. Since the foci are $(\pm 2, 0)$, then $c = 2$. Also $\frac{b}{a} = \frac{1}{\sqrt{3}}$, so $a = b\sqrt{3}$. So:
$$c^2 = a^2 + b^2$$
$$4 = \left(b\sqrt{3}\right)^2 + b^2$$
$$4 = 4b^2$$
$$1 = b^2$$
$$1 = b$$

So $a = 1\sqrt{3} = \sqrt{3}$, thus the equation is $\frac{x^2}{3} - \frac{y^2}{1} = 1$.

9. We plot points in polar form:

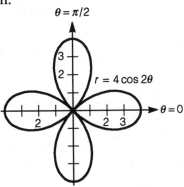

10. (a) Substituting $P(6, 5)$ into the equation $5x^2 - 4y^2 = 80$, we have:
$$5(6)^2 - 4(5)^2 = 5(36) - 4(25) = 180 - 100 = 80$$
So $P(6, 5)$ lies on the hyperbola.

(b) The quantity $(F_1P - F_2P)^2$ can be computed without determining the coordinates of

F_1 and F_2. By definition we have $|F_1P - F_2P| = 2a$ for any point P on the
hyperbola. Squaring both sides here yields $(F_1P - F_2P)^2 = 4a^2$. Now to compute a,
we convert the equation $5x^2 - 4y^2 = 80$ to standard form. The result is

$\frac{x^2}{4^2} - \frac{y^2}{(2\sqrt{5})^2} = 1$. Therefore $a = 4$ and we obtain $(F_1P - F_2P)^2 = 4a^2 = 4(16) = 64$.

11. Dividing by 100 yields the standard form $\frac{x^2}{25} + \frac{y^2}{4} = 1$. The length of the major axis is 10
and the length of the minor axis is 4. Using $c^2 = a^2 - b^2$, we find:
$$c = \sqrt{a^2 - b^2} = \sqrt{25 - 4} = \sqrt{21}$$
So the foci are $(\pm\sqrt{21}, 0)$.

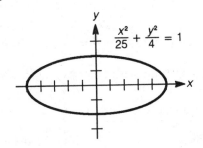

12. Using the double-angle formula for $\cos 2\theta$ and multiplying by r^2 yields:
$$r^2 = \cos^2 \theta - \sin^2 \theta$$
$$r^4 = r^2 \cos^2 \theta - r^2 \sin^2 \theta$$
$$\left(x^2 + y^2\right)^2 = x^2 - y^2$$

13. Here $(x_0, y_0) = (-1, 0)$, $A = 2$, $B = -1$ and $C = -1$, so using the distance formula from a
point to a line yields:
$$d = \frac{|Ax_0 + By_0 + C|}{\sqrt{A^2 + B^2}} = \frac{|-2 + 0 - 1|}{\sqrt{4 + 1}} = \frac{3}{\sqrt{5}} = \frac{3\sqrt{5}}{5}$$

14. We complete the square to convert the equation to standard form:
$$16x^2 + y^2 - 64x + 2y + 65 = 0$$
$$16(x^2 - 4x) + (y^2 + 2y) = -65$$
$$16(x^2 - 4x + 4) + (y^2 + 2y + 1) = -65 + 64 + 1$$
$$16(x - 2)^2 + (y + 1)^2 = 0$$

Since the only solution to this equation is the point $(2, -1)$, the graph consists of a single point.

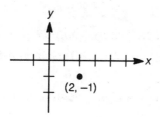

15. We plot points in polar form:

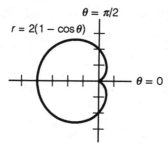

16. The equation represents a hyperbola with center $(-4, 4)$. We graph the hyperbola:

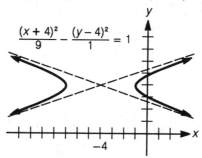

17. Comparing the given equation with the four basic types, it appears this is the type associated with Figure 3 (Section 7.8). We divide both numerator and denominator by 5 to obtain:

$$r = \frac{9}{5 - 4\cos\theta} = \frac{\frac{9}{5}}{1 - \frac{4}{5}\cos\theta}$$

Since the eccentricity is $e = \frac{4}{5} < 1$, this conic is an ellipse. Computing the values of r when $\theta = 0, \frac{\pi}{2}, \pi$ and $\frac{3\pi}{2}$, we have:

θ	0	$\frac{\pi}{2}$	π	$\frac{3\pi}{2}$
r	9	$\frac{9}{5}$	1	$\frac{9}{5}$

So the vertices on the major axis are $(9, 0)$ and $(-1, 0)$, and thus the center is $(4, 0)$ and the length of the major axis is $2a = 10$, so $a = 5$. Since a focus is $(0, 0)$, then $c = 4$ and thus:

$$b = \sqrt{a^2 - c^2} = \sqrt{25 - 16} = \sqrt{9} = 3$$

We graph the ellipse:

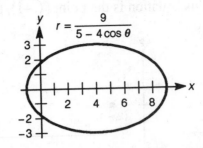

$$r = \frac{9}{5 - 4\cos\theta}$$

18. Multiplying the second equation by 2 yields $x = 4\sin t$ and $2y = 4\cos t$, so:

$$x^2 + (2y)^2 = 16\sin^2 t + 16\cos^2 t$$
$$x^2 + 4y^2 = 16$$
$$\frac{x^2}{16} + \frac{y^2}{4} = 1$$

So the given parametric equations determine an ellipse.

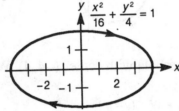

$$\frac{x^2}{16} + \frac{y^2}{4} = 1$$

19. Since $4p = 8$, then $p = 2$. The focal width is 8 and the vertex is $(1,2)$.

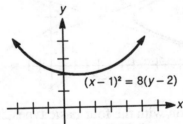

$$(x-1)^2 = 8(y-2)$$

20. (a) Since $a = 6$ and $b = 5$, we find c:

$$c = \sqrt{a^2 - b^2} = \sqrt{36 - 25} = \sqrt{11}$$

So the eccentricity is $e = \frac{c}{a} = \frac{\sqrt{11}}{6}$. Thus the directricies are given by:

$$x = \pm\frac{a}{e} = \pm\frac{6}{\frac{\sqrt{11}}{6}} = \pm\frac{36}{\sqrt{11}} = \pm\frac{36\sqrt{11}}{11}$$

(b) The focal radii are given by:

$$F_1 P = a + ex = 6 + \frac{\sqrt{11}}{6}(3) = 6 + \frac{\sqrt{11}}{2} = \frac{12 + \sqrt{11}}{2}$$
$$F_2 P = a - ex = 6 - \frac{\sqrt{11}}{6}(3) = 6 - \frac{\sqrt{11}}{2} = \frac{12 - \sqrt{11}}{2}$$